Pitman Research Notes in Mathematics Series

Submission of proposals for consideration
Suggestions for publication, in the form of outlines and representative samples, are invited by the Editorial Board for assessment. Intending authors should approach one of the main editors or another member of the Editorial Board, citing the relevant AMS subject classifications. Alternatively, outlines may be sent directly to the publisher's offices. Refereeing is by members of the board and other mathematical authorities in the topic concerned, throughout the world.

Preparation of accepted manuscripts
On acceptance of a proposal, the publisher will supply full instructions for the preparation of manuscripts in a form suitable for direct photo-lithographic reproduction. Specially printed grid sheets can be provided and a contribution is offered by the publisher towards the cost of typing. Word processor output, subject to the publisher's approval, is also acceptable.

Illustrations should be prepared by the authors, ready for direct reproduction without further improvement. The use of hand-drawn symbols should be avoided wherever possible, in order to maintain maximum clarity of the text.

The publisher will be pleased to give any guidance necessary during the preparation of a typescript, and will be happy to answer any queries.

Important note
In order to avoid later retyping, intending authors are strongly urged not to begin final preparation of a typescript before receiving the publisher's guidelines. In this way it is hoped to preserve the uniform appearance of the series.

Addison Wesley Longman Ltd
Edinburgh Gate
Harlow, Essex, CM20 2JE
UK
(Telephone (0) 1279 623623)

Titles in this series. A full list is available from the publisher on request.

Gerhard Dangelmayr

Universität Tübingen, Germany and Colorado State University, USA

Bernold Fiedler

Freie Universität Berlin, Germany

Klaus Kirchgässner

Universität Stuttgart, Germany

and

Alexander Mielke

Universität Hannover, Germany

Dynamics of nonlinear waves in dissipative systems: reduction, bifurcation and stability

 LONGMAN

Addison Wesley Longman Limited
Edinburgh Gate, Harlow
Essex CM20 2JE, England
and Associated Companies throughout the world.

Published in the United States of America
by Addison Wesley Longman Inc.

First published 1996

AMS Subject Classifications: (Main) 35B40, 35C20, 58F14
                             (Subsidiary) 34C20, 47H20, 76E30

ISSN 0269-3674

ISBN 0 582 22929 4

British Library Cataloguing in Publication Data

A catalogue record for this book is
available from the British Library

Printed and bound in Great Britain
by Biddles Ltd, Guildford and King's Lynn

Contents

Preface

The time evolution of wave patterns in unbounded continua comprises the themes of the four contributions of this book. Their study has originated as a joint research project between the universities of Stuttgart and Tübingen and has been supported by the Land Baden–Württemberg during 1991 – 1993. In the meantime, the authors were dispersed in various parts of the world, but their common interest in this area has prevailed. Their cooperation has now found support from the Deutsche Forschungsgemeinschaft within the Schwerpunktprogramm "Ergodentheorie, Analysis und effiziente Simulation Dynamischer Systeme".

Following the character of these Research Notes we present a number of new results on evolution equations in unbounded domains which should illuminate some of the features which are in contrast to the well developed theories for spatially bounded domains. The most obvious one is the change in the character of spectra, and their inseparability from critical regions due to the continuous spectrum. Other aspects concern the role which the breaking of the translational invariance has on travelling waves when an infinite cylinder is shortened to a large, but finite one. Finally, patterns such as pulses and fronts, also spatially periodic solutions, generate an inhomogeneous structure in space for which neither its global appearance nor its effect on local bifurcations is known on a general level. The reader will find some answers to these questions in this book. We abstain from describing the general research background of our endeavour here. In an active field such an effort would be bound to fail. Instead, the individual contributions will take on this task in a more localized sense.

April 1996

Gerhard Dangelmayr, Tübingen and Fort Collins
Bernold Fiedler, Berlin
Klaus Kirchgässner, Stuttgart
Alexander Mielke, Hannover

Introduction and Overview

The time evolution of nonlinear wave patterns in unbounded continua comprises the themes of the four contributions of this book. Following the character of these Research Notes we present a number of new results on evolution equations in unbounded domains which should illuminate some of the features which are in contrast to the well developed theories for spatially bounded domains. The most obvious one is the change in the character of spectra, and their inseparability from critical regions due to the continous spectrum. Others concern the role which the breaking of the translational invariance has on travelling waves when an infinite cylinder is shortened to a large, but finite one. Finally, patterns such as pulses - and fronts, also spatially periodic solutions, generate an inhomogeneous structure in space for which neither its global appearance nor its effect on local bifurcations is known on a general level. Some answers to these questions can be found in the following chapters.

The subject of the first chapter is the derivation and analysis of envelope equations of the Ginzburg-Landau type. These equations describe the propagation of waves in systems of large spatial extent near an oscillatory instability of a basic, spatially and temporally uniform state. A formal procedure, based on asymptotic expansions in Fourier space, is developed which allows the computation of the Ginzburg-Landau equations to any desired order. Since any real physical system is of finite extent, special emphasis is laid on the effect of distant side walls. Depending on the critical group velocity, the Ginzburg-Landau equations reduce either to a single equation with a global nonlinear term or to a system of locally coupled partial differential equations for the envelopes of left and right traveling waves. In the latter case, the presence of side walls is revealed by Robin boundary conditions that induce a breaking of the translation symmetry of the infinitely extended system. Analytical and numerical studies of these equations show a variety of modes of dynamical behaviours. In particular, the spatio-temporal patterns derived near a homoclinic orbit resemble much of the so called "blinking state" observed in convection experiments with binary fluids.

1

In general, homoclinic orbits play an important role for parabolic systems of partial differential equations. In particular, spatially localized patterns like fronts and pulses can often be described by homoclinic orbits of a system of ordinary differential equations derived from the underlying pde-system. Typically, these ode-systems depend on parameters, including those of the original equation as well as the propagation speeds of the localized patterns.

The main goal of the second chapter is to obtain a global continuation theorem for homoclinic solutions of autonomous ordinary differential equations with two real parameters. In one-parameter flows, Hopf bifurcation serves as a starting point for global paths of periodic orbits. B-points, alias Arnol'd-Bogdanov-Takens points, play an analogous role for paths of homoclinic orbits in two-parameter flows. In fact, a path of homoclinic orbits emanating from a B-point can be continued in phase space until it terminates at another B-point, or becomes unbounded, or approaches a region with chaotic dynamics. This result is obtained via a new topological invariant for homoclinic orbits, based on an approximation of the homoclinic orbit by nearby periodic orbits. Several local bifurcation results for homoclinic and heteroclinic orbits are reviewed, along the way, to illustrate scope, significance, and limitations of the global approach. The chapter concludes with an extensive discussion, including "nongeneric" aspects like equivariance, Hamiltonian and reversible systems, and topics like numerical continuation and partial differential equations.

In the third chapter, a semilinear parameter dependent parabolic system in an infinitely extended strip in space is investigated. Near a critical parameter-value the system possesses a family of travelling fronts on a two-dimensional spacelike- manifold on which the flow can be derived from a KPP-equation. The fronts connect two equilibria. In a space with suitably weighted norms, which suppress the eigenvalue zero, each individual front can be shown to be asymptotically stable even in some semiglobal sense. The problem arises, whether the long-time asymptotics of local - and semiglobal perturbations can be determined by reducing properly the full system to the two-dimensional subspace on which the two-dimensional invariant manifold is living, which contains the front. This question is answered affirmatively for the case, that the front considered has not the minimal speed. However, the result is also true in the case excluded here, as a recently completed paper shows. The reduction procedure is nonlocal in nature, robust and can be extended to cover systems of infinite dimensions. The stability results are presented for two types of initial conditions which differ in the degree of regularity assumed for the initial conditions.

The main emphasis of the fourth chapter is to develop the functional analytical framework concerning the Floquet theory for elliptic systems. In contrast to ODEs and parabolic PDEs, the Floquet decomposition is not always possible, since the number of Floquet multipliers outside the unit circle is infinite as well as the number inside. In addition to these general questions, several typical examples are treated, such as reaction-diffusion systems, water waves, spatial period doubling and stationary bifurcations associated to the Eckhaus instability.

Chapter 1
Ginzburg–Landau Description of Waves in Extended Systems

by **G DANGELMAYR**

Contents

1 Introduction

There has been much recent interest in the dynamics of wave patterns in dissipative, spatially extended systems. This interest was particularly motivated by the observation of traveling waves in binary fluid convection [WKPS85] and the subsequent discovery of the so-called "confined" and "blinking" states [FMS88, KoS88, KoSW89, SFMR89]. In these experiments the term traveling wave is used to refer to a state of approximately parallel convection rolls drifting in a direction perpendicular to their axes. The traveling wave states differ from the usual convection rolls in a moving frame of reference in having well-developed phase differences between the velocity, thermal and concentration fields, which decrease with the phase velocity as the applied Rayleigh number is raised [DKT87]. In a finite geometry, however, a pure traveling wave cannot exist because there is no moving reference frame in which they become a time-independent spatially periodic state. Instead, the experiments described in [FMS88, KoS88, KoSW89, SFMR89] reveal that, when a roll reaches the sidewall, it shrinks and a new roll is born at the opposite end. It is evident, therefore, that the ratio of the length of the container to that of the roll wavelength is an important parameter in the problem. The so-called confined traveling waves are states that do not fill the whole container, but are confined to some bounded region near a sidewall. Finally, the blinking states are those in which the pattern oscillates, often irregularly, between a confined left-traveling wave near the left wall and a confined right-traveling wave near the right wall. The mathematical description of phenomena of this kind forms the subject of this contribution.

Mathematically, wave states are expected to appear in parameter regimes which are close to an oscillatory instability, or Hopf bifurcation, of a basic uniform solution of the governing evolution equation. In an idealized description one commonly assumes that the system is of infinite extent, and imposes periodic boundary conditions in order to obtain a discrete spectrum of the linearized operator. The invariance of the governing equation under spatial translation and reflection then induces an action of the circle group $O(2)$ on spatially periodic states. The most simple and straightforward approach to describing the dynamics of waves in dissipative, infinitely extended systems is, therefore, based on the normal form for a Hopf bifurcation with $O(2)$-symmetry. In a rotating frame of reference this normal form is given by [GSS88]

$$\frac{dA}{dT} = (a_0 \Lambda + a|A|^2 + b|B|^2)A \qquad (1.1)$$

6

$$\frac{dB}{dT} = (a_0\Lambda + a|B|^2 + b|A|^2)B, \tag{1.2}$$

and describes the temporal evolution of the complex amplitudes A and B of left and right propagating traveling waves of the form $Ae^{\imath(\omega_c t + k_c x)}$, $Be^{\imath(\omega_c t - k_c x)}$. Here, T is a slow time, Λ is the bifurcation parameter describing the onset of instability when Λ increases and a_0, a, b are complex coefficients in which the details of the original evolution equation are compressed. The parameters ω_c, k_c are the critical frequency and wave number at the stability threshold. The o.d.e.-system (1.1), (1.2) exhibits three invariances, two rotation invariances

$$\mathbf{S}^1 : (A, B) \to (e^{\imath\varphi}A, e^{\imath\varphi}B), \quad \mathbf{SO}(2) : (A, B) \to (e^{\imath\Psi}A, e^{-\imath\Psi}B),$$

and a reflection invariance

$$\mathbf{Z}_2 : (A, B) \to (B, A).$$

The $\mathbf{S}^1$- and $\mathbf{SO}(2)$-symmetries result from the temporal and spatial translation invariances of the original system, respectively, whereas $\mathbf{Z}_2$ corresponds to interchanging left and right going traveling waves and is a consequence of the spatial reflection invariance. The system (1.1), (1.2) possesses two basic states, namely traveling waves of the form $(A, 0)$ and $(0, A)$, referred to as TW, and standing waves $(A, e^{\imath\phi}A)$, referred to as SW, which are constant in time. The stability properties of these states are easily characterized in terms of the nonlinear coefficients a, b. Generically the SW and TW solutions are the only non-wandering solutions of (1.1), (1.2).

The system (1.1), (1.2) may be derived from the original evolution equation by means of a center manifold reduction followed by a normal form transformation [GSS88, GH90]. This procedure is possible near an oscillatory instability (Hopf bifurcation) if the evolution equation is supplemented by periodic boundary conditions, where the spatial period is chosen in accordance with the critical wave number. The existence and stability properties of ideal wave patterns can then be inferred from the coefficients a, b. The normal form approach describes, however, only the stability of the waves against spatially periodic perturbations, spatially modulated patterns are not available. The main objective here is to discuss the modifications that are necessary for describing spatial modulations as well as the effects of distant sidewalls.

When spatial modulations are admitted, the o.d.e.'s (1.1), (1.2) have to be replaced by spatio-temporal evolution equations. The amplitudes A, B play now the role of wave envelopes with slow variation in space and time, similarly as the envelopes introduced in the pioneering work of Newell and Whitehead [NW69] and Segel [Se69]

7

serve for modulations of spatially periodic patterns. This kind of approach to the description of amplitude modulations near instabilities is commonly referred to as the Ginzburg-Landau formalism. In contrast to the stationary patterns discussed in [NW69, Se69], however, wave propagation is combined with energy transport. The rate of this transport is measured by the group velocity which plays a crucial role for the resulting envelope equations. For small group velocities the evolution equations for (A, B) are p.d.e's, i.e., the system (1.1), (1.2) is supplemented by diffusion terms whereas the couplings of the envelopes remain still local. For large group velocity we have again additional diffusion terms, however, the couplings of A and B are non-local, as was first recognized by Knobloch and deLuca [KD90].

The second modification of (1.1), (1.2) results from the presence of distant sidewalls in the original evolution system. Clearly, the main effect of sidewalls is to break the spatial translation invariance which in turn results in the breaking of the $\mathbf{SO}(2)$-symmetry of the envelope equations. Again, there is a strong difference between the cases of small and large group velocities. For small group velocity we have to supplement the envelope-p.d.e.'s by boundary conditions at the ends of a normalized, bounded interval. The effect of sidewalls is then simply to change their type from ideal (Dirichlet) to mixed (Robin) type which breaks the $\mathbf{SO}(2)$-symmetry. For finite group velocities instead, the envelope equations reduce to a single equation for one of the envelopes subject to periodic boundary conditions.

We begin, in Section 2, by introducing a general class of evolution equations on which the subsequent discussion relies, and describe the basic steps of the linear stability analysis. In Section 3 we represent the solution of the evolution equation in terms of slowly varying envelopes and show how to compute asymptotic equations for these. In this section we still work with the original unscaled space-time variables. The computations rely on a Fourier-space formulation and involve Taylor expansions in wave number and frequency space, combined with an application of the Liapunov-Schmidt method. Explicit expressions for the coefficients are summarized in the appendix. Section 4 is devoted to suitable rescalings of the envelope equations and to a derivation of their boundary conditions for both, finite and small group velocities. The evolution equations resulting from this procedure are referred to as the non-local and the local Ginzburg-Landau equations, respectively. The exposition of Sections 3 and 4 is formal and focuses mainly on computational aspects.

Recently the Ginzburg-Landau formalism obtained a rigorous basis by a series of papers. The time-independent and the time-periodic cases for certain classes of hydro-dynamic instabilities was treated in [IoMD89] and [IoM91], respectively. At the same time, the general time-dependent case was studied in [CE93, VH91, KSM92, Sch94a], but only for scalar problems. By now a general theory for p.d.e.'s, like Navier-Stokes in cylindrical domains, is available in [Sch94b]. It is expected that the methods devel-oped in these papers may also be applied to give a rigorous justification of the formal approach to oscillatory instabilities utilized here.

The techniques of Sections 3 and 4 are applied in Section 5 to two-dimensional Boussi-nesq convection in a vertical magnetic field. In Section 6 we review recent results on the dynamics of the local Ginzburg-Landau equations. In particular it is shown how a perturbed version of the o.d.e.-system (1.1), (1.2) can be extracted from the Ginzburg-Landau equations and can be used to explain much of the observed exper-imental behavior. In the final Section 7 we summarize and discuss the methods and results described in the previous sections.

2 Basic evolution equation and linear stability analysis

Consider a function $u(t, x)$ describing a perturbation of some basic state in a physical problem, e.g., of the heat conduction state in a convection system. Here, t is time and $x \in (-l, l)$ is a distinguished spatial variable which varies in a "large" domain of size $2l$. Of particular interest is the limit $l \to \infty$ in which case we have an idealized, infinitely extended system. The meaning of "large l" if l is finite is that the system inherits a basic wavelength $l_c = 2\pi/k_c$ and that $l/l_c \gg 1$. The presence of possibly other spatial variables, which are assumed to vary in a bounded domain, is incorporated by viewing u as a mapping from $\mathbf{R} \times (-l, l)$ to an appropriate function space $\mathcal{H}$, a real Hilbert or Banach space, whatever setting is best suited. Technical questions concerning the functional analytic foundations will not be dealt with here (see, e.g., Mielke's article in this volume and [IoMD89, Sch94b] for related problems), we confine ourselves to a purely formal treatment. In the case of two dimensional magnetoconvecton considered in Section 5, $\mathcal{H}$ is a space of vector valued functions of the vertical coordinate z with appropriate boundary conditions at the top $z = 1$ and the bottom $z = 0$.

Assume that $u(t, x)$ obeys the following evolution equation,

$$\mathcal{L}(\partial_t, \partial_x, R)u + \mathcal{N}(u, R) = 0, \tag{2.1}$$

where $\mathcal{L}$ is polynomial in (∂_t, ∂_x) with R-dependent coefficients that are linear operators in $\mathcal{H}$. The second term $\mathcal{N}$ is strictly nonlinear and may be expanded as usual in a "Taylor series",

$$\mathcal{N}(u, R) = \mathcal{N}_2[u, u; R] + \mathcal{N}_3[u, u, u; R] + \ldots, \tag{2.2}$$

where the subscripts $2, 3$ etc. refer to bilinear, trilinear etc. operators which may act as local differential operators or as global operators with respect to t and x. We specify these operators in more detail in Section 3 using a Fourier space formulation; here we merely assume that they allow the problem to be posed in any spatial domain $x \in (-l, l)$, including the unbounded limit $l \to \infty$. The parameter $R \in \mathbf{R}$ can be varied freely and characterizes the transition from linear stability to instability of the trivial solution $u = 0$ when R increases. In the bounded case we have to impose boundary conditions which are assumed to be homogeneous,

$$\mathcal{B}_{\pm}^{(\nu)}(\partial_t, \partial_x)u = 0 \text{ at } x = \pm l; 1 \leq \nu \leq M. \tag{2.3}$$

The $\mathcal{B}_{\pm}^{(\nu)}$ are also polynomials in (∂_t, ∂_x) with coefficients that are linear operators from $\mathcal{H}$ to some suitable space $\mathcal{H}_B^{(\nu)}$. We do not specify the boundary conditions at this stage (see Section 4), but assume that they are compatible with the operators in (2.1), i.e., (2.1) and (2.3) should constitute a well-posed initial-boundary value problem. In the unbounded limit $l \to \infty$ (2.3) is disregarded and we only require boundedness of the solutions of (2.1) in the limit $x \to \infty$ for finite t.

We assume that for any domain $(-l, l)$ the problem (2.1) with (2.3) admits a temporal translation symmetry and a spatial reflection symmetry. Letting $\tau \in \mathbf{R}$, time translation invariance means that

$$\mathcal{N}(u_\tau, R)(t, x) = \mathcal{N}(u, R)(t + \tau, x) \tag{2.4}$$

if $u_\tau(t, x) := u(t + \tau, x)$. Concerning reflection invariance, let S be a linear operator in $\mathcal{H}$ with $S^2 = \text{id}$ and associate to $u(t, x)$ the reflected function $u_-(t, x) := u(t, -x)$. Invariance of (2.1) under simultaneous space inversion and operation via S requires that

$$\mathcal{L}(\partial_t, \partial_x, R)S = S\mathcal{L}(\partial_t, -\partial_x, R), \quad \mathcal{N}(Su_-, R) = S\mathcal{N}(u, R)_-,$$
$$\mathcal{B}_+^{(\nu)}(\partial_t, \partial_x) = \mathcal{B}_-^{(\nu)}(\partial_t, -\partial_x)S \quad (1 \leq \nu \leq M), \tag{2.5}$$

10

i.e., if $u(t,x)$ is a solution of (2.1) with (2.3), then $Su(t,-x)$ is also a solution. In the unbounded limit $l = \infty$ ($x \in \mathbf{R}$), we assume in addition to the symmetries (2.4) and (2.5) also invariance under spatial translation. Letting $y \in \mathbf{R}$, this means that

$$\mathcal{N}(u_y, R)(t,x) = \mathcal{N}(u, R)(t, x+y) \tag{2.6}$$

if $u_y(t,x) := u(t, x+y)$. Clearly, (2.6) is not a symmetry for (2.1) with (2.3) in the bounded case $l < \infty$: the presence of sidewalls unavoidably breaks the spatial translation invariance.

The remainder of this section is devoted to a linear stability analysis of (2.1) for the infinitely extended case $l = \infty$. The linear stability of the trivial solution is determined by the linearized problem,

$$\mathcal{L}(\partial_t, \partial_x, R)u = 0. \tag{2.7}$$

If for a given R (2.7) possesses only exponentially decaying solutions with respect to t, then u is linearly stable. The temporal and spatial translation invariances suggest an ansatz for (2.7) in the form of plane waves, $u(t,x) = \exp(\mu t + \imath k x)v$ with $v \in \mathcal{H}_c = \mathcal{H} \oplus \imath \mathcal{H}$, yielding

$$\mathcal{L}(\mu, \imath k, R)v = 0. \tag{2.8}$$

The operator in (2.8) is a linear operator in $\mathcal{H}_c$ that depends polynomially on the yet undetermined parameters μ and k. We assume that $\mathcal{H}_c$ can be decomposed into finite dimensional subspaces $\mathcal{H}_n$ ($n = 1, 2, \ldots$), $\mathcal{H}_c = \bigoplus_{n \geq 1} \mathcal{H}_n$, and that $\mathcal{L}$ and S leave each $\mathcal{H}_n$ invariant. The requirement that (2.8) possesses a solution v in $\mathcal{H}_n$ leads to the solvability condition

$$g_n(\mu, k^2, R) := \det \mathcal{L}_n(\mu, \imath k, R) = 0 \quad (n = 1, 2, \ldots), \tag{2.9}$$

where $\mathcal{L}_n := \mathcal{L}\,|_{\mathcal{H}_n}$ and, as a consequence of the spatial reflection symmetry, $\det \mathcal{L}_n$ is a polynomial in (μ, k^2) with real, R-dependent coefficients. If (2.9) has a solution (μ, k^2) with $\Re\mu > 0$ for some n, then the trivial solution $u = 0$ of (2.1) is linearly unstable. The transition to instability is characterized by stability boundaries in the (R, k^2)-plane which are determined by the equations

$$h_{ns}(k^2, R) := g_n(0, k^2, R) = 0 \text{ and } h_{no}(k^2, R) = 0. \tag{2.10}$$

The first equation in (2.10) leads to nontrivial stationary solutions, $\mu = 0$, of (2.6). The second equation is derived by setting $\mu = \imath \omega$ ($\omega \in \mathbf{R}$), writing g_n as

$$g_n(\imath\omega, k^2, R) = g_{nr}(\omega^2, k^2, R) + \imath g_{ni}(\omega^2, k^2, R)$$

11

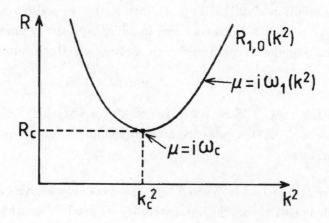

Figure 2.1: Neutral oscillatory stability curve $R_{1,o}(k^2)$ with minimum R_c at $k^2 = k_c^2$.

and eliminating ω^2 from the two polynomial equations $g_{nr} = 0 = g_{ni}$, e.g., via calculation of the resultant of these polynomials. The solutions (R, k^2) must be restricted here so that $\omega^2 \geq 0$.

The typical situation is that the equations (2.10) define smooth functions $R_{ns}(k^2)$ and $R_{no}(k^2)$ such that for fixed k^2 there is no solution μ of (2.9) with $0 < \mu \in \mathbb{R}$ and $\Im\mu \neq 0, \Re\mu > 0$ if $R < R_{ns}(k^2)$ and $R < R_{no}(k^2)$, respectively, whereas above these values positive growth rates occur. The graphs of the functions $R_{ns}(k^2)$ and $R_{no}(k^2)$ are referred to as neutral stationary and neutral oscillatory stability curves [Ma90]. Assuming that $R_{ns}(k^2)$ and $R_{n0}(k^2)$ have unique minima $R_{n,cs} = R_{ns}(k_{n,cs}^2)$ and $R_{n,co} = R_{no}(k_{n,co}^2)$ that occur for certain critical wavenumbers $k_{n,cs}^2$ and $k_{n,co}^2$, these minima define the critical stationary and critical oscillatory parameter values for the subspace $\mathcal{H}_n$, i.e., if R is below these values all solutions of (2.7) in $\mathcal{H}_n$ decay exponentially. In applications R is usually a measure for the rate of energy fed into the system, e.g., heat supply in the case of convection systems. Thus the sets of critical values $R_{n,cs}$ and $R_{n,co}$ typically have lower bounds $R_{cs} := \min_n R_{n,cs}$ and $R_{co} := \min_n R_{n,co}$. Then, if R is increased from below, the trivial solution $u = 0$ of (2.1) encounters the first instability at $R_c := \min\{R_{cs}, R_{co}\}$ with a certain critical wavenumber $k_c > 0$.

In this contribution we are exclusively interested in oscillatory instabilities (Hopf bifurcations), so let us assume that $R_c = R_{co} \equiv R_{1,co}$. The passage from linear stability to linear instability when R increases through R_c is then characterized by the neutral oscillatory stability curve $R = R_{1,o}(k^2)$ which is typically a parabola-like

12

curve as indicated in Figure 2.1. For R near R_c the linearized equation (2.7) possesses solutions in $\mathcal{H}_1$ of the form (*cc* refers to the complex conjugate expression)

$$u(t, x) = \exp[\mu(k^2, R)t \pm \imath kx]v_\pm(k, R) + cc, \qquad (2.11)$$

where $v_-(k, R) = S v_+(-k, R)$ by virtue of the reflection symmetry and both,

$$\mu(k^2, R) = \mu_r(k^2, R) + \imath\omega(k^2, R) \qquad (2.12)$$

and $\bar{\mu}(k^2, R)$ solve the determinantal equation $g_1(\mu, k^2, R) = 0$, with the bar denoting complex conjugation. Below the neutral stability curve μ_r is negative and above this curve μ_r is positive. Along $R = R_{1,o}(k^2)$ we have

$$\mu(k^2, R_{1,o}(k^2)) = \imath\omega(k^2, R_{1,o}(k^2)) \equiv i\omega_1(k^2), \qquad (2.13)$$

i.e., solutions oscillating in time. The frequency $\omega_c := \omega_1(k_c^2)$ occurring at the minimum R_c is referred to as the critical frequency and is assumed to be positive so that $\omega(k^2, R)$ is positive as well for (k^2, R) sufficiently close to (k_c^2, R_c).

We end this section by relating the local form of $\omega(k^2, R)$, Eq. (2.12), to its defining equation $g_1(\mu, k^2, R) = 0$. Due to our basic assumptions we have $g_1(\imath\omega_c, k_c^2, R_c) = 0$ and, assuming further that $\partial g_1(\imath\omega_c, k_c^2, R_c)/\partial\mu \neq 0$, the real and imaginary parts of $\mu(k^2, R)$ have unique Taylor expansions about (k^2, R_c) of the form

$$\mu_r(k^2, R) = d(R - R_c) - c(k^2 - k_c^2)^2 + \cdots \qquad (2.14)$$

$$\omega(k^2, R) = \omega_c + \frac{c_g}{2k_c}(k^2 - k_c^2) + e(R - R_c) + \cdots \qquad (2.15)$$

A term proportional to $k^2 - k_c^2$ is not present in μ_r because R_c is a minimum of $R_{1,o}(k^2)$. The local form of this curve follows from $\mu_r(k^2, R) = 0$ as

$$R_{1,o}(k^2) = R_c + \frac{c}{d}(k^2 - k_c^2)^2 + O((k^2 - k_c^2)^3),$$

whence dc must be positive in order to have a non-degenerate minimum. The condition for $\partial\mu_r(k_c^2, R_c)/\partial k^2 = 0$ is

$$\Re\left(\frac{\partial g_1}{\partial k^2}\frac{\overline{\partial g_1}}{\partial\mu}\right)\Bigg|_c = 0, \qquad (2.16)$$

where the "c" indicates that the expression is to be evaluated at $(\mu, k^2, R) = (i\omega_c, k_c^2, R_c)$. The quantity c_g in (2.15) is the group velocity at the instability point,

$$c_g = \frac{\partial\omega_1}{\partial k}\Big|_c = -2k_c\Im(\frac{\partial g_1}{\partial k^2}/\frac{\partial g_1}{\partial\mu})\Bigg|_c. \qquad (2.17)$$

13

Here the second relation follows from differentiating the identity $g_1(\mu(k^2, R), k^2, R) = 0$ with respect to k^2. Similarly we find

$$d + \imath e = -\left(\frac{\partial g_1}{\partial R}\bigg/\frac{\partial g_1}{\partial \mu}\right)\bigg|_c,\tag{2.18}$$

$$c = \frac{1}{2}\Re\left\{\left(\frac{\partial^2 g_1}{(\partial k^2)^2} - \frac{c_g^2}{4k_c^2}\frac{\partial^2 g_1}{\partial \mu^2}\right)\bigg/\frac{\partial g_1}{\partial \mu}\right\}\bigg|_c - \frac{c_g}{2k_c}\Im\left(\frac{\partial^2 g_1}{\partial \mu \partial k^2}\bigg/\frac{\partial g_1}{\partial \mu}\right)\bigg|_c.\tag{2.19}$$

3   Derivation of envelope equations

In this section we derive a system of coupled envelope or modulation equations for (2.1) that describes the dynamics of wave patterns for R slightly above the stability threshold R_c of an oscillatory instability. The basic idea behind the formal derivation is the following. At the instability point and in the infinitely extended case $l = \infty$ the linearized problem, Eq. (2.7), possesses two solutions of the form $e^{\imath(\omega_c t + k_c x)}v_0 + cc$ and $e^{\imath(\omega_c t - k_c x)}\, S\, v_0 + cc$, that correspond to left and right traveling waves, with $v_0 \in \mathcal{H}_c$ being a nontrivial solution of $\mathcal{L}(\imath\omega_c, \imath k_c, R_c)v_0 = 0$. The existence of such solutions in the linearized problem suggests a solution ansatz $u = U$ for the full nonlinear problem as a superposition of these wave patterns, i.e.

$$U(t, x) = e^{\imath\omega_c t}\{e^{\imath k_c x}A(t, x)v_0 + e^{-\imath k_c x}B(t, x)\, S\, v_0\} + cc + h.o.t.,\tag{3.1}$$

where "h.o.t." refers to terms of higher order in an sense that will be specified below. The complex functions $A(t, x)$ and $B(t, x)$ are envelopes of the two waves and are considered as small quantities which vary on slow scales in both the spatial variable x and the temporal variable t. Thus rapid (oscillatory) variations are solely captured by the wave patterns, whereas A and B play the role of small modulation functions. In writing (3.1) it is implicitly assumed that A, B depend also on the local bifurcation parameter

$$\lambda := (R - R_c)/R_c\tag{3.2}$$

and should tend to zero when $\lambda \to +0$. The slow variation of A and B means that $\partial_t A$, $\partial_x A$ and $\partial_t B$, $\partial_x B$ are much smaller than A and B themselves. Thus one is tempted to try asymptotic expansions in $(\partial_t, \partial_x, A, B, \lambda)$ with the understanding that $\partial_t^{m+1}f(A, \bar{A}, B, \bar{B}, \lambda)$ is considered as a term of higher order than $\partial_t^m f(A, \bar{A}, B, \bar{B}, \lambda)$ if f is a polynomial in $(A, \bar{A}, B, \bar{B}, \lambda)$, and similarly for spatial derivatives. This ordering in powers of $(\partial_t, \partial_x, A, B, \lambda)$ can be justified by introducing slow variables

$T = \varepsilon^p t, X = \varepsilon^q x$ with $0 < \varepsilon \ll 1$ and properly chosen exponents p, q, setting $\lambda = \varepsilon^2 \Lambda$ with $\Lambda = 0(1)$ and expanding A as $A(t, x; \varepsilon) = \varepsilon A_0(T, X) + \varepsilon^2 A_1(T, X) + \dots$. We wish, however, to carry out such a scaling at the last stage of the analysis and therefore postpone ε-expansions to Section 4. In this section we make a formal "Taylor expansion" of the solution of (2.1) with respect to $(\partial_t, \partial_x, A, B, \lambda)$ and do not yet specify distinct orders of these quantities. The "h.o.t." in (3.1) then indicates terms which are at least quadratic in $(\partial_t, \partial_x, A, B, \lambda)$. These terms involve also higher harmonics of the basic left and right travelling wave patterns. The boundary conditions (2.3) are disregarded in this section, whence the asymptotic representation of the solutions of (2.1) derived here is valid either for $l = \infty$ or, in the case of $1 \ll l < \infty$ in the core of the interval $(l, -l)$, excluding certain boundary layers near the two ends. Boundary conditions for the envelope functions will be derived in the next section by means of boundary correction terms.

3.1 Fourier space formulation of the basic equations

The notion of "Taylor expansions" with respect to the differential operators (∂_t, ∂_x) becomes more transparent in a Fourier space formulation. Let

$$v(\omega, k) = (\mathcal{F}u)(\omega, k) := \int dt \int dx \, u(t, x) e^{-i(\omega t + kx)} \qquad (3.3)$$

be the formal Fourier transform of $u(t, x)$, with the inverse operation

$$u(t, x) = (\mathcal{F}^{-1}v)(t, x) := \frac{1}{4\pi^2} \int d\omega \int dk \, v(\omega, k) e^{i(\omega t + kx)}. \qquad (3.4)$$

For notational convenience integral domains are not written down explicitly. Formally all integrals in (3.3), (3.4) and in the expansions below are extended from $-\infty$ to ∞. Convergence questions will not be dealt with here; convergences may be assured by choosing slightly distorted paths in the complex plane or simply by viewing $u(t, x)$ and $v(\omega, k)$ as generalized functions in an appropriate setting. Some care must be laid on the case of finite l, here we assume that $u(t, x)$ is extended beyond $(-l, l)$ in a way so that the contributions from this domain to (3.3) are negligible and boundary singularities are avoided. Let

$$V_+(\omega, k) := \int dt \int dx \, A(t, x) e^{-i(\omega t + kx)} \qquad (3.5)$$

$$V_-(\omega, k) := \int dt \int dx \, B(t, x) e^{-i(\omega t + kx)} \qquad (3.6)$$

15

be the Fourier transforms of the envelopes introduced in (3.1). Because A, B are slowly varying in space and time, the $V_\pm(\omega, k)$ are centered in a small neighborhood of $(\omega, k) = (0, 0)$ and should have a strong decay in the (ω, k)-plane with increasing distance from the origin. We consider these functions as smeared out versions of weighted Delta functions $\hat{V}_\pm \delta(\omega) \delta(k)$ with small weights $\hat{V}_\pm$. Due to the correspondences $\partial_t \leftrightarrow \imath\omega$, $\partial_x \leftrightarrow \imath k$, the Fourier transform of $\partial_t^m \partial_x^n A(t, x)$ is given by $(\imath\omega)^m (\imath k)^n V_+(\omega, k)$. Thus an expansion with respect to (∂_t, ∂_x) corresponds to a power series in (ω, k) around $(0, 0)$ which is justified because the $V_\pm$ are centered near the origin.

By virtue of (3.1), the leading term of $v(\omega, k)$ for $\omega \geq 0$ is a superposition of the Fourier transform of the basic wave patterns,

$$A(t, x) e^{\imath(\omega_c t + k_c x)} v_0 \overset{\mathcal{F}}{\mapsto} V_+(\omega - \omega_c, k - k_c) v_0,$$
$$B(t, x) e^{\imath(\omega_c t - k_c x)} S v_0 \overset{\mathcal{F}}{\mapsto} V_-(\omega - \omega_c, k + k_c) S v_0,$$

which are centered around $\omega = \omega_c$, $k = \pm k_c$. These leading terms will now be extended to an asymptotic series in terms of the still unknown functions $V_\pm(\omega, k)$. Due to the nonlinear part of (2.1), the higher order terms are centered near multiples $(n\omega_c, mk_c)$ $(n, m \in \mathbb{Z})$ of the critical frequency and wavelength. They involve convolutions between $V_+(\omega - \omega_c, k - k_c)$, $V_-(\omega - \omega_c, k + k_c)$ and their complex conjugates as well as power series with respect to (ω, k) around $(n\omega_c, mk_c)$. Equations for $V_\pm(\omega, k)$ follow then from a Fredholm alternative and lead finally via Fourier inversion to the desired system of coupled equations for A and B.

Due to the spatial and temporal translation invariances, the general forms of the nonlinear terms $\mathcal{N}_l$ $(l = 2, 3, \dots)$ in (2.2) can be represented by integral kernels as

$$\mathcal{N}_l[u, \cdots, u; R](t, x) = \left(\prod_{\mu=1}^{l} \int dt_\mu \int dx_\mu \right) F_l[u(t - t_1, x - x_1), \dots, u(t - t_l, x - x_l)].$$

(3.7)

Here, $F_l \equiv F_l(t_1, x_1 | \dots | t_l, x_l; R)$ is an l-linear operator in $\mathcal{H}$ that depends on the $2l$ variables (t_μ, x_μ) $(1 \leq \mu \leq l)$ and R. This notation allows us to treat both local differential operators as well as global nonlinearities, where in the former case the (t_μ, x_μ)-dependence should be expressed in terms of derivatives of Delta functions. We work mainly with the Fourier transforms

$$G_l(\omega_1, k_1 | \dots | \omega_l, k_l; R) := (4\pi^2)^{1-l} \left(\prod_{\mu=1}^{l} \int dt_\mu \int dx_\mu \, e^{-\imath(\omega_\mu t_\mu + k_\mu x_\mu)} \right) F_l, \quad (3.8)$$

16

which are assumed to depend smoothly on (ω_j, k_j) and R. Specifically, when $\mathcal{N}_l$ is a differential operator, the kernel G_l has the form $G_l = \sum_\nu P_{l\nu} A_{l\nu}$ with complex polynomial functions $P_{l\nu}(\omega_1, k_1 | \ldots | \omega_l, k_l)$ and l-linear operators $A_{l\nu} : \mathcal{H}^l \to \mathcal{H}$. After Fourier transforming (2.1) we obtain then an equation for $v(\omega, k)$,

$$\mathcal{L}(\imath\omega, \imath k, R)v + \mathcal{M}_2[v, v; R] + \mathcal{M}_3[v, v, v; R] + \ldots = 0. \tag{3.9}$$

The nonlinear terms $\mathcal{M}_l$ are represented by the transformed kernels G_l,

$$\mathcal{M}_l[v, \ldots, v; R](\omega, k) = \left(\prod_{\mu=1}^{l} \int d\omega_\mu \int dk_\mu \right) \delta\left(\omega - \sum_{\mu=1}^{l} \omega_\mu \right) \delta\left(k - \sum_{\mu=1}^{l} k_\mu \right)$$
$$\times G_l(\omega_1, k_1 | \ldots | \omega_l, k_l; R) [v(\omega_1, k_1), \ldots, v(\omega_l, k_l)], \tag{3.10}$$

with $\delta(y)$ $(y \in \mathbb{R})$ denoting Dirac's Delta function.

As was explained before, we seek a solution of (3.9) that is centered near multiples $(n\omega_c, mk_c)$ $(n, m \in \mathbb{Z})$ of the critical frequency and wavenumber. This suggests an ansatz of the form

$$v(\omega, k) = \sum_{n, m \in \mathbb{Z}} \tilde{v}_{nm}(\omega, k), \tag{3.11}$$

where $\tilde{v}_{nm}$ has a sharp peak at $(n\omega_c, mk_c)$ and is negligibly small outside some neighborhood of this point. We represent this local variation in terms of local variables (σ, κ) defined by

$$v_{nm}(\sigma, \kappa) := \tilde{v}_{nm}(n\omega_c + \sigma, mk_c + \kappa) \tag{3.12}$$

and denote by

$$\mathcal{L}(n, m)(\sigma, \kappa; \lambda) := \mathcal{L}(\imath n\omega_c + \imath\sigma, \imath mk_c + \imath\kappa; R_c(1 + \lambda)) \tag{3.13}$$

the corresponding local forms of the linearized operator. Neglecting the contribution of $\tilde{v}_{nm}(\omega, k)$ to multiples $(n'\omega_c, m'k_c)$ with $(n', m') \neq (n, m)$, the equation (3.9) is split into a set of equations for the v_{nm},

$$\mathcal{L}(n, m)(\sigma, \kappa; \lambda)v_{nm} + \sum_{l \geq 2} \sum_{[n,m]} \mathcal{M}_l[v_{n_1 m_1}, \ldots, v_{n_l m_l}; R_c(1 + \lambda)] = 0, \tag{3.14}$$

where the symbol $[n, m]$ in the second sum denotes that this sum is extended over l-tuples $((n_1, m_1), \ldots, (n_l, m_l))$ which satisfy $n_1 + \cdots + n_l = n$, $m_1 + \cdots + m_l = m$, as a consequence of the δ-functions in (3.9). Each of the equations (3.14) labelled

17

by (n, m) is considered for $(\sigma, \kappa, \lambda)$ in a small neighborhood of $(0, 0, 0)$. Noting that $v(-\omega, k) = \bar{v}(\omega, -k)$ because $u(t, x)$ is real, i.e.,

$$v_{-n,m}(\sigma, \kappa; \lambda) = \bar{v}_{n,-m}(-\sigma, -\kappa; \lambda), \tag{3.15}$$

it is sufficient to solve (3.14) for $n \geq 0$, $m \in \mathbb{Z}$.

By assumption, the operators

$$\mathcal{L}_0(n, m) := \mathcal{L}(\imath n \omega_c, \imath m k_c; R_c) = \mathcal{L}(n, m)(0, 0, 0) \tag{3.16}$$

are invertible for $(n, m) \neq (1, \pm 1)$. (Invertibility is understood here in the sense that the domain of these operators is compactly embedded in $\mathcal{H}_c$.) Hence the corresponding equations for v_{nm} can be solved using a regular perturbation expansion for $(\sigma, \kappa, \lambda)$ close to $(0, 0, 0)$. On the other hand, the operators

$$\mathcal{L}_{\pm} := \mathcal{L}(\imath \omega_c, \pm \imath k_c; R_c) \tag{3.17}$$

have one-dimensional kernels spanned by v_0 and $S v_0$.

We therefore split the equations for v_{11} and $v_{1,-1}$ into their regular and singular parts. Let $\mathcal{L}_{\pm}^*$ be the adjoint operator of $\mathcal{L}_{\pm}$ and let $\mathcal{L}_{+}^* v_0^* = 0$ with the normalization $\langle v_0, v_0^* \rangle = 1$, where the bracket denotes the pairing between vectors in $\mathcal{H}_c$ and dual vectors in $\mathcal{H}_c^*$ (inner product if $\mathcal{H}$ is a Hilbert space). Due to the reflection symmetry, $S^* v_0^*$ is the corresponding coeigenvector of $\mathcal{L}_{-}^*$. As usual in bifurcation theory we introduce projections $Q_{\pm}$ in $\mathcal{H}_c$ onto subspaces $\mathcal{H}_c^{\pm}$ of codimension one by setting

$$Q_+ v := v - \langle v, v_0^* \rangle v_0, \quad Q_- v := v - \langle S v, v_0^* \rangle S v_0, \quad (v \in \mathcal{H}_c) \tag{3.18}$$

and defining $\mathcal{H}_c^{\pm} = Q_{\pm} \mathcal{H}_c$. The equations (3.14) for $(n, m) = (1, \pm 1)$ are then projected via $Q_{\pm}$,

$$Q_{\pm} \mathcal{L}(1, \pm 1) v_{1,\pm 1} + \sum_{l \geq 2} \sum_{[1, \pm 1]} Q_{\pm} \mathcal{M}_l [v_{n_1 m_1}, \ldots, v_{n_l m_l}] = 0 \tag{3.19}$$

and the $v_{1, \pm 1}$ are decomposed into

$$v_{11}(\sigma, \kappa) = V_+(\sigma, \kappa) v_0 + \hat{v}_{1,1}(\sigma, \kappa), \quad v_{1,-1}(\sigma, \kappa) = V_-(\sigma, \kappa) v_0 + \hat{v}_{1,-1}(\sigma, \kappa), \tag{3.20}$$

with $\hat{v}_{1, \pm 1}(\sigma, \kappa) \in \mathcal{H}_c^{\pm}$

18

3.2 Asymptotic expansions in Fourier space

Our approach to reducing (3.14) and (3.19) for $v_\pm(\sigma, \kappa)$ relies on an extension of the classical Liapunov-Schmidt reduction procedure. We first solve the coupled equations (3.14) and (3.19) for v_{nm} with $(n, m) \neq (1, \pm 1)$ and $\hat{v}_{1,\pm 1}$ in terms of the still unknown functions $V_\pm(\sigma, \kappa)$. Equations for $V_\pm$ are then obtained by scalar multiplication of (3.14) with $\langle \cdot, v_0^* \rangle$ for $(n, m) = (1, 1)$ and with $\langle \cdot, S^* v_0^* \rangle$ for $(m, n) = (1, -1)$. Contrary to the standard Liapunov-Schmidt reduction, however, the v_{nm} and $\hat{v}_{1,\pm 1}$ are expressed as functionals of $V_\pm(\sigma, \kappa)$ that involve convolution integrals. These functionals have to be determined in a way so that the essential contributions of the leading terms of $\tilde{v}_{\pm 1,\pm 1}(\omega, k)$, which are centered near $(\pm \imath \omega_c, \pm \imath k_c)$, to the higher harmonics $\tilde{v}_{nm}(\omega, k)$ are taken into account.

To present the general form of the solution ansatz for v_{nm}, we introduce four-fold indices $\boldsymbol{p} = (p_1, p_2, p_3, p_4)$ with nonnegative integers p_j, and define

$$n(\boldsymbol{p}) = p_1 + p_2 - p_3 - p_4, \quad m(\boldsymbol{p}) = p_1 - p_2 - p_3 + p_4,$$

$$I(n, m) = \{\boldsymbol{p} : n(\boldsymbol{p}) = n, m(\boldsymbol{p}) = m\}, \quad |\boldsymbol{p}| := \sum_{j=1}^{4} p_j \geq 1,$$

To each fixed $\boldsymbol{p}$ we associate variables $(\sigma_{j\nu}, \kappa_{j\nu}) \in I\!\!R^2$ and multiindices $\boldsymbol{\alpha} = (\alpha_{j\nu})$, $\boldsymbol{\beta} = (\beta_{j\nu})$ with nonnegative integers $\alpha_{j\nu}$, $\beta_{j\nu}$ and arbitrary degrees $|\boldsymbol{\alpha}|$, $|\boldsymbol{\beta}|$, where $1 \leq j \leq 4$ and ν runs for each j from 1 to p_j. The ansatz for v_{nm} is then written as

$$v_{nm}(\sigma, \kappa) = \sum_{\boldsymbol{p} \in I(n,m)} \sum_{\boldsymbol{\alpha}, \boldsymbol{\beta}, \gamma} v(\boldsymbol{p}; \boldsymbol{\alpha}, \boldsymbol{\beta}, \gamma) V(\boldsymbol{p}; \boldsymbol{\alpha}, \boldsymbol{\beta}, \gamma)(\sigma, \kappa), \tag{3.21}$$

with $v(\boldsymbol{p}; \boldsymbol{\alpha}, \boldsymbol{\beta}, \gamma) \in \mathcal{H}_c$ and further nonnegative integers γ. The expression $V(\boldsymbol{p}; \boldsymbol{\alpha}, \boldsymbol{\beta}, \gamma)$ denotes the $2|\boldsymbol{p}|$-fold integral

$$V(\boldsymbol{p}; \boldsymbol{\alpha}, \boldsymbol{\beta}, \gamma)(\sigma, \kappa) = \lambda^\gamma \left[\prod_{j=1}^{4} \prod_{\nu=1}^{p_j} \int d\sigma_{j\nu} \int d\kappa_{j\nu} (\sigma_{j\nu})^{\alpha_{j\nu}} (\kappa_{j\nu})^{\beta_{j\nu}} V_j(\sigma_{j\nu}, \kappa_{j\nu}) \right]$$
$$\times \delta(\sigma - \sum_{j,\nu} \sigma_{j\nu}) \delta(\kappa - \sum_{j,\nu} \kappa_{j\nu}), \tag{3.22}$$

where $(V_1, V_2, V_3, V_4) = (V_+, V_-, V_+^\dagger, V_-^\dagger)$ and $V_\pm^\dagger(\sigma, \kappa) := \overline{V}_\pm(-\sigma, -\kappa)$. In accordance with (3.20) we set $v(\boldsymbol{p}; 0, 0, 0) = v_i$ if $p_j = \delta_{ij}$, with $(v_1, v_2, v_3, v_4) = (v_0, Sv_0, \bar{v}_0, S\bar{v}_0)$ and $\delta_{ij} = 1$ for $i = j$, $\delta_{ij} = 0$ for $i \neq j$. Moreover, to meet (3.15) we have to impose the condition

$$v(\boldsymbol{p}^\dagger; \boldsymbol{\alpha}^\dagger, \boldsymbol{\beta}^\dagger, \gamma) = (-1)^{|\boldsymbol{\alpha}|+|\boldsymbol{\beta}|} \bar{v}(\boldsymbol{p}; \boldsymbol{\alpha}, \boldsymbol{\beta}, \gamma), \tag{3.23}$$

19

with $\boldsymbol{p}^\dagger$ and $\boldsymbol{\alpha}^\dagger$, $\boldsymbol{\beta}^\dagger$ defined by permuting the indices j and $j\nu$ in $\boldsymbol{p}$ and $\boldsymbol{\alpha}$, $\boldsymbol{\beta}$ according to $(1,2,3,4) \to (3,4,1,2)$. Finally, the restriction that $\hat{v}_{1,\pm 1}(\sigma,\kappa) \in \mathcal{H}_c^\pm$ means that

$$v(\boldsymbol{p};\boldsymbol{\alpha},\boldsymbol{\beta},\gamma) \in \mathcal{H}_c^\pm \text{ if } n(\boldsymbol{p}) = 1, \ m(\boldsymbol{p}) = \pm 1 \text{ and } |\boldsymbol{p}| > 1. \tag{3.24}$$

A further restriction on the vectors $v(\boldsymbol{p};\boldsymbol{\alpha},\boldsymbol{\beta},\gamma)$ follows by taking the reflection symmetry of (2.1) into account. In the Fourier space formulation this means that the l.h.s. of (3.9) is equivariant under the operation $v(\omega,k) \to Sv(\omega,-k)$ which becomes $v_{nm}(\sigma,\kappa) \to Sv_{n,-m}(\sigma,\kappa)$ in the representation (3.11) and (3.12). In particular this induces the replacement $V_\pm \to V_\pm^{(-)}$ in the leading terms of $v_{1,\pm 1}(\sigma,\kappa)$, Eq. (3.20), where $V_\pm^{(-)}$ denotes the reflected functions $V_\pm^{(-)}(\sigma,\kappa) = V_\pm(\sigma,-\kappa)$. Hence, viewing (3.21) as a functional representation $v_{nm}[V_+, V_-](\sigma,\kappa)$, we have

$$v_{n,-m}[V_+, V_-](\sigma,\kappa) = Sv_{nm}[V_-^{(-)}, V_+^{(-)}](\sigma,-\kappa).$$

This relation implies in turn that

$$v(\boldsymbol{p}_-;\boldsymbol{\alpha}_-,\boldsymbol{\beta}_-,\gamma) = (-1)^{|\boldsymbol{\beta}|} Sv(\boldsymbol{p};\boldsymbol{\alpha},\boldsymbol{\beta},\gamma), \tag{3.25}$$

where $\boldsymbol{p}_-$ and $\boldsymbol{\alpha}_-$, $\boldsymbol{\beta}_-$ are defined by permuting the indices j and $j\nu$ in $\boldsymbol{p}$ and $\boldsymbol{\alpha}$, $\boldsymbol{\beta}$ according to $(1,2,3,4) \to (2,1,4,3)$. By virtue of (3.23) and (3.25) we need only to consider the equations for v_{nm} with $n \geq 0$, $m \geq 0$.

When the ansatz (3.21) is substituted into (3.14) and (3.19) if $(n,m) \neq (1,1)$ and $(n,m) = (1,1)$, respectively, a recursive system of equations for the vectors $v(\boldsymbol{p};\boldsymbol{\alpha},\boldsymbol{\beta},\gamma) \in \mathcal{H}_c$ is obtained as follows. We write the expression (3.22) as a multiple convolution

$$V(\boldsymbol{p};\boldsymbol{\alpha},\boldsymbol{\beta},\gamma) = (\sigma^{\alpha_{11}}\kappa^{\beta_{11}}V_+) * \cdots * (\sigma^{\alpha_{4,p_4}}\kappa^{\beta_{4,p_4}}V_-) \tag{3.26}$$

where the asterisk denotes the convolution operation $f * g$ between two complex functions $f(\sigma,\kappa)$, $g(\sigma,\kappa)$,

$$(f * g)(\sigma,\kappa) := \int d\sigma' \int d\kappa' \ f(\sigma - \sigma', \kappa - \kappa')g(\sigma',\kappa') = (g * f)(\sigma,\kappa).$$

Noting that multiplication by σ or κ acts like a derivation on the convolution product,

$$\sigma(f * g) = (\sigma f) * g + f * (\sigma g), \quad \kappa(f * g) = (\kappa f) * g + f * (\kappa g),$$

we find by induction that

$$\left[\sigma^{\alpha_1}\kappa^{\beta_1}V(\boldsymbol{p}_1;\boldsymbol{\alpha}_1,\boldsymbol{\beta}_1,\gamma_1)\right] * \cdots * \left[\sigma^{\alpha_l}\kappa^{\beta_l}V(\boldsymbol{p}_l;\boldsymbol{\alpha}_l,\boldsymbol{\beta}_l,\gamma_l)\right] = \sum_{\boldsymbol{\alpha}\boldsymbol{\beta}} q_{\boldsymbol{\alpha}\boldsymbol{\beta}}V(\boldsymbol{p};\boldsymbol{\alpha},\boldsymbol{\beta},\gamma),$$

$$\tag{3.27}$$

20

where $\boldsymbol{p} = \sum_\mu \boldsymbol{p}_\mu$, $\gamma = \sum_\mu \gamma_\mu$. The $q_{\alpha\beta}$ in (3.27) are certain combinatorial numbers that depend on $(\boldsymbol{\alpha}_\mu, \boldsymbol{\beta}_\mu, \alpha_\mu, \beta_\mu)$ and are nonzero only if $|\boldsymbol{\alpha}| = \sum_\mu (|\boldsymbol{\alpha}_\mu| + \alpha_\mu)$ and $|\boldsymbol{\beta}| = \sum_\mu (|\boldsymbol{\beta}_\mu| + \beta_\mu)$. The linear parts are now expanded in Taylor series,

$$\mathcal{L}(n,m)(\sigma, \kappa; \lambda) = \mathcal{L}_0(n,m) + \sum_{\alpha+\beta+\gamma \geq 1} (\imath\sigma)^\alpha (\imath\kappa)^\beta \lambda^\gamma \mathcal{L}_{\alpha\beta\gamma}(n,m). \tag{3.28}$$

When $(\imath\sigma)^\alpha (\imath\kappa)^\beta \lambda^\gamma \mathcal{L}_{\alpha\beta\gamma}(n(\boldsymbol{p}), m(\boldsymbol{p}))$ is applied to a particular term $vV(\sigma, \kappa)$ in the sum (3.21), labelled by $(\boldsymbol{p}, \boldsymbol{\alpha}, \boldsymbol{\beta}, \gamma')$, this results in an expression of the form

$$[\mathcal{L}_{\alpha\beta\gamma}(n,m)v]\, V'(\sigma, \kappa).$$

The function $V'(\sigma, \kappa)$ is, by virtue of (3.27) with $l = 1$, a superposition of functions $V(\boldsymbol{p}; \boldsymbol{\alpha}', \boldsymbol{\beta}', \gamma+\gamma')$, where $(\boldsymbol{\alpha}', \boldsymbol{\beta}')$ runs through a finite set of indices with total degrees $|\boldsymbol{\alpha}'| = |\boldsymbol{\alpha}|+\alpha$, $|\boldsymbol{\beta}'| = |\boldsymbol{\beta}|+\beta$. A nonlinear term $\mathcal{M}_l$ in (3.14) with fixed (n,m) involves l double integrals with respect to (ω_μ, k_μ) $(1 \leq \mu \leq l)$ over a function which is centered near $(n_\mu \omega_c, m_\mu k_c)$. These integrals are first transformed to local integration variables (σ_μ, κ_μ) via $(\omega_\mu, k_\mu) = (n_\mu \omega_c + \sigma_\mu, m_\mu k_c + \kappa_\mu)$ and then the integral kernel is expanded in a Taylor series about $(\sigma_\mu, \kappa_\mu, \lambda) = (0,0,0)$. A particular Taylor term in this series leads to a contribution to (3.14) of the form

$$\left(\prod_{\mu=1}^l \int d\sigma_\mu \sigma_\mu^{\alpha_\mu} \int d\kappa_\mu \kappa_\mu^{\beta_\mu} \right) \delta\left(\sigma - \sum_{\mu=1}^l \sigma_\mu \right) \delta\left(\kappa - \sum_{\mu=1}^l \kappa_\mu \right)$$
$$\times \lambda^\gamma G_l'[v_{n_1 m_1}(\sigma_1, \kappa_1), \ldots, v_{n_l m_l}(\sigma_l, \kappa_l)], \tag{3.29}$$

where G_l' is now an l-linear operator that does not depend on the integration variables (σ_μ, κ_μ). Substituting $v_{n_\mu m_\mu}(\sigma_\mu, \kappa_\mu)$ according to (3.21) into (3.29) results again in a sum of terms with each of these involving l index combinations $(\boldsymbol{p}_\mu; \boldsymbol{\alpha}_\mu, \boldsymbol{\beta}_\mu, \gamma_\mu)$ $(1 \leq \mu \leq l)$. Such a term is given by the l.h.s of (3.27) times the vector

$$G_l'[v(\boldsymbol{p}_1; \boldsymbol{\alpha}_1, \boldsymbol{\beta}_1, \gamma_1), \ldots, v(\boldsymbol{p}_l; \boldsymbol{\alpha}_l, \boldsymbol{\beta}_l, \gamma_l)]$$

and can, therefore, again be written as a finite superposition of functions $V(\boldsymbol{p}; \boldsymbol{\alpha}, \boldsymbol{\beta}, \gamma+\sum_\mu \gamma_\mu)$ with fixed $\boldsymbol{p} = \sum_\mu \boldsymbol{p}_\mu$, i.e., $n(\boldsymbol{p}) = n$, $m(\boldsymbol{p}) = m$, and with the $\boldsymbol{\alpha}, \boldsymbol{\beta}$ satisfying $|\boldsymbol{\alpha}| = \sum_\mu (|\boldsymbol{\alpha}_\mu|+\alpha_\mu)$ and $|\boldsymbol{\beta}| = \sum_\mu (|\boldsymbol{\beta}_\mu|+\beta_\mu)$. In summary, the substitution of (3.21) into the l.h.s of (3.14) yields the same kind of expansion as (3.21), but with new vectors $v'(\boldsymbol{p}; \boldsymbol{\alpha}, \boldsymbol{\beta}, \gamma)$. Moreover, each of these vectors can be decomposed into

$$v'(\boldsymbol{p}; \boldsymbol{\alpha}, \boldsymbol{\beta}, \gamma) = \mathcal{L}_0\left(n(\boldsymbol{p}), m(\boldsymbol{p}) \right) v(\boldsymbol{p}; \boldsymbol{\alpha}, \boldsymbol{\beta}, \gamma) + w_r(\boldsymbol{p}; \boldsymbol{\alpha}, \boldsymbol{\beta}, \gamma),$$

21

where the remainder term w_r depends on a set of vectors $v(\boldsymbol{p}'; \boldsymbol{\alpha}', \boldsymbol{\beta}', \gamma')$ for which the indices are restricted by $p'_j \leq p_j$ $(1 \leq j \leq 4)$ and $|\boldsymbol{\alpha}'| \leq |\boldsymbol{\alpha}|$, $|\boldsymbol{\beta}'| \leq |\boldsymbol{\beta}|$, $\gamma' \leq \gamma$ with at least one true inequality holding in these relations. Owing to (3.23) and (3.25) it suffices to consider $\boldsymbol{p}$ with $n(\boldsymbol{p}) \geq 0$, $m(\boldsymbol{p}) \geq 0$. Hence, setting $v'(\boldsymbol{p}; \boldsymbol{\alpha}, \boldsymbol{\beta}, \gamma) = 0$ and $Qv'(\boldsymbol{p}; \boldsymbol{\alpha}, \boldsymbol{\beta}, \gamma) = 0$ if $(n(\boldsymbol{p}), m(\boldsymbol{p})) \neq (1,1)$ and $(n(\boldsymbol{p}), m(\boldsymbol{p})) = (1,1)$, respectively, we obtain the coupled system of equations

$$\mathcal{L}_0(n(\boldsymbol{p}), m(\boldsymbol{p}))v(\boldsymbol{p}; \boldsymbol{\alpha}, \boldsymbol{\beta}, \gamma) + w_r(\boldsymbol{p}; \boldsymbol{\alpha}, \boldsymbol{\beta}, \gamma) = 0 \quad \text{if} \quad (n(\boldsymbol{p}), m(\boldsymbol{p})) \neq (1,1) \quad (3.30)$$

$$\hat{\mathcal{L}}_+ v(\boldsymbol{p}; \boldsymbol{\alpha}, \boldsymbol{\beta}, \gamma) + Q_+ w_r(\boldsymbol{p}; \boldsymbol{\alpha}, \boldsymbol{\beta}, \gamma) = 0 \quad \text{if} \quad n(\boldsymbol{p}) = m(\boldsymbol{p}) = 1, \quad (3.31)$$

where $\hat{\mathcal{L}}_+ := Q_+ \mathcal{L}_+|_{\mathcal{H}_c^+}$, which is invertible as well as $\mathcal{L}_0(n, m)$ for $(n, m) \neq (1, 1)$. These equations can be solved recursively, order by order. In practice one will truncate the expansion (3.21) by restricting $\boldsymbol{p}$ to $|\boldsymbol{p}| \leq p < \infty$ and for each $\boldsymbol{p}$ one fixes upper bounds for $|\boldsymbol{\alpha}|, |\boldsymbol{\beta}|, \gamma$. The bound $|\boldsymbol{p}| \leq p$ selects then a finite set of relevant harmonics (n, m) which have to be considered, for example $(0, 0)$, $(2, 0)$, $(0, 2)$, $(2, 2)$ and the critical $(1, 1)$ if $|\boldsymbol{p}| \leq 2$.

The remaining equations for $V_\pm(\sigma, \kappa)$ are obtained by replacing the projections $Q_\pm$ in (3.14) by $\langle \cdot, v_0^* \rangle$ if $m = 1$ and by $\langle \cdot, S v_0^* \rangle$ if $m = -1$. Because $\langle \mathcal{L}_+ v_{11}(\sigma, \kappa), v_0^* \rangle = 0$, these equations can be written as

$$\tilde{\mathcal{G}}[V_+, V_-] = 0 \quad \text{for } (1, 1), \quad \tilde{\mathcal{G}}[V_-^{(-)}, V_+^{(-)}] = 0 \quad \text{for } (1, -1), \quad (3.32)$$

with the functional $\tilde{\mathcal{G}}$ given by

$$\tilde{\mathcal{G}}[V_+, V_-] = \sum_{\boldsymbol{p} \in I(1,1)} \sum_{\boldsymbol{\alpha}, \boldsymbol{\beta}, \gamma} \langle v_r(\boldsymbol{p}; \boldsymbol{\alpha}, \boldsymbol{\beta}, \gamma), v_0^* \rangle V(\boldsymbol{p}; \boldsymbol{\alpha}, \boldsymbol{\beta}, \gamma), \quad (3.33)$$

and $V_\pm^{(-)}(\sigma, \kappa) = V_\pm(\sigma, -\kappa)$. The second equation in (3.32) follows by virtue of the symmetry (3.25) and w_r is the remainder term in (3.31). We note that if we set $V_\pm(\sigma, \kappa) = 4\pi^2 \hat{V}_\pm \delta(\sigma - \sigma_0)\delta(\kappa)$, i.e., $(A, B) = e^{i\sigma_0 t}(\hat{V}_+, \hat{V}_-)$, (3.33) reduces to coupled algebraic equations for $(\sigma_0, \hat{V}_+, \hat{V}_-)$. These equations coincide with the result of a Liapunov Schmidt reduction [MC76] of (2.1), viewed as an o.d.e. in the space of functions which are periodic in x with spatial period $2\pi/k_c$. The temporal period $2\pi/(\omega_c + \sigma_0)$ has then to be determined from (3.32). It is also worth noting that in (3.30), (3.31) the equations for vectors v with $|\boldsymbol{\alpha}| = |\boldsymbol{\beta}| = 0$ decouple from vectors with $|\boldsymbol{\alpha}| \neq 0$ or $|\boldsymbol{\beta}| \neq 0$. Thus the $v(\boldsymbol{p}; 0, 0, \gamma)$ can also be determined from the Liapunov Schmidt reduction for $\sigma_0 = 0$.

22

3.3 Asymptotic expansions in physical space

We now reformulate the expansions in Fourier space in terms of the physical variables (t, x). Setting

$$U(\boldsymbol{p}; \boldsymbol{\alpha}, \boldsymbol{\beta}, \gamma)(t, x) := (4\pi^2)^{-|\boldsymbol{P}|} \int d\sigma \int d\kappa \, V(\boldsymbol{p}; \boldsymbol{\alpha}, \boldsymbol{\beta}, \gamma)(\sigma, \kappa) e^{i(\sigma t + \kappa x)} \qquad (3.34)$$

we find, using the fact that Fourier inversion of a convolution yields a product,

$$U(\boldsymbol{p}; \boldsymbol{\alpha}, \boldsymbol{\beta}, \gamma)(t, x) = \prod_{j=1}^{4} \prod_{\nu=1}^{p_j} \left[(-i\partial_t)^{\alpha_{j\nu}} (-i\partial_x)^{\beta_{j\nu}} U_j(t, x) \right], \qquad (3.35)$$

where $(U_1, U_2, U_3, U_4) := (A, B, \bar{A}, \bar{B})$. By virtue of (3.11), (3.12) and (3.21) the desired representation of the solution $u(t, x)$ of (2.1) as a series expansion in $(\partial_t, \partial_x, A, B, \lambda)$ is then given by

$$
\begin{aligned}
u(t, x) = \sum_{n, m \geq 0} \sum_{\boldsymbol{p} \in I(n,m)} \sum_{\boldsymbol{\alpha}, \boldsymbol{\beta}, \gamma} \Big\{ & e^{in\omega_c t} \Big[e^{imk_c x} v(\boldsymbol{p}; \boldsymbol{\alpha}, \boldsymbol{\beta}, \gamma) U(\boldsymbol{p}; \boldsymbol{\alpha}, \boldsymbol{\beta}, \gamma)(t, x) \\
& + (1 - \delta_{m0})(-1)^{|\boldsymbol{\beta}|} e^{-imk_c x} S v(\boldsymbol{p}; \boldsymbol{\alpha}, \boldsymbol{\beta}, \gamma) U(\boldsymbol{p}_-; \boldsymbol{\alpha}_-, \boldsymbol{\beta}_-, \gamma)(t, x) \Big] \\
& + (1 - \delta_{n0}\delta_{m0}) cc \Big\},
\end{aligned}
\qquad (3.36)
$$

where cc refers to the complex conjugate expression and $(\boldsymbol{p}_-; \boldsymbol{\alpha}_-, \boldsymbol{\beta}_-)$ is defined below (3.25). The leading term $U_1(t, x)$ in (3.36) is a superposition of modulated left and right travelling waves associated with the critical wave number and frequency,

$$u_1(t, x) = e^{i\omega_c t} \left\{ e^{ik_c x} A(t, x) v_0 + e^{-ik_c x} B(t, x) S v_0 \right\} + cc. \qquad (3.37)$$

Writing the Fourier inverted versions of (3.32) as

$$\mathcal{G}[A, B] = 0, \quad \mathcal{G}[B_-, A_-] = 0 \qquad (3.38)$$

with $A_-(t, x) = A(t, -x)$, $B_-(t, x) = B(t, -x)$, the functional $\mathcal{G}$ is represented by

$$\mathcal{G}[A, B] = \sum_{\boldsymbol{p} \in I(1,1)} \sum_{\boldsymbol{\alpha}, \boldsymbol{\beta}, \gamma} \langle v_r(\boldsymbol{p}; \boldsymbol{\alpha}, \boldsymbol{\beta}, \gamma), v_0^* \rangle U(\boldsymbol{p}; \boldsymbol{\alpha}, \boldsymbol{\beta}, \gamma). \qquad (3.39)$$

Because the indices $\boldsymbol{p}$ in the expansion (3.39) satisfy $n(\boldsymbol{p}) = 1$ and $m(\boldsymbol{p}) = 1$, $\mathcal{G}$ is equivariant under two independent rotation symmetries, denoted by

$$\mathbf{S}^1 : \quad \mathcal{G}[e^{i\phi} A, e^{i\phi} B] = e^{i\phi} \mathcal{G}[A, B] \qquad (3.40)$$

$$\mathbf{SO}(2) : \quad \mathcal{G}[e^{i\Psi} A, e^{-i\Psi} B] = e^{i\Psi} \mathcal{G}[A, B]. \qquad (3.41)$$

The first symmetry (3.40) is a consequence of the temporal translation invariance of (2.1). In the leading term (3.37) a temporal translation $t \to t + \tau$ results in the phase factor $e^{i\omega_c\tau}$, whereas $A(t + \tau, x) - A(t, x) = O(\partial_t A)$ is considered as a term of higher order. Similarly, the second symmetry (3.41) corresponds to spatial phase shifts $e^{\pm ik_c x} \to e^{\pm ik_c(x+\xi)}$ and is a consequence of the spatial translation invariance of (2.1) in the infinitely extended limit. In the next section, the $\mathbf{SO}(2)$-equivariance will be broken by supplementing (3.38) by suitable boundary conditions in the finitely extended case; however, the equations (3.38) themselves still preserve this symmetry. In addition to these rotation symmetries, (3.38) is equivariant under the reflection

$$\mathbf{Z}_2 : \quad (A(t, x), B(t, x)) \to (B(t, -x), A(t, -x)) \qquad (3.42)$$

which is a consequence of the spatial reflection symmetry $u(t, x) \to Su(t, -x)$ of (2.1) and has already been incorporated in the presentation of the coupled system (3.38) of equations for (A, B). This symmetry is preserved also in the finite case $l < \infty$ because the boundary conditions (2.3) respect the reflection equivariance of (2.1).

The calculation of the first few terms $v(\mathbf{p}; \alpha, \beta, \gamma)$ from (3.30), (3.31) is straightforward following the procedure outline before. We present here the expansion of (3.21) up tp second order, assuming that σ, κ, V_+, V_- are considered as first order terms whereas λ is regarded as a term of second order. The corresponding terms for the relevant harmonics $v_{nm}(\sigma, \kappa; \lambda)$ are written as

$$\begin{aligned}
v_{11} &= (v_0 + v_\sigma\sigma + v_\kappa\kappa)V_+ \\
v_{00} &= v_{00}^{(0)}V_+ * V_+^\dagger + (Sv_{00}^{(0)})V_- * V_-^\dagger + \cdots \\
v_{20} &= v_{20}^{(0)}V_+ * V_- + \cdots \\
v_{02} &= v_{02}^{(0)}V_+ * V_-^\dagger + \cdots \\
v_{22} &= v_{22}^{(0)}V_+ * V_+ + \cdots,
\end{aligned} \qquad (3.43)$$

To simplify the notation we have labelled the vectors in (3.43) in a way that differs from the general form in terms of multiindices $(\mathbf{p}; \alpha, \beta, \gamma)$; the correspondences between these labellings should be obvious. The condition $v(\omega, k) = \bar{v}(-\omega, -k)$ implies $v_{00}^{(0)} = \bar{v}_{00}^{(0)}$ and by virtue of the reflection symmetry we have $v_{20}^{(0)} = Sv_{20}^{(0)}$, $v_{02}^{(0)} = S\bar{v}_{02}^{(0)}$. Higher order terms can be written down in the same way; here some care must be laid on which local Fourier variable belongs to which function in a convolution product. For example, a typical third order term occuring in v_{00} has the form $(v_{00}^{(\sigma)}\sigma_1 - \bar{v}_{00}^{(\sigma)}\sigma_2)V_- * V_-^\dagger$ with the convention $\sigma_1(V_- * V_-^\dagger) = (\sigma V_-) * V_-^\dagger$ and $\sigma_2(V_- * V_-^\dagger) = V_- * (\sigma V_-^\dagger)$. The

24

vectors summarized in (3.43) are those which are needed for computing the vectors w_r occurring in (3.31) up to third order. Setting

$$w_{r,11}(\sigma,\kappa) = \sum_{\boldsymbol{p}\in I(1,1)} \sum_{\boldsymbol{\alpha},\boldsymbol{\beta},\gamma} w_r(\boldsymbol{p};\boldsymbol{\alpha},\boldsymbol{\beta},\gamma)V(\boldsymbol{p};\boldsymbol{\alpha},\boldsymbol{\beta},\gamma)(\sigma,\kappa), \qquad (3.44)$$

we rewrite the third order truncation of $w_{r,11}$ in the form

$$w_{r,11} = \left[\sum_{\nu+\mu=1}^{2} w_{\nu\mu}(\imath\sigma)^\nu(\imath\kappa)^\mu - \lambda w_1\right] V_+ - 16\pi^4 \sum_{s=\pm} w_s V_+ * V_s * V_s^\dagger + \cdots. \quad (3.45)$$

In the Appendix explicit expressions for the vectors occurring in (3.43) and (3.45) are summarized in terms of the Taylor coefficients of the linear part, Eq. (3.28), and analogous coefficients of the kernels G_2, G_3. The expansion of the first equation in (3.38) follows then via Fourier inversion and scalar multiplication with $\langle\cdot,v_0^*\rangle$ from (3.45). We write the third order truncation of this equation in the form

$$A_t - c_g A_x = \left(\sum_{\nu+\mu=2} D_{\nu\mu}\partial_t^\nu\partial_x^\mu + a_0\lambda + a|A|^2 + b|B|^2\right) A + h.o.t., \qquad (3.46)$$

where, assuming that $q_{10} := \langle v_{10}, v_0^*\rangle \neq 0$,

$$D_{\nu\mu} = -\langle w_{\nu\mu}, v_0^*\rangle/q_{10}, \quad a_0 = \langle w_1, v_0^*\rangle/q_{10}, \quad a = \langle w_+, v_0^*\rangle/q_{10}, \quad b = \langle w_-, v_0^*\rangle/q_{10}$$

and h.o.t. refers to higher order terms. The equation for B follows by applying the operation $(A, B, \partial_x) \to (B, A, -\partial_x)$ to (3.46).

The quantity $c_g = -q_{10}/q_{01}$, where $q_{01} = \langle v_{01}, v_0^*\rangle$, is real and is just the critical group velocity introduced in Section 2. To see this, consider the linearized operator $\mathcal{L}(\imath\omega, \imath k, R)$ along the neutral stability curve $\omega = \omega_1(k^2)$, $R = R_{1,o}(k^2)$. Along this curve we have, for k near k_c, a solution $v_o(k)$ with $v_o(k_c) = v_0$, of the homogeneous problem

$$\mathcal{L}(\imath\omega_1(k^2), \imath k, R_{1,o}(k^2))v_o(k) = 0.$$

Differentiating this identity with respect to k at $k = k_c$ yields, recalling that $\partial R_{1,o}(k^2)/\partial k^2 = 0$,

$$\imath c_g \mathcal{L}_{10}^+ v_0 + \imath \mathcal{L}_{01}^+ v_0 + \mathcal{L}_+\tilde{v}_{01} = 0,$$

where $\mathcal{L}_{10}^+ = \mathcal{L}_{100}(1,1)$, $\mathcal{L}_{01}^+ = \mathcal{L}_{010}(1,1)$, and $\tilde{v}_{01} = \partial v_o(k_c)/\partial k$. By scalar multiplication with v_0^* we obtain, using (A.1),

$$\langle(c_g\mathcal{L}_{10}^+ + \mathcal{L}_{01}^+)v_0, v_0^*\rangle = c_g q_{10} + q_{01} = 0.$$

25

Together with (A.2), this relation tells us further that

$$\mathcal{L}_+(c_g v_\sigma + v_\kappa) = -\imath(c_g \mathcal{L}_{10}^+ + \mathcal{L}_{01}^+)v_0$$

and $Q_+ \tilde{v}_{01} = c_g v_\sigma + v_\kappa$. Analogously we find

$$
\begin{aligned}
D &\equiv D_{02} + c_g D_{11} + c_g^2 D_{20} = -\frac{\imath c_g}{2k_c} - k_c^2 \frac{\partial^2 \mu(k_c^2, R_c)}{(\partial k^2)^2} \\
&= -\frac{\imath c_g}{2k_c} + \left(\frac{\partial g_1}{\partial \mu}\right)^{-1} \left\{ 2k^2 \frac{\partial^2 g_1}{(\partial k^2)^2} + 2\imath k c_g \frac{\partial^2 g_1}{\partial \mu \partial k^2} - \frac{1}{2} c_g^2 \frac{\partial^2 g_1}{\partial \mu^2} \right\}\bigg|_c \quad (3.47)
\end{aligned}
$$

$$
a_0 = R_c \frac{\partial \mu}{\partial R}\bigg|_c = -R_c \left(\frac{\partial g_1}{\partial R} \bigg/ \frac{\partial g_1}{\partial \mu}\right)\bigg|_c, \quad (3.48)
$$

where $\mu(k^2, R)$ is the solution of $g_1(\mu, k^2, R) = 0$ with $\mu(k_c^2, R_c) = \imath \omega_c$. Hence these linear coefficients are entirely determined from the determinant $g_1(\mu, k^2, R)$.

We conclude this section by a comment on the higher order terms in (3.46). As mentioned before, this equation represents the expansion of the first functional equation of (3.38) up to third order with the understanding that λ is considered as a second order quantity. In principle it is possible to extend this expansion to any desired order. The general form for a given order can be easily written down by taking the invariances (3.40)–(3.42) into account. For example, up to fourth order we have

$$
\begin{aligned}
h.o.t. = &\left(\sum_{\nu+\mu=2} D_{\nu\mu} \partial_t^\nu \partial_x^\mu\right) A + \lambda(a_1 A_t + a_2 A_x) + (s_1 A_t + s_2 A_x)|B|^2 \\
&+ (s_3 B_t + s_4 B_x)A\bar{B} + (s_5 \bar{B}_t + s_6 \bar{B}_x)AB \\
&+ (c_1 A_t + c_2 A_x)|A|^2 + (c_3 \bar{A}_t + c_4 \bar{A}_x)A^2 + \cdots, \quad (3.49)
\end{aligned}
$$

with further complex coefficients D_{30}, D_{21}, a_1, a_2 etc. These coefficients can also be represented in the form $\langle w, v_0^* \rangle / q_{10}$ with the vectors w taken from the fourth order expansion of (3.44). Writing these vectors down explicitly by extending the computations in the appendix to third order is not a complicated task, but needs much space, and has therefore been omitted. Moreover, as we will see in the next section, generically the third order terms are sufficient for obtaining the rescaled envelope equations which govern the dominant behaviour.

26

4 Coupled Ginzburg-Landau equations and their boundary conditions

In the last section we have represented the solution of (2.1) by the series (3.36), in terms of the functions $A(t, x)$, $B(t, x)$ which are slowly varying in space and time. The functional occurring in the coupled system of equations (3.38) for (A, B) is represented in (3.39) also as an asymptotic series in $(\partial_t, \partial_x, A, B, \lambda)$, with its third order truncation given by (3.45). In the finitely extended case, $l < \infty$, the envelope equations (3.38) must be supplemented by boundary conditions for (A, B) at $x = \pm l$ which are to be determined from a boundary layer expansion of the solution of (2.1) in which the bulk solution (3.36) is extended by further terms. These terms are negligibly (i.e., exponentially) small in the interior of the interval $(l, -l)$, but have to be respected in certain boundary layers near the two ends in order that (2.3) be satisfied. As will become apparent in Subsection 4.2, in the case of finite group velocity it suffices to work with the leading order,

$$u_1(t, x) = e^{i(\omega_c t + k_c x)} A(t, x) v_0 + e^{i(\omega_c t - k_c x)} B(t, x)(S v_0) + cc, \qquad (4.1)$$

for deriving a reflection coefficient that relates A and B to each other at the boundaries. This in turn will lead to a single, nonlocal amplitude equation from which both envelopes can be obtained. In the case $c_g \approx 0$, instead, we use

$$u_2(t, x) = u_1(t, x) + \{i e^{i(\omega_c t - k_c x)} B_x(t, x)(S v_\kappa) - i e^{i(\omega_c t + k_c x)} A_x(t, x) v_\kappa + cc\}, \qquad (4.2)$$

since, as will become evident in Subsection 4.3, the second order is needed to break the translation invariance of the truncated pde-system. The omission of A_t and λA is justified in this case because $A_t = O(A_{xx}) = \lambda A$ are third order terms.

The approximations (4.1) and (4.3), respectively, must now be supplemented by suitable boundary layer correction terms. Before we pusue this, we have to specify the boundary operators in more detail. Recall from Section 2 that the linearized operator $\mathcal{L}$ leaves the finite dimensional subspace $\mathcal{H}_1$, $\dim \mathcal{H}_1 = M_1$, invariant and that all critical quantities have been derived from the restriction $\mathcal{L}_1 = \mathcal{L}|_{\mathcal{H}_1}$. In fact, noting that $v_0, v_\kappa, v_\sigma \in \mathcal{H}_1$, the approximation $u_2(t, x)$ can be viewed as the leading term of an asymptotic solution of the linearized equation $\mathcal{L}_1 u = 0$. It suffices, therefore, to supplement (4.1) by a solution $u_B(t, x)$ of this equation which is of the same order as u_2 near the boundary, but exponentially small in the bulk. Consider then a left

27

boundary operator $\mathcal{B}_-^{(\nu)}(\partial_t, \partial_x)$ which is explicitly written as

$$\mathcal{B}_-^{(\nu)}(\partial_t, \partial_x) = \sum_{i,j} \partial_t^i \partial_x^j \mathcal{B}_{ij}^{(\nu)}, \tag{4.3}$$

with a finite set of operators $\mathcal{B}_{ij}^{(\nu)}$ acting on $u \in \mathcal{H}$. We assume that for all (i,j) the range of the restriction $\mathcal{B}_{ij}^{(\nu)}|_{\mathcal{H}_1}$ is a finite dimensional space $\mathcal{H}_1^{(\nu)}$ with $\dim\mathcal{H}_1^{(\nu)} = N_\nu$. Choose a basis $(u_n)_{1 \leq n \leq M_1}$ for $\mathcal{H}_1$ and let $(u_n^*)_{1 \leq n \leq M_1}$ be the dual basis, $\langle u_n, u_m^* \rangle = \delta_{nm}$. Represent a solution of $\mathcal{L}_1 u = 0$ as

$$u(t,x) = \sum_{m=1}^{M_1} \langle u(t,x), u_m^* \rangle u_m,$$

and let

$$\mathcal{B}_{ij}^{(\nu)} u_m = \sum_{n=1}^{M_1} b_{ij,nm}^{(\nu)} u_n^{(\nu)},$$

where $(u_n^{(\nu)})_{1 \leq n \leq M_1}$ is a basis for $\mathcal{H}_1^{(\nu)}$. The boundary conditions (2.3) for $u(t,x)$ at $x = -l$ can then be reformulated in terms of a collection of $N = \sum_{\nu=1}^M N_\nu$ scalar equations

$$\sum_{m=1}^{M_1} \mathcal{B}_{nm}^{(\nu)}(\partial_t, \partial_x)\langle u(t,x), u_m^* \rangle|_{x=-l} = 0 \quad (1 \leq \nu \leq M, 1 \leq n \leq N_\nu) \tag{4.4}$$

where the

$$\mathcal{B}_{nm}^{(\nu)}(\partial_t, \partial_x) = \sum_{i,j} b_{ij,nm}^{(\nu)} \partial_t^i \partial_x^j \tag{4.5}$$

are polynomial differential operators with real coefficients. In order that the linear problem $\mathcal{L}_1 u = 0$ is compatible with the boundary conditions we have to assume further that

$$g_1(\imath\omega, k^2, R) := \det\mathcal{L}_1(\imath\omega, \imath k, R) \tag{4.6}$$

is a polynomial of degree N with respect to k^2.

The following subsection is devoted to the derivation of boundary conditions for the envelope functions (A, B). We work entirely with the restriction of the linearized problem to $\mathcal{H}_1$ for $R = R_c$ so that the equations (4.4) and (4.6) can be applied. In the case of $c_g = O(1)$ we confine the boundary analysis to the derivation of a reflection coefficient, whereas for $c_g \approx 0$ the second order leads to boundary conditions of the

Robin–type. While the analysis in Subsection 4.1 relies essentially on the slow variation of the envelopes, it is still in terms of the original variables (x, t). We make this
slow variation more explicit in Subsection 4.2 by introducing rescaled, characteristic
wave variables for the case $c_g = O(1)$. This leads to a system of partial differential equations for (A, B) with nonlocal coupling terms which was first introduced by
Knobloch and DeLuca [KD90] and was later more systematically derived by Knobloch
[Kn92]. When combined with the boundary conditions, this coupled system of equations reduces to an integro-differential equation for a single envelope that is periodic
with respect to a characteristic variable. Loosely speaking, the appearance of nonlocal terms is a consequence of the fact that in the case of large group velocities global
transport processes have to be balanced by dissipation. The perturbation expansion
for $c_g = O(1)$ is, however, only valid if the parameters occurring in the boundary
conditions satisfy certain conditions.

In Subsection 4.3 we consider the case of small group velocity. In this case the rescaled
equation system coincides more or less with the cubic truncation of (3.45) and there are
no problems with the boundary conditions. All couplings are local in both the partial
differential equations as well as in the boundary conditions. These equations have
been suggested by Cross [Cr86, Cr88] without relating them explicitly to evolution
equations. Their structure reflects, in a sense, a balance between local transport and
dissipation.

We regard the case $c_g = O(1)$ as the generic case, i.e., the underlying instability at
$R = R_c$ is viewed as a bifurcation of codimension one. In contrast, the case of small c_g
is viewed as an unfolding of a codimension-two bifurcation that occurs only if a further
parameter in the basic evolution equation is varied. Denoting this parameter at the
moment by Q, we assume that at (R_c, Q_c) an oscillatory instability occurs with $c_g = 0$.
Varying Q near Q_c induces then a family of oscillatory instabilities with critical values
$\tilde{R}_c(Q)$ and group velocities $c_g(Q)$ such that $\tilde{R}_c(Q_c) = R_c$ and $c_g(Q_c) = 0$. Generically
$dc_g(Q_c)/dQ$ will be non-zero, so that we may identify the additional parameter $Q - Q_c$
with the group velocity c_g itself. Hence, in the case of small group velocity, c_g is
considered as a free parameter which varies in a neighborhood of $c_g = 0$.

4.1   Boundary conditions for the envelopes

In order to obtain the desired boundary conditions for A, B, we have to investigate
the general solution of the linearized problem $\mathcal{L}_1 u = 0$. For $R = R_c$, all linearly

independent solutions with temporally oscillating behavior $e^{\imath\omega_c t}$ follow from

$$P(k^2) := g_1(\imath\omega_c, k^2, R_c) = 0, \tag{4.7}$$

which is a polynomial equation in k^2 of degree N with complex coefficients. One solution of (4.7) is $k^2 = k_c^2$ and there are, by assumption, no further real, positive solutions. Differentiating the identity

$$g_1(\imath\omega_1(k^2), k^2, R_{1,o}(k^2)) = 0$$

with respect to k^2 and evaluating the result at $k^2 = k_c^2$ yields

$$-\frac{c_g}{2k_c}\frac{\partial g_1}{\partial\imath\omega}|_c = \frac{\partial g_1}{\partial k^2}|_c = \frac{dP(k_c^2)}{dk^2}.$$

It follows that if $c_g \neq 0$, there are generically $N-1$ further zeroes $k_j^2 \neq k_c^2$ ($1 \leq j \leq N-1$) of $P(k^2)$ with either $k_j^2 < 0$ or $\Im k_j^2 \neq 0$. Contrary to that we have for $c_g = 0$ only $N-2$ further zeroes k_j^2 ($1 \leq j \leq N-2$) which are negative or have nonzero imaginary parts. We assume genericity in the sense that all $k_j^2 \neq k_c^2$ are distinct, i.e., they do not induce secular terms in the solution of $\mathcal{L}_1(\imath\omega_c, \partial_x, R_c)u = 0$. Note, however, that $xe^{\imath k_c x}v_{01}$ is an exact solution of this equation provided $c_g = 0$.

To each k_j^2 with $\Im k_j^2 \neq 0$ or $k_j^2 < 0$ there correspond two complex wave numbers, one with a positive and the other with a negative imaginary part. For the boundary correction of (4.1) near the left boundary we choose the square root k_j of k_j^2 such that $\Im k_j > 0$. Let w_j, w_j^* with $\langle w_j, w_j^* \rangle = 1$ be the vectors associated with k_j according to

$$\mathcal{L}_1^{(j)}w_j = 0, \quad \mathcal{L}_1^{*(j)}w_j^* = 0, \tag{4.8}$$

where a superscript (j) indicates that the corresponding expression is evaluated at $(\imath\omega_c, \imath k_j, R_c)$. Assuming that

$$s_j := -\langle \partial_{\imath\omega}\mathcal{L}_1^{(j)}w_j, w_j^* \rangle / \langle \partial_{\imath k}\mathcal{L}_1^{(j)}w_j, w_j^* \rangle = \frac{1}{2\imath k}\frac{\partial g_1/\partial\mu}{\partial g_1/\partial k^2}\bigg|_j \tag{4.9}$$

is finite, the boundary correction term can be written as

$$u_B(t, x) = \sum_{j=1}^{\tilde{N}} e^{\imath(\omega_c t + k_j \tilde{x})}F_j(\xi_j)w_j, \tag{4.10}$$

where

$$\xi_j = t + s_j\tilde{x}, \quad \tilde{x} = x + l, \tag{4.11}$$

30

$$\tilde{N} = N - 1 \text{ if } c_g = O(1), \quad \tilde{N} = N - 2 \text{ if } c_g \approx 0, \tag{4.12}$$

and $F_j(z)$ is an arbitrary complex analytic function that is slowly varying with respect to $z \in \mathbb{C}$. Here the usual interpretation that $F_j \gg dF_j/dz \gg d^2F_j/dz^2$ was made which, in particular, means that $dF_j/dz = O(A_t)$ and is thus a third order term in the case $c_g \approx 0$. Moreover, because $\Im k_j > 0$, F_j is exponentially small in the interior of the interval $(-l, l)$, but cannot be neglected near the left boundary $\tilde{x} = 0$. Inserting the boundary layer solution $u_2 + u_B$ into the left boundary conditions (4.4) yields a system of N linear equations for A, A_x, B, B_x, F_j at $\tilde{x} = 0$,

$$a_{n0}^{(\nu)} A + a_{n1}^{(\nu)} A_x + b_{n0}^{(\nu)} B + b_{n1}^{(\nu)} B_x + \sum_{j=1}^{\tilde{N}} c_{nj}^{(\nu)} F_j = 0, \tag{4.13}$$

for $1 \leq \nu \leq M$, $1 \leq n \leq N_\nu$, with the coefficients given by

$$a_{n0}^{(\nu)} = e^{-\imath k_c l} \sum_{m=1}^{M_1} \langle v_0, u_m^* \rangle \mathcal{B}_{nm}^{(\nu)}(\imath \omega_c, \imath k_c)$$

$$b_{n0}^{(\nu)} = e^{\imath k_c l} \sum_{m=1}^{M_1} \langle S v_0, u_m^* \rangle \mathcal{B}_{nm}^{(\nu)}(\imath \omega_c, -\imath k_c)$$

$$c_{nj}^{(\nu)} = \sum_{m=1}^{M_1} \langle w_j, u_m^* \rangle \mathcal{B}_{nm}^{(\nu)}(\imath \omega_c, \imath k_j)$$

$$a_{n1}^{(\nu)} = e^{-\imath k_c l} \sum_{m=1}^{M_1} (\langle v_0, u_m^* \rangle \frac{\partial}{\partial \imath k} - \imath \langle v_\kappa, u_m^* \rangle) \mathcal{B}_{nm}^{(\nu)}(\imath \omega_c, \imath k_c)$$

$$b_{n1}^{(\nu)} = -e^{\imath k_c l} \sum_{m=1}^{M_1} (\langle S v_0, u_m^* \rangle \frac{\partial}{\partial \imath k}) + \imath \langle S v_\kappa, u_m^* \rangle) \mathcal{B}_{nm}^{(\nu)}(\imath \omega_c, -\imath k_c) \tag{4.14}$$

In the case $c_g = O(1)$ the terms involving A_x, B_x in (4.13) will be omitted, since here we confine ourselfes to the leading order which is determined already from $u_1 + u_B$. More precisley, a second order expansion would involve here also A_t, B_t and dF_j/dz. In contrast, $u_2 + u_B$ with u_2 as defined in (4.2) (A_t, B_t neglected) is sufficient for the second order expansion in the case $c_g \approx 0$. We now discuss the cases $c_g = O(1)$ and $c_g \approx 0$ in succession.

(1) The case $c_g = O(1)$. The derivation of a full set of boundary conditions in this case is much more complicated than for $c_g \approx 0$. Nevertheless, it is obvious from a physical point of view that the dominant behaviour near the boundaries should be governed by

31

a reflection coefficient relating the amplitudes of the left and right traveling waves to each other at the side walls. We focus here on a derivation of the reflection coefficient which involves only the leading terms of the boundary layer expansion.

Omitting the derivatives A_x, B_x in (4.13), we write this equation in a compact matrix–vector form as

$$Aa_0 + Bb_0 + \mathbf{CF} = 0,\tag{4.15}$$

where the N coefficients $a_{n0}^{(\nu)}$ and $b_{n0}^{(\nu)}$ have been combined to complex column vectors $\mathbf{a}_0 = (a_{n0}^{(\nu)})$ and $\mathbf{b}_0 = (b_{n0}^{(\nu)})$ in $\mathcal{C}^N$, respectively. The matrix $\mathbf{C}$ denotes the $N \times (N-1)$-matrix whose columns are formed by the $\tilde{N} = N - 1$ vectors $\mathbf{c}_j = (c_{nj}^{(\nu)})$, and in $\mathbf{F}(t) \in \mathcal{C}^{N-1}$ the functions $F_j(t)$ are collocated. We assume that the matrix $\mathbf{C}$ has full rank $N-1$, whence for a solution $\mathbf{F}$ to exist $A\mathbf{a}_0 + B\mathbf{b}_0$ must satisfy the solvability condition

$$(\mathbf{a}_0, \mathbf{c}_0)A + (\mathbf{b}_0, \mathbf{c}_0)B + \cdots = 0,\tag{4.16}$$

where $\mathbf{c}_0 \in \mathcal{C}^N$ spans the kernel of $\bar{\mathbf{C}}^T$ and $(.,.)$ denotes here the standard unitary product in $\mathcal{C}^N$. Assuming that $(\mathbf{b}_0, \mathbf{c}_0) \neq 0$, (4.16) describes the interaction of the left and right traveling waves at the left boundary, i.e.,

$$B(t, -l) = rA(t, -l) + \cdots .\tag{4.17}$$

Here, $r = -(\mathbf{a}_0, \mathbf{c}_0)/(\mathbf{b}_0, \mathbf{c}_0) \in \mathcal{C}$ is referred to as the reflection coefficient and the dots denote as usual terms of higher order involving derivatives and nonlinear terms in (A, B). The relected version of (4.17) may be written as

$$A(t, l) = rB(t, l) + \cdots .\tag{4.18}$$

In the next subsection (4.17) and (4.18) will be combined with the evolution equations for A and B, yielding a single, nonlocal equation that governs both envelopes. In this subsection we also address the question for further boundary conditions.

(2) The case $c_g \approx 0$. As will be shown in Subsection 4.3, the case $c_g \approx 0$ leads to a system of local, coupled partial diffential equations for the leading terms of the envelopes A, B. These equations exhibit a full translation symmetry, thus the breaking of the translational invariance has to be incorporated through its boundary conditions. It turns, however, that to leading order the boundary conditions still preserve translation invariance so that here an expansion up to second order is required. We use, therefore, the full equation (4.13), which is a consistent second order expansion

in this case because A_t is of higher order than A_x. We write (4.13) again as

$$A\mathbf{a}_0 + A_x\mathbf{a}_1 + B\mathbf{b}_0 + B_x\mathbf{B}_1 + \mathbf{CF} = 0, \tag{4.19}$$

where $\mathbf{a}_1, \mathbf{b}_1 \in \mathbb{C}^N$ are defined in the same way in terms of the a_{n1}^ν, b_{n1}^ν as $\mathbf{a}_0$ and $\mathbf{b}_0$. The vector $\mathbf{F}(t)$ is now in $\mathbb{C}^{N-2}$ and $\mathbf{C}$ is a $N \times (N-2)$-matrix which we assume again to have full rank. Let the kernel of $\bar{\mathbf{C}}^T$ be spanned by two linearly independent vectors $\mathbf{c}_1, \mathbf{c}_2 \in \mathbb{C}^N$. The equation (4.19) can be solved for $\mathbf{F}$, provided the solvability conditions

$$(\mathbf{a}_0, \mathbf{c}_j)A + (\mathbf{b}_0, \mathbf{c}_j)B + (\mathbf{a}_1, \mathbf{c}_j)A_x + (\mathbf{b}_1, \mathbf{c}_j)B_x = 0 \quad (j = 1, 2) \tag{4.20}$$

are satisfied. Assuming that $(\mathbf{a}_0, \mathbf{c}_1)(\mathbf{b}_0, \mathbf{c}_2) \neq (\mathbf{b}_0, \mathbf{c}_1)(\mathbf{a}_0, \mathbf{c}_2)$, we can rewrite (4.20) in the form

$$A(t, -l) = \alpha A_x(t, -l) + \beta B_x(t, -l), \quad B(t, -l) = \gamma A_x(t, -l) + \delta B_x(t, -l), \tag{4.21}$$

where $(\alpha, \beta, \gamma, \delta)$ are genuine complex numbers which can be calculated from the coefficients occurring in (4.20). By virtue of the reflection symmetry, the analogous boundary conditions at the right end are given by

$$B(t, l) = -\alpha B_x(t, l) - \beta A_x(t, l), \quad A(t, l) = -\gamma B_x(t, l) - \delta A_x(t, l). \tag{4.22}$$

As usual, (4.21) and (4.22) are to be interpreted as the first two leading terms of the boundary conditions associated with the harmonic $e^{iw_c t}$. Higher orders involve also t-derivatives and nonlinear terms.

4.2 Nonlocal Ginzburg-Landau equations

We discuss now the case in which $c_g = O(1)$ in more detail using suitable rescalings. Our exposition is closely related to that of Knobloch and DeLuca [KD90], but focuses more on the implications that follow from the boundary conditions in the bounded case $l < \infty$. If c_g is a finite and nonzero quantity of order $O(1)$, the first order truncation of (3.45) reads $A_t - c_g A_x \approx 0$; thus the dominant spatio-temporal variation of A is captured by its dependence on the characteristic wave variable $c_g t + x$. Following [KD90] we introduce two slow time variables

$$\tau = \epsilon t, \quad T = \epsilon^2 t, \tag{4.23}$$

33

where $0 < \epsilon \ll 1$ and associate to τ the scaled characteristic variables

$$X = c_g\tau + \epsilon x, \quad Y = c_g\tau - \epsilon x. \tag{4.24}$$

The solution $(A(t, x), B(t, x))$ of (3.45) and the corresponding reflected equation is expanded as

$$
\begin{align}
A(t,x) &= \epsilon A_1(X,T) + \epsilon^2 A_2(X,Y,T) + \epsilon^3 A_3(X,Y,T) + \ldots \tag{4.25}\\
B(t,x) &= \epsilon B_1(Y,T) + \epsilon^2 B_2(Y,X,T) + \epsilon^3 B_3(Y,X,T) + \ldots, \tag{4.26}
\end{align}
$$

i.e., A_1 and B_1 depend only on (X,T) and (Y,T), respectively, so that $A_t - c_g A_x = O(\epsilon^3) = B_t + c_g B_x$. We require that the functions A_2, A_3 etc. are periodic with respect to Y and, similarly, that B_2, B_3 etc. are periodic with respect to X with the same period $2L$ which will be determined from the boundary conditions. When (4.25), (4.26) are substituted into (3.45) and its reflected equation, we obtain at order $O(\epsilon^3)$,

$$
\begin{align}
2c_g A_{2Y} &= -A_{1T} + DA_{1XX} + a_0\Lambda A_1 + (a|A_1|^2 + b|B_1|^2)A_1 \tag{4.27}\\
2c_g B_{2X} &= -B_{1T} + DB_{1YY} + a_0\Lambda B_1 + (a|B_1|^2 + b|A_1|^2)B_1 \tag{4.28}
\end{align}
$$

where $D = D_{20}c_g^2 + D_{11}c_g + D_{02}$ can be calculated from g_1, Eq. (3.47), and Λ is the rescaled bifurcation parameter,

$$\lambda = \epsilon^2\Lambda$$

which is considered as an $O(1)$-quantity. The equations (4.27), (4.28) provide the starting point for deriving coupled nonlinear Ginzburg-Landau equations with non-local coupling terms. Namely, the condition that (4.27) admits a solution A_2 that is periodic in Y requires that the average of the r.h.s. of (4.27) with respect to Y vanishes. Denoting by

$$\langle f \rangle_\xi := \frac{1}{2L}\int_{-L}^{L} f(\xi)d\xi$$

the average of any function that is $2L$-periodic in the variable ξ, the solvability conditions for A_2 and B_2 induce the nonlocal system of coupled equations,

$$
\begin{align}
A_{1T} &= \left\{D\partial_X^2 + a_0\Lambda + b\langle|B_1|^2\rangle_Y + a|A_1|^2\right\} A_1 \tag{4.29}\\
B_{1T} &= \left\{D\partial_Y^2 + a_0\Lambda + b\langle|A_1|^2\rangle_X + a|B_1|^2\right\} B_1. \tag{4.30}
\end{align}
$$

Clearly, in order that the averages are well defined, we must assume that $|A_1|^2$ and $|B_1|^2$ are $2L$-periodic with respect to X and Y, respectively. This does not, however, necessarily mean that A_1 and B_1 themselves are periodic. That asymptotic expansions of nonlinear wave problems lead to nonlocal couplings between left and right going waves has already been observed earlier (see [CK72, Ec75] and Chapter 4.4 in [KC81]). The description of oscillatory instabilities in terms of the nonlocal system of Ginzburg-Landau-type equations (4.29), (4.30) has been first established by Knobloch and DeLuca [KD90] and was further analyzed later by Vega in [Ve93].

Consider now the boundary conditions (4.17) and (4.18). If we insert (4.25) into these equations, the $O(\epsilon)$-order leads here to

$$A_1(c_g\tau + \epsilon l, T) = rB_1(c_g\tau - \epsilon l, T), \quad B_1(c_g\tau + \epsilon l, T) = rA_1(c_g\tau - \epsilon l, T),$$

i.e.,

$$A_1(X, T) = rB_1(X - 2\epsilon l, T), \quad B_1(Y, T) = rA_1(Y - 2\epsilon l, T).$$

In the following we assume that the boundaries are perfectly reflecting, i.e., there is no energy loss when an incoming wave is reflected at a side wall. This means that $|r|^2 = 1$ so that r may be written as

$$r = e^{2\imath K\epsilon l}, \ K \in \mathbb{R},$$

whereby ϵl is regarded as $O(1)$. To $O(\epsilon)$ the boundary conditions then become

$$A_1(c_g\tau + \epsilon l, T) = e^{2\imath K\epsilon l}B_1(c_g\tau - \epsilon l, T), \quad B_1(c_g\tau + \epsilon l, T) = e^{2\imath K\epsilon l}A_1(c_g\tau - \epsilon l, T),$$

which is equivalent to

$$A_1(X + 4\epsilon l, T) = e^{4\imath K\epsilon l}A_1(X, T), \quad A_1(X + 2\epsilon l, T) = e^{2\imath K\epsilon l}B_1(X, T),$$

$$B_1(Y + 4\epsilon l, T) = e^{4\imath K\epsilon l}B_1(Y, T), \quad B_1(Y + 2\epsilon l, T) = e^{2\imath K\epsilon l}A_1(Y, T).$$

It follows that

$$A_1(X, T) = e^{\imath KX}G(X, T), \quad B_1(Y, T) = e^{\imath KY}G(Y + L, T) \qquad (4.31)$$

where $G(X, T)$ is $2L$-periodic in X, $G(X + 2L, T) = G(X, T)$, with $L = 2\epsilon l$. Both equations (4.27), (4.28) reduce then to the single, nonlocal Ginzburg-Landau equation

$$G_T = \left\{ D(\partial_X + \imath K)^2 + a_0\Lambda + a|G|^2 + b\langle|G|^2\rangle \right\} G, \qquad (4.32)$$

where the subscript X at the average has now been suppressed because we have only one periodic function at leading order.

35

A similar self–consistent system of equations can be also obtained at the next order in the perturbation expansion, provided the full expansion of (4.17) (including the dots) does not contain terms which are linear in the first order derivatives of A and B. We omit here the details of the derivatin, but state the resulting equations for A_2 and B_2. The solvability conditions for A_3 and B_3 induce here a representation of A_2 and B_2 in the form

$$
\begin{aligned}
A_2(X,Y,T) &= H(Y+L,T)e^{\imath KX}G(X,T) + A_{2m}(X,T), & (4.33) \\
B_2(Y,X,T) &= H(X,T)e^{\imath KY}G(Y+L,T) + B_{2m}(Y,T), & (4.34)
\end{aligned}
$$

where

$$
H(X,T) = \frac{b}{2c_g}\left\{\frac{1}{2L}\int_{-L}^{L}X'|G(X',T)|^2 dX' + (L-X)\langle|G|^2\rangle - \int_{X}^{L}|G(X',T)|^2 dX'\right\}
$$

is $2L$-periodic in X with zero mean, $\langle H\rangle = 0.$, i.e., we have isolated the Y-average of A_2 in the function $A_{2m}(X,T)$. This function and its reflected version $B_{2m}(Y,T)$ can then be expressed by a single $2L$-periodic function $G_2(X,T)$ as

$$
A_{2m}(X,T) = e^{\imath KX}G_2(X,T), \quad B_{2m}(Y,T) = e^{\imath KY}G_2(Y+L,T),
$$

which satisfies the linear, nonlocal and inhomogeneous equation

$$
\begin{aligned}
G_{2T} =\; & \left\{D(\partial_X + \imath K)^2 + a_0\Lambda - K^2 + 2a|G|^2 + b\langle|G|^2\rangle\right\}G_2 \\
& + aG^2\bar{G}_2 + b\langle\bar{G}G_2 + G\bar{G}_2\rangle G + 2b_r\langle|G|^2\rangle HG \\
& + e^{-\imath KX}\left\{D_1\partial_X\partial_T + D_3\partial_X^3 + b_1(\Lambda + d_1|G|^2 + e_1\langle|G|^2\rangle)\partial_X\right\}e^{\imath KX}G_2 \\
& + d_2 G^2 e^{\imath KX}\partial_X(e^{-\imath KX}\bar{G}) + (e_2 - e_3)\langle\bar{G}G_X\rangle G,
\end{aligned}
$$

where $b_r = \Re b$, $D_1 = 2D_{20}c_g + D_{11}$, $D_2 = D_{20}c_g^2 - D_{02}$, and

$$
D_3 = D_{30}c_g^3 + D_{21}c_g^2 + D_{12}c_g + D_{03}, \quad b_1 = a_1 c_g + a_2, \quad d_1 = c_1 c_g + c_2,
$$

$$
d_2 = c_3 c_g + c_4, \quad e_1 = s_1 c_g + s_2, \quad e_2 = (s_3 - s_5)c_g + s_6 - s_4,
$$

are determined by the fourth order coefficients in the equations for A and B (see (3.49)).

So far the analysis has been confined to perfectly reflecting boundaries. The general case, $|r|^2 \neq 1$, is rather complicated. In a recent paper Martel and Vega [MV96] derive, by means of a matching procedure, an additional set of boundary conditions

36

for a restricted class of evolution equations. These boundary conditions involve both derivatives and nonlinear terms of the envelopes. Similarly as here, the authors embed the boundary value problems for A_1, A_2, B_1, B_2 into nonlocal equations for functions that are periodic in the rescaled wave variables. For $|r|^2 \neq 1$, they are then forced to take a discontinuity in the derivative of the second order function (G_2 in our notation) into account. Moreover, they do not incorporate the next order in the first set of boundary conditions (Eq. (4.17) which also should have an effect on the second order terms. There are, therefore, still some open questions in the general case that deserves further study.

4.3 Local Ginzburg-Landau equations

If c_g varies in a neighborhood of $c_g = 0$, one introduces the slow variables X, T and the rescaled quantities s and Λ by

$$X = \epsilon x, \quad t = \epsilon^2 T, \quad c_g = \epsilon s, \quad \lambda = \epsilon^2 \Lambda, \tag{4.35}$$

where $L = \epsilon l = O(1)$ and expands

$$\begin{align}
A(t,x) &= \epsilon A_1(T, X) + \epsilon^2 A_2(T, X) + \cdots \tag{4.36}\\
B(t,x) &= \epsilon B_1(T, X) + \epsilon^2 B_2(T, X) + \cdots . \tag{4.37}
\end{align}$$

The orders $O(\epsilon^3)$ in (3.45) and $O(\epsilon)$ in the boundary conditions (4.21), (4.22) become then

$$A_{1T} - sA_{1X} = \left\{ D_{02}\partial_X^2 + a_0\Lambda + a|A_1|^2 + b|B_1|^2 \right\} A_1, \tag{4.38}$$

$$A_1(T, -L) = A_1(T, L) = 0, \tag{4.39}$$

and the B_1-equations follow from (4.38), (4.39) as usual by the replacements

$$(A_1, B_1, \partial_X) \to (B_1, A_1, -\partial_X)$$

. At the next order we find

$$\begin{align}
A_{2T} - sA_{2X} =\ & \left\{ D_{02}\partial_X^2 + a_0\Lambda + 2a|A_1|^2 + b|B_1|^2 \right\} A_2 + aA_1^2\bar{A}_2\\
&+ b(B_1\bar{B}_2 + \bar{B}_1 B_2)A_1 + D_{11}A_{1XT} + D_{03}A_{1XXX} + c_4 A_1^2\bar{A}_{1X}\\
&+ (a_2\Lambda + s_2|B_1|^2 + c_2|A_1|^2)A_{1X}\\
&+ s_4 A_1\bar{B}_1 B_{1X} + s_6 A_1 B_1\bar{B}_{1X} \tag{4.40}
\end{align}$$

$$A_2 = \alpha A_{1X} + \beta B_{1X} \text{ at } X = -L; \quad A_2 = -\gamma B_{1X} - \delta A_{1X} \text{ at } X = L, \qquad (4.41)$$

with the additional coefficients following from the fourth order terms (see (3.49)). The equations (4.38) through (4.41) and their reflected versions are the correct asymptotic equations up to second order. Alternatively one may write

$$A(t, x) = \tilde{A}(T, X; \epsilon) = \epsilon A_1(T, X) + \epsilon^2 A_2(T, X) + \cdots$$

$$B(t, x) = \tilde{B}(T, X; \epsilon) = \epsilon B_1(T, X) + \epsilon^2 B_2(T, X) + \cdots$$

and work directly with the asymptotic expansion (3.45). Omitting the tilde in $(\tilde{A}, \tilde{B})$, this yields

$$
\begin{aligned}
A_T - s A_x &= \left\{ D_{02} \partial_X^2 + a_0 \Lambda + a|A|^2 + b|B|^2 \right\} A + O(\epsilon) & (4.42) \\
B_T + s B_x &= \left\{ D_{02} \partial_X^2 + a_0 \Lambda + a|B|^2 + b|A|^2 \right\} B + O(\epsilon) & (4.43)
\end{aligned}
$$

with the boundary conditions

$$A = \epsilon \alpha A_X + \epsilon \beta B_X + O(\epsilon^2), \quad B = \epsilon \gamma A_X + \epsilon \delta B_X + O(\epsilon^2) \text{ at } X = -L \qquad (4.44)$$

$$B = -\epsilon \alpha B_X - \epsilon \beta A_X + O(\epsilon^2), \quad A = -\epsilon \gamma B_X - \epsilon \delta A_X + O(\epsilon^2) \text{ at } X = L. \qquad (4.45)$$

A natural approximation for A, B is then given by the omission of the $O(\epsilon)$-terms in (4.42), (4.43) and of the $O(\epsilon^2)$-terms in (4.44), (4.45). The reason why we keep the $O(\epsilon)$-terms in the boundary conditions, but neglect them in the p.d.e.s, is the following. To any order in ϵ the p.d.e-system (4.42), (4.43) inherits the two rotation symmetries $\mathbf{S}^1$ and $\mathbf{SO}(2)$ and a reflection symmetry $\mathbf{Z}_2$, which are defined by the operations

$$\mathbf{S}^1: \quad A \to e^{\iota\phi} A, \ B \to e^{\iota\phi} B \qquad (4.46)$$

$$\mathbf{SO}(2): \quad A \to e^{\iota\psi} A, \ B \to e^{-\iota\psi} B \qquad (4.47)$$

$$\mathbf{Z}_2: \quad (A, B, X) \to (B, A, -X). \qquad (4.48)$$

The $\mathbf{S}^1$-invariance corresponds to temporal translations $t \to t + \phi/\omega_c$ and the $\mathbf{SO}(2)$-invariance corresponds to spatial translations $x \to x + \psi/k_c$ in the basic harmonics $e^{\iota(\omega_c t \pm k_c x)}$. Both invariances hold for the original evolution equation in the unbounded limit $l = \infty$. The "presence of sidewalls", $l < \infty$, unavoidably breaks the spatial translation symmetry, although this breaking is not revealed in the p.d.e-system (4.42), (4.43). These p.d.e.s exhibit all three invariances (4.46)–(4.48) at any finite order in ϵ. The first breaking of $\mathbf{SO}(2)$ occurs at $O(\epsilon)$ in the boundary conditions (4.44), (4.45); thus these terms should have a much stronger effect than the $O(\epsilon)$-terms in the p.d.e.s (4.42), (4.43). The implications of this symmetry breaking on the behavior of the solutions will be discussed in more detail in Section 6.

5   Convection in a vertical magnetic field

Double diffusive convection systems, which are characterized by the presence of a second diffusion process besides the heat diffusion, provide a number of analytically tractable bifurcation problems of physical interest [BLN86, CFT85, ND88, NLHGD87]. A common feature of these systems is the possibility of an oscillatory instability of the heat conduction state when the Rayleigh number reaches its critical value [Ch61, KnP81]. Such an instability may occur if a stabilizing effect is present that competes with the destabilizing heating and if the second diffusion process is slower than the heat diffusion. In binary mixtures, for example, a sufficiently negative Soret coefficient induces a stabilizing force, because then the component with the lower density tends to migrate to the lower boundary. Similarly, in thermohaline convection a solute gradient imposed at the boundaries acts stabilizing if the bottom concentration is larger than the top concentration. Another example is provided by convection of an electrically conducting fluid in the presence of an external magnetic field which will be discussed below. Here again the Lorentz force associated with a vertically oriented field acts as a stabilizing effect on the heat conduction state.

When attention is restricted to two dimensional motions with periodic boundary conditions in the horizontal, the partial differential equations describing a double diffusive convection system acquire $\mathbf{O}(2)$-symmetry. Near an oscillatory instability threshold a center manifold reduction and normal form transformation [GH90] then leads to the spatially independent version of the envelope equations (3.45) with the leading terms characterized by the complex coefficients a_0, a and b. In this framework a and b have been calculated analytically for both binary mixtures [Kn86b] and the thermohaline problem [DKT87] for idealized, i.e. stress free, boundary conditions at the top and bottom. The problem with these coefficients is that the real part of a (denoted b in [Kn86b] and [CK91]) vanishes in both cases. This creates a degeneracy which should be understood as an artifact of the Boussinesq approximation combined with idealized boundary conditions. Another degeneracy that occurs in these equations is that the neutral stationary and the neutral oscillatory stability curves attain their minima at the same wave number $k_c^2 = \pi^2/2$. For a distinguished value of another parameter the two minima may then coalesce, which results in a degenerate bifurcation of the Takens-Bogdanov type [DaK87b, Gu86, KnP81]. The appearance of this bifurcation as a codimension two phenomenon is also non-generic. Generically one expects the coincidence of the two minima only if three parameters are fixed.

Both degeneracies are removed if the stress free boundary conditions are replaced by non-slip (rigid) ones. Numerical calculations for the rigid case show indeed that $\Re a < 0$ [SW89] and that the critical wave numbers are distinct when the minima of the two neutral stability curves occur at the same height ($R_{c,o} = R_{c,s}$). However, for these realistic boundary conditions even the linearized Boussinesq equations are not tractable analytically.

In order to illustrate the general techniques developed so far, we investigate in this section the Boussinesq equations for two dimensional convection in a vertical magnetic field. The analytic approach to this problem is made possible by the imposition of stress free boundary conditions at the top and bottom. In contrast to binary mixtures and thermohaline convection, these boundary conditions do not lead here to non-generic degeneracies. We begin, in Subsection 5.1, by introducing the basic system of equations and its Fourier transform. The linear stability analysis is carried out in Subsection 5.2. This analysis shows in particular that the critical wave numbers vary with the physical parameters and that the minima of the two neutral stability curves cannot coincide. The boundary conditions for the envelopes and the nonlinear coefficients are considered in Subsections 5.3 and 5.4, respectively. For small group velocity we find here a certain kind of non-genericity in the boundary conditions for the envelopes if the *horizontal* boundary conditions for the fluid are also stress free.

Previous nonlinear investigations [DaK86, KnWD81, KnP81, Na86] of two dimensional convection in a vertical magnetic field rely on center manifold and normal form techniques. In [KnWD81] a five mode system for a finite box is studied analytically and numerically. In particular, conditions are set up which guarantee the existence of a Hopf bifurcation from the basic state. The results of [KnWD81] are extended in [KnP81] by a thorough study of a Takens-Bogdanov bifurcation. In [Na86] and [DaK86] a center manifold reduction and normal form analysis for an infinitely extended system with periodic boundary conditions in the horizontal has been performed for, respectively, a Hopf bifurcation and a Takens-Bogdanov bifurcation. In these papers the spatial period, i.e., the wave number has been chosen arbitrarily, so that minimality of the critical Rayleigh number was not assured.

40

5.1 Basic equations

In non dimensional units the Boussinesq equations describing two dimensional convection in a vertical, external magnetic field are [We77]

$$\frac{1}{\sigma} \left[\nabla^2 \psi_t + J(\psi, \nabla^2 \psi) \right] = \pi^4 R \Theta_x + \nabla^4 \psi + \pi^2 \xi Q J(x + \mathcal{A}, \nabla^2 \mathcal{A})$$
$$\Theta_t + J(\psi, \Theta) = \psi_x + \nabla^2 \Theta$$
$$\mathcal{A}_t + J(\psi, \mathcal{A}) = \psi_y + \xi \nabla^2 \mathcal{A}. \tag{5.1}$$

Here, $\psi(t, x; y)$ is the stream function, and $\Theta(t, x; y)$ and $\mathcal{A}(t, x; y)$ denote departures of the temperature and magnetic potential from the basic state. The bilinear operator J refers to the Jacobian, $J(f, g) = f_x g_y - f_y g_x$ of two functions f and g depending on the horizontal coordinate x and the vertical coordinate y, $0 \leq y \leq 1$. The actual magnetic field $\mathbf{B}$ is given by $\mathbf{B} = B_0(-\mathcal{A}_y, 1 + \mathcal{A}_x)$, where B_0 is the externally imposed magnetic field strength, and $(-\psi_y, \psi_x)$ is the velocity field. The equations (5.1) contain four dimensionless parameters, the Rayleigh number $\pi^4 R$, the Chandrasekhar number $\pi^2 Q$ measuring the strength of the applied field ($Q \sim B_0^2$), and the Prandtl numbers σ and ξ which are, respectively, the ratios of viscous and magnetic diffusivities to the thermal diffusivity. All four parameters have to be positive. The scaling factors π^4 and π^2 in the Rayleigh and Chandrasekhar numbers have been introduced for convenience in the calculation. The height of the fluid conveying box is rescaled to unity so that boundary conditions at $y = 0$ and $y = 1$ have to be imposed. These are chosen as stress free,

$$\psi = \psi_{yy} = \Theta = \mathcal{A}_y = 0 \text{ at } y = 0, 1. \tag{5.2}$$

In the finitely extended case, $-l \leq x \leq l$, we have also to impose horizontal boundary conditions. We consider both stress free boundary conditions,

$$\psi = 0, \ \psi_{xx} = 0, \ \Theta_x = 0, \ \mathcal{A} = 0 \text{ at } x = \pm l, \tag{5.3}$$

and non-slip boundary conditions in which case the first two equations in (5.3) have to be replaced by

$$\psi_y = 0, \ \psi_x = 0 \text{ at } x = \pm l, \tag{5.4}$$

whereas the other two equations remain the same. The condition for Θ means in particular that no heat flux takes place across the horizontal boundaries, i.e., the box

41

is perfectly isolating. For both types of horizontal boundary conditions the system is invariant under the reflection

$$(x, \psi, \Theta, \mathcal{A}) \to (-x, -\psi, \Theta, -\mathcal{A}).$$

We now rewrite (5.1) with the boundary conditions (5.2) in the form used in Sections 2–4. Our basic space $\mathcal{H}$ consists of vector valued functions

$$u : [0, 1] \to I\!\!R^3, \quad u(y) = (\psi(y), \Theta(y), \mathcal{A}(y))^T,$$

which satisfy the vertical boundary conditions (5.2), and

$$\langle u_1, u_2 \rangle = \int_0^1 dz \, (\psi_1 \bar{\psi}_2 + \Theta_1 \bar{\Theta}_2 + \mathcal{A}_1 \bar{\mathcal{A}}_2),$$

if $u_1 \in \mathcal{H}_c = \mathcal{H} + \imath \mathcal{H}$ and $u_2 \in \mathcal{H}_c^*$. The reflection operator S acts in this space according to

$$Su = (-\psi, \Theta, -\mathcal{A})^T.$$

The Fourier transformed equation (3.9) becomes

$$\mathcal{L}(\imath \omega, \imath k, R)v + \mathcal{M}_2[v, v] = 0, \tag{5.5}$$

with the linearized part

$$\mathcal{L}(\imath \omega, \imath k, R) = \begin{pmatrix} \Delta^2(k^2) - \frac{\imath \omega}{\sigma} \Delta(k^2) & \pi^4 R \imath k & \pi^2 \xi Q \Delta(k^2) \partial_y \\ -\imath k & \imath \omega - \Delta(k^2) & 0 \\ -\partial_y & 0 & \imath \omega - \xi \Delta(k^2) \end{pmatrix}. \tag{5.6}$$

Here,

$$\Delta(k^2) = \partial_y^2 - k^2 \tag{5.7}$$

is the transformed Laplacian. Formally an appropriate Sobolev space, consistent with the boundary conditions (5.2), may be chosen as the domain of $\mathcal{L}$. The nonlinear part $\mathcal{M}_2$ is written as

$$\mathcal{M}_2[v_1, v_2] = \int d\omega_1 \int dk_1 \int d\omega_2 \int dk_2 \, \delta(\omega - \omega_1 - \omega_2) \delta(k - k_1 - k_2)$$
$$\times G_2(k_1|k_2)[v_1(\omega_1, k_1), v_2(\omega_2, k_2)], \tag{5.8}$$

whose kernel G_2 does not depend on the frequencies (ω_1, ω_2). Letting

$$v_j = (\psi_j, \Theta_j, \mathcal{A}_j)^T \quad (j = 1, 2)$$

42

we use the following form for the bilinear operator G_2,

$$-4\imath\pi^2 G_2(k_1|k_2)[v_1, v_2] = k_1\psi_1 K(k_2^2)\partial_y v_2 - (\partial_y\psi_1)k_2 K(k_2^2)v_2$$
$$+\pi^2\xi Q\left[k_1\mathcal{A}_1\Delta(k_2^2)\partial_y\mathcal{A}_2 - (\partial_y\mathcal{A}_1)k_2\Delta(k_2^2)\mathcal{A}_2\right]e_1, \tag{5.9}$$

where e_j $(1 \leq j \leq 3)$ is the j-th column vector of the 3×3-unit matrix, and

$$K(k^2) = \text{diag}(-\frac{1}{\sigma}\Delta(k^2), 1, 1). \tag{5.10}$$

The space $\mathcal{H}_c$ can be decomposed as $\mathcal{H}_c = \bigoplus_{n\geq 0}\mathcal{H}_n$, with $\mathcal{H}_0 = \text{span}\{e_3\}$ and

$$\mathcal{H}_n = \text{span}\{e_1 \sin n\pi y, e_2 \sin n\pi y, e_3 \cos n\pi y\}.$$

Functions in $\mathcal{H}_{n\geq 1}$ may thus be represented in terms of their complex coordinates $z = (z_1, z_2, z_3)^T \in \mathbb{C}^3$ as

$$(v[z])(y) = \sum_{j=1}^{2} z_j e_j \sin n\pi y + z_3 e_3 \cos n\pi y. \tag{5.11}$$

The linearized operator $\mathcal{L}$ acts through the 3×3-matrix $\mathcal{L}_n$ in $\mathcal{H}_n$,

$$\mathcal{L}v[z] = v[\mathcal{L}_n z], \tag{5.12}$$

where

$$\mathcal{L}_n(\imath\omega, \imath k, R) = \begin{pmatrix} \Delta_n^2 - \frac{\imath\omega}{\sigma}\Delta_n & \pi^4 R\imath k & -\pi^3\xi Qn\Delta_n \\ -\imath k & \imath\omega - \Delta_n & 0 \\ -n\pi & 0 & \imath\omega - \xi\Delta_n \end{pmatrix}, \tag{5.13}$$

with $\Delta_n(k^2) = -(n^2\pi^2 + k^2)$. In $\mathcal{H}_0$, $\mathcal{L}$ acts simply as scalar multiplication,

$$\mathcal{L}v = (\imath\omega + \xi k^2)v \text{ if } v \in \mathcal{H}_0. \tag{5.14}$$

Concerning the nonlinear part, we note that

$$G_2(k_1|k_2)[v_1, v_2] \in \mathcal{H}_{|n-m|} \oplus \mathcal{H}_{n+m} \tag{5.15}$$

if $v_1 \in \mathcal{H}_n$ and $v_2 \in \mathcal{H}_m$.

In Subsection 5.4 we will need the restrictions of the boundary operators occurring in (5.3) or (5.4) to the space $\mathcal{H}_1$. In the basis introduced above the non-zero components among the $B_m^{(\nu)}$ $(1 \leq m \leq 3)$ for $1 \leq \nu \leq 4$ are (after rescaling)

$$B_1^{(1)}(\imath k) = -k^2/\pi^4 \text{ (stress free) }, \quad B_1^{(1)}(\imath k) = \imath k/\pi^3 \text{ (non-slip)}$$
$$B_1^{(2)}(\imath k) = 1/\pi^2, \quad B_2^{(3)}(\imath k) = \imath k/\pi^2, \quad B_3^{(4)}(\imath k) = 1/\pi, \tag{5.16}$$

where the first subscript in (4.5) has been suppressed because $N_\nu = 1$ $(1 \leq \nu \leq 4)$.

5.2 Linear stability analysis

In Boussinesq systems with stress free boundary conditions the critical Rayleigh numbers and wave numbers follow from the restriction of the linearized system to the first vertical modes [Ch61]. In terms of

$$\rho = 1 + k^2/\pi^2, \quad \mu = \imath\omega\pi^2, \tag{5.17}$$

the determinantal equation (2.9) for $n = 1$ becomes

$$g_1 \equiv \mu^3 + \rho(1 + \sigma + \xi)\mu^2 + \left[\rho^2(\sigma + \xi + \sigma\xi) - R\frac{\sigma}{\rho}(\rho - 1) + \sigma\xi Q\right]\mu$$
$$+\sigma\xi\rho^3 - R\sigma\xi(\rho - 1) + \sigma\xi Q = 0. \tag{5.18}$$

A zero growth rate, $\mu = 0$, exists along the neutral stationary stability curve

$$R_{1,s}(\rho) = \frac{1}{\rho - 1}(\rho^3 + Q\rho), \tag{5.19}$$

whose minimum is determined by the equation $2\rho^3 - 3\rho^2 = Q$. An imaginary growth rate, $\mu = \imath\Omega$, leads to

$$R_{1,o} = \frac{1}{\rho - 1}\left[\frac{1}{\sigma}(\sigma + \xi)(1 + \xi)\rho^3 + \frac{\xi(\sigma + \xi)}{1 + \sigma}Q\rho\right], \tag{5.20}$$

with the frequency

$$\Omega^2 = -\xi^2\rho^2 + \frac{1 - \xi}{1 + \sigma}\sigma\xi Q = \frac{\sigma\xi(\rho - 1)}{(1 + \sigma + \xi)\rho}(R_{1,s} - R_{1,o}). \tag{5.21}$$

This quantity is positive if

$$\xi < 1 \text{ and } Q > \frac{\xi(1 + \sigma)}{\sigma(1 - \xi)}\rho^2, \tag{5.22}$$

which is consistent with the condition derived in [KnWD81]. The minimum of $R_{1,o}$ is determined by the equation

$$2\rho^3 - 3\rho^2 = \frac{\sigma\xi Q}{(1 + \sigma)(1 + \xi)}. \tag{5.23}$$

and thus can not coincide with the minimum of $R_{1,s}$.

Some calculation shows that the relative position of the two neutral stability curves (5.19) and (5.20) in the $(R \geq 0, \rho \geq 1)$-half plane changes when Q passes through the values

$$Q_1 = \frac{\xi(1 + \sigma)}{\sigma(1 - \xi)}, \quad Q_o = \frac{\xi(1 + \sigma)(3 - 2\xi^2)^2}{4\sigma(1 - \xi)^3(1 + \xi)^2}, \quad Q_s = \frac{\xi(1 + \sigma)(3\sigma + \xi - 2\sigma\xi)^2}{4\sigma^3(1 - \xi)^3}.$$

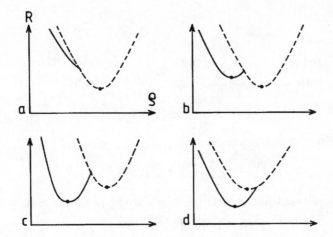

Figure 5.1: Neutral stationary (dashed) and oscillatory (full) stability curves for $\xi < 1$ and (a) $Q_1 < Q < Q_o$, (b) $Q_o < Q < Q_+$, (c) $Q_+ < Q < Q_s$, (d) $Q_s < Q$.

When $\xi > 1$ or when $\xi < 1$ and $Q < Q_1$, $R_{1,o}(\rho)$ is above $R_{1,s}(\rho)$ throughout and hence does not lead to real frequencies Ω. For $\xi < 1$ and $Q > Q_1$ the qualitative behaviour of these curves is as sketched in Figure 5.1. From this figure we infer that there exists a value $Q = Q_+(\sigma, \xi)$ between $Q_1(\sigma, \xi)$ and $Q_2(\sigma, \xi)$ at which the transition from Figure 5.1(b) to Figure 5.1(c) occurs. For $Q = Q_+$ the minima of the two neutral stability curves have the same height, but occur at different values of ρ. When $Q > Q_+$ we obtain an oscillatory instability as defined in Section 2 (Figs. 2(c), (d)), i.e., $R_c = \min_\rho R_{1,o} < \min_\rho R_{1,s}$ (cf., also [ClK94]). Some calculation leads to the following expression for Q_+,

$$Q_+(\sigma, \xi) = \frac{27}{4} \frac{(1+\sigma)(\delta^{1/3} - 1)}{(1-\xi)^3(1+\sigma+\xi)^3} \left[(1+\sigma)\delta^{1/3} - \xi(\sigma+\xi)^2 \right], \qquad (5.24)$$

where $\delta = \frac{1}{\sigma}(\sigma + \xi)(1 + \xi)$. For $Q = Q_+$, the minimum of $R_{1,o}$ occurs at

$$\rho_+(\sigma, \xi) = \frac{1}{2} \left\{ 1 + \left[2\sqrt{P + P^2} + 1 + 2P \right]^{1/3} - \left[2\sqrt{P + P^2} - 1 - 2P \right]^{1/3} \right\},$$

where $P = \sigma\xi Q_+/[(1+\sigma)(1+\xi)]$.

Since the minimum $\rho = \rho_c(Q)$ of $R_{1,o}(\rho; Q)$ varies monotonically with Q ($d\rho_c/dQ > 0$ for $\rho > 1$), we use in the following ρ_c as independent parameter and express Q in terms of ρ_c as

$$Q = \frac{1}{\sigma\xi}(1+\sigma)(1+\xi)\rho_c^2(2\rho_c - 3), \qquad (5.25)$$

where ξ is restricted to $\xi < 1$. In this parametrization an oscillatory instability occurs if $\rho_c > \rho_+(\sigma,\xi)$, with the critical Rayleigh number given by

$$R_c = \frac{2}{\sigma}(\sigma + \xi)(1+\xi)\rho_c^3, \qquad (5.26)$$

and with the critical frequency $\omega_c = \pi^2\Omega_c$, where

$$\Omega_c^2 = \rho_c^2\left[2\rho_c(1-\xi^2) + 2\xi^2 - 3\right]. \qquad (5.27)$$

From (5.16) the relevant linear coefficients occurring in (3.45) may be determined as functions of ρ_c. The critical group velocity is given by

$$c_g = -\frac{2k_c}{\Omega_c}\rho_c\xi^2 \quad (k_c^2 = \pi^2(\rho_c - 1)). \qquad (5.28)$$

The coefficient a_0 and the relevant diffusion constant $D = D_{02} + c_g D_{11} + c_g^2 D_{20}$ take the form

$$\frac{a_0}{R_c} = \frac{\pi^2\sigma(\rho_c - 1)}{2\rho_c\Omega_c\Omega_1^2}\left\{(1+\sigma)\rho_c\Omega_c - i\left[(1+\sigma+\xi)\xi\rho_c^2 + \Omega_c^2\right]\right\} \qquad (5.29)$$

$$\begin{aligned} D = {} & \frac{3(\sigma+\xi)(1+\sigma)(\rho_c^2 + \Omega_c^2)}{(1-\xi)\rho_c\Omega_1^2} - \frac{3i(\rho_c^2 + \Omega_c^2)\Omega_2^2}{(1-\xi^2)\rho_c^2\Omega_c\Omega_1^2} \\ & - \frac{i\xi^2(3\Omega_c^4 + \rho_c^2\Omega_c^2 - 2\xi^2\rho_c^4)}{2(1-\xi^2)\rho_c^2\Omega_c^3}, \end{aligned} \qquad (5.30)$$

where

$$\Omega_1 = (1+\sigma+\xi)^2\rho_c^2 + \Omega_c^2, \quad \Omega_2 = \sigma\xi(1+\sigma+\xi)\rho_c^2 + (\sigma+\xi+\sigma\xi)\Omega_c^2.$$

The coordinate vectors z_0 and z_0^* corresponding to the eigenfunction $v_0 = v[z_0]$ and $v_0^* = 2v[z_0^*]$ span the kernels of the matrices $\mathcal{L}_1(i\omega_c, ik_c, R_c)$ and $\mathcal{L}_1^*(i\omega_c, ik_c, R_c)$. We choose

$$z_0 = \left(\pi^2(\rho_c + i\Omega_c)(\xi\rho_c + i\Omega_c), ik_c(\xi\rho_c + i\Omega_c), \pi(\rho_c + i\Omega_c)\right)^T \qquad (5.31)$$

$$z_0^* = C\left((\rho_c - i\Omega_c)(\xi\rho_c - i\Omega_c), i\pi^2 k_c R_c(\xi\rho_c - i\Omega_c), \pi^3\xi Q\rho_c(i\Omega_c - \rho_c)\right)^T \qquad (5.32)$$

where C serves for the normalization $z_0^T \bar{z}_0^* = 1$. Noting that

$$v_\sigma = v[\mathrm{diag}(\pi^2\rho_c/\sigma, 1, 1)z_0],$$

the basic linear coefficient q_{10} is obtained as

$$q_{10} = \frac{2}{\sigma}\pi^4\bar{C}i\omega\rho(\rho + i\omega)(\xi\rho + i\omega)\left[(1+\sigma+\xi)\rho + i\omega\right]. \qquad (5.33)$$

46

5.3 Boundary conditions

There are four scalar boundary conditions to consider, hence (4.15) and (4.19) reduce to systems of linear equations in $\mathbb{C}^4$. The relevant vectors $\mathbf{a}_0$, $\mathbf{b}_0 \in \mathbb{C}^4$ follow from (5.16). In the stress free case they are related to each other by $\mathbf{a}_0 = -e^{2\imath k_c l}\mathbf{b}_0$ so that the reflection coefficient has the simple form $r = e^{-2\imath k_c l}$. In the non-slip case $\mathbf{a}_0$ is given by

$$
\begin{aligned}
e^{\imath k_c l}\mathbf{a}_0 \;=\; (&\frac{\imath k_c}{\pi}(\rho_c + \imath\Omega_c)(\xi\rho_c + \imath\Omega_c), (\rho_c + \imath\Omega_c)(\xi\rho_c + \imath\Omega_c), \\
&-(\xi\rho_c + \imath\Omega_c)(\rho_c - 1), \rho_c + \imath\Omega_c)^T,
\end{aligned}
\tag{5.34}
$$

and $\mathbf{b}_0 = e^{2\imath k_c l}\mathrm{diag}(-1, 1, 1, 1)\mathbf{a}_0$. There is no such simple relation as in the stress free case. One has, therefore, to consider the remaining roots of $g_1(\imath\Omega_c, \rho, R_c) = 0$, Eq. (5.16), which satisfy the cubic equation

$$
(\rho^2 + \rho_c\rho)(\rho + \imath\delta_1\Omega_c) - 2\delta_2\rho_c^2(\xi\rho + \imath\Omega_c) = 0,
\tag{5.35}
$$

with $\delta_1 = 1 + (\sigma + \xi)/\sigma\xi$, $\delta_2 = (\sigma + \xi)(1 + \xi)/\sigma\xi$. Denote the (complex) roots of (5.35) by ρ_j ($1 \le j \le 3$). The vectors $\mathbf{c}_j \in \mathbb{C}^4$ which form the columns of the 4×3-matrix $\mathbf{C}$ defined below (4.15) are then given by the right hand side of (5.3), with ρ_c replaced by ρ_j and k_c replaced by $k_j := \pi\sqrt{\rho_j - 1}$, where the square root is chosen such that $\Im k_j > 0$. Proceeding as explained in Section 4.4 we may then express, after some calculation, the reflection coefficient in terms of the roots ρ_j as

$$
r = \frac{r_+ - r_-}{r_+ + r_-}e^{-2\imath k_c l},
$$

where

$$
r_+ = \sum_{\nu=0}^{2} r_\nu \sum_{j=1}^{3} \imath k_j(\rho_{j+2} - \rho_{j+1})\rho_j^\nu \quad (\rho_{j+3} := \rho_j)
$$

$$
r_- = k_c\Omega_c\xi(1 - \xi)(\rho_c + \imath\Omega_c)(\xi\rho_c + \imath\Omega_c)\sum_{j=1}^{3}(\rho_{j+1} - \rho_{j+2})\rho_j^2,
$$

and

$$
\begin{aligned}
r_0 \;=\; & (1 - \xi)(\xi\rho_c + \imath\Omega_c)\rho_c\Omega_c^2\,[2\delta_2\xi\rho_c + \imath\Omega_c\rho_c(1 + \delta_1\xi)] - 2\delta_2\xi^3\rho_c^2(\rho_c + \imath\Omega_c) \\
& -(1 - \delta_1\xi)(\xi\rho_c\imath\Omega_c + \Omega_c^2)(\rho_c + \imath\Omega_c + \xi\rho_c\imath\Omega_c + \imath\Omega_c^3) \\
r_1 \;=\; & \xi(\rho_c + \imath\Omega_c)(\xi\rho_c - \imath\Omega_c + \delta_1\xi\imath\Omega_c) - \Omega_c^2(1 - \delta_1\xi)(\rho_c^2\xi^2 + \Omega_c^2) \\
& -\imath\Omega_c(1 - \xi)(\xi\rho_c + \imath\Omega_c)\,[2\xi\rho_c(\delta_2\xi\rho_c + \imath\Omega_c) + \imath\Omega_c(\rho_c + \delta_1\xi\imath\Omega_c)] \\
r_2 \;=\; & \xi(\rho_c + \imath\Omega_c)\,[\xi - (1 - \xi)\imath\Omega_c(\xi\rho_c + \imath\Omega_c)].
\end{aligned}
$$

Further analysis requires explicit determination of the roots ρ_j. This is rather complicated and will not be pursued here. In particular, the question whether $|r|^2 = 1$, i.e., $r_+\bar{r}_- + \bar{r}_+r_- = 0$, must be left open at this stage.

According to (5.28), the group velocity vanishes in the limit $\xi \to 0$. For small ξ, therefore, the local Ginzburg-Landau theory applies. This limit is particularly interesting for astrophysical applications, for example $\xi \approx 10^{-2}$ in the sun. We consider the limits $\xi \to 0$, $Q \to \infty$ in a way so that $\xi Q = Q'$ remains finite. The expressions derived in the preceding subsection reduce then to

$$Q' = \frac{1+\sigma}{\sigma}(2\rho_c^3 - 3\rho_c^2), \quad R_c = 2\rho_c^3, \quad \Omega_c^2 = 2\rho_c^3 - 3\rho_c^2,$$

$$D = \frac{6\sigma(\rho_c - 1)}{\Omega_1^2}[(1+\sigma)\rho_c - \imath\Omega_c], \quad \frac{a_0}{R_c} = \frac{\pi^2 D}{12\rho_c},$$

and $Q'_+(\sigma) \equiv 0$, $\rho_+(\sigma) = 3/2$.

The boundary conditions for the local Ginzburg-Landau equations, which are relevant in the limit $\xi \to 0$, are again very easy to obtain in the stress free case. Here,

$$\mathbf{a}_\nu = e^{-\imath k_c l}\mathbf{p}_\nu, \quad \mathbf{b}_\nu = -e^{\imath k_c l}\mathbf{p}_\nu, \quad (\nu = 0,1),$$

with certain vectors $\mathbf{p}_\nu \in \mathbb{C}^4$, whose explicit form is not needed. Namely, form (4.20) we obtain

$$A - e^{2\imath k_c l}B = -\frac{(\mathbf{p}_1, \mathbf{c}_j)}{(\mathbf{p}_0, \mathbf{c}_j)}(A_x - e^{2\imath k_c l}B_x) \quad (x = -l)$$

for $j = 1,2$, where $\mathbf{c}_1$, $\mathbf{c}_2$ span the kernel of the 2×4-matrix $\bar{\mathbf{C}}_T$. Obviously, this system of equations cannot be transformed to the form (4.21). Instead, we find the boundary conditions

$$A - e^{2\imath k_c l}B = 0, \quad A_x - e^{2\imath k_c l}B_x = 0 \quad \text{at } x = -l, \tag{5.36}$$

and those at $x = l$ are obtained in the usual way by reflection. It follows that the leading order Ginzburg-Landau equations (4.38) (with $s = -2(\xi^2/\epsilon)k_c\rho_c/\Omega_c$, $\xi^2 = O(\epsilon)$) have to be supplemented by the boundary conditions (5.36) (with $A \to A_1$, $B \to B_1$, $l \to L$) instead of (4.39) An important consequence of these boundary conditions is that the $\mathbf{O}(2)$-symmetry is already broken to $\mathbf{Z}_2$ at leading order.

In the non-slip case, explicit calculations must be performed. For $\xi \to 0$ the two remaining roots of (5.35) are

$$\rho_1 = -2\rho_c + O(\xi), \quad \rho_2 = -\frac{\imath\Omega_c}{\xi}\left(1 + \frac{1+\sigma}{\sigma}\xi + O(\xi^2)\right),$$

i.e., one of the root diverges for $\xi \to 0$ which does not, however, cause problems. The vectors $z_j \in \mathbb{C}^3$ corresponding to $w_j = v[z_j] \in \ker \mathcal{L}(\imath \omega_c, \imath k_j, R_c)$ are given by

$$z_1 = \left(-\pi^2(\Omega_c^2 + 2\imath\Omega_c\rho_c), -\imath\pi\Omega_c\sqrt{2\rho_c+1}, \pi\imath\Omega_c\right)^T + O(\xi)$$

$$z_2 = \left(-\pi\imath\Omega_c\xi(1+\sigma)/\sigma + O(\xi^2), \frac{\sqrt{\Omega_c}}{\sigma}(1+\sigma)e^{5\pi\imath/4}\xi^{3/2} + O(\xi^{5/2}), 1 + O(\xi)\right)^T,$$

and lead to the following columns $\hat{\mathbf{c}}_1$, $\hat{\mathbf{c}}_2$ of the 4×2-matrix $\mathbf{C}$,

$$\hat{\mathbf{c}}_1 = \left(\imath\Omega_c - 2\rho_c, \sqrt{2\rho_c+1}(2\rho_c - \imath\Omega_c), 2\rho_c + 1, 1\right)^T + O(\xi)$$

$$\hat{\mathbf{c}}_2 = \left(O(\xi), O(\sqrt{\xi}), O(\sqrt{\xi}), 1 + O(\xi)\right)^T.$$

The kernel of $\bar{\mathbf{C}}^T$ is spanned by

$$\mathbf{c}_1 = (-(2\rho_c + 1), 1, 0, 0)^T + O(\sqrt{\xi})$$
$$\mathbf{c}_2 = (0, 1, 2\rho_c + \imath\Omega_c, 0)^T + O(\sqrt{\xi}),$$

so that (4.20) can be written as

$$\begin{pmatrix} p + p_1 & p - p_1 \\ p + p_2 & p - p_2 \end{pmatrix} \begin{pmatrix} A \\ e^{2\imath k_c l}B \end{pmatrix} + \begin{pmatrix} q + q_1 & q - q_1 \\ q + q_2 & q - q_2 \end{pmatrix} \begin{pmatrix} A_x \\ e^{2\imath k_c l}B_x \end{pmatrix} = 0, \quad (5.37)$$

where

$$p = -\frac{k_c}{\pi}\Omega_c(\rho_c + \imath\Omega_c) \qquad q = \frac{\imath}{\pi}\Omega_c(\rho_c + \imath\Omega_c) + k_c z_{01,1}/\pi^3$$

$$p_1 = -\pi\imath\Omega_c(\rho_c + \imath\Omega_c)(2\rho_c + 1) \quad q_1 = \frac{\imath}{\pi}(2\rho_c + 1)z_{01,1}$$

$$p_2 = \pi\imath\Omega_c(1 - \rho_c)(2\rho_c - \imath\Omega_c) \quad q_2 = \frac{k_c}{\pi}(z_{01,2} - \Omega_c)(2\rho_c - \imath\Omega_c).$$

Here, $z_{01,1}$ and $z_{01,2}$ are the first two components of the vector $z_{01} \in \mathbb{C}^3$ corresponding to $v_\kappa = v[z_{01}]$ (cf. (3.43)) which has to be calculated in order to find the final form of the boundary conditions. For genuine σ this is still very tedious, and we therefore confine ourselves to the additional limit $\sigma \to 0$ that is also of astrophysical interest. In this limit we assume that $\sigma\xi Q$ tends to a finite quantity Q'' of order 1 so that $\Omega_c^2 = Q''$. Setting $w_{01} = v[\tilde{z}_{01}]$ (cf. (A.1)), we find

$$\tilde{z}_{01} = \imath(\rho_c + \imath\Omega_c)z_{01} + O(\sigma) = \left(0, \Omega_c^2 + (\rho_c - 2)\imath\Omega_c, 0\right)^T.$$

The boundary conditions for the envelopes can then be written in the form of (4.21), with

$$\alpha = \frac{\imath}{2k_c} + \mu, \quad \beta = \left(\frac{\imath}{2k_c} - \mu\right)e^{2\imath k_c l}, \quad (\gamma, \delta) = e^{-2\imath k_c l}(\alpha, \beta), \quad (5.38)$$

49

where

$$\mu = \frac{\imath\Omega_c - 2\rho_c}{3\rho_c(\imath\Omega_c + \rho_c)}\left(\frac{\imath k_c}{\pi^2} + 2\rho_c + 1\right).$$

5.4 Nonlinear coefficients

In principle it is straightforward to calculate the nonlinear coefficients a, b that occur in both, the global and the local Ginzburg-Landau equations. Applying the bilinear operator (5.9) to $(v_0, \bar{v}_0)$ etc. as stated in equation (A.4) leads to $v_{00}^{(0)}$, $v_{00}^{(0)} \in \mathcal{H}_2$ and $v_{02}^{(0)}$, $v_{22}^{(0)} \in \mathcal{H}_0$. When these are substituted into (A.5) we get w_+, $w_- \in \mathcal{H}_1 \oplus \mathcal{H}_3$ from which only the $\mathcal{H}_1$-component is needed. Projection with v_0^* and dividing by q_{10} leads finally to the desired coefficients. Although the procedure is clear, the explicit calculations are very tedious, even with the aid of computer algebra.

Recall that a and b coincide with the "normal form coefficients" which would be obtained from a standard Liapunov Schmidt reduction, in which the problem is posed as a nonlinear operator equation for spatially and temporally periodic solutions. Alternatively one may pursue a center manifold reduction and normal form transformation in which the basic equations (5.1), (5.2) (with $l = \infty$) are considered as a dynamical system in the space of spatially periodic functions. From that point of view, a and b have already been calculated by Nagata [Na86] for arbitrary wave numbers (see, also, [MR93]). His general expressions for a and b are complicated and will not be written down in full. Instead, we consider again the limit $\xi \to 0$. In our normalization and after specializing the wave number to k_c, the coefficients a, b take the form

$$a = \frac{\pi^2(\rho_c - 1)}{8\Omega_c^2\left[(1 + \sigma)\rho_c + \imath\Omega_c\right]}\left\{2\sigma\imath\Omega_c^3\rho_c(\rho - \imath\Omega_c) - \sigma\Omega_c^2\rho_c^2\right.$$

$$\left. +2(\rho_c + \imath\Omega_c)\left[(1 + \sigma)(4\rho_c - 3)\imath\Omega_c - 8\sigma\right]\right\} \tag{5.39}$$

$$a + b = \frac{\imath\pi^2\rho_c(\rho_c - 1)}{2\Omega_c\left[(1 + \sigma)\rho_c + \imath\Omega_c\right]}\left\{\frac{\sigma\imath\Omega_c\left[3\rho_c + \imath\Omega_c(\rho_c - 1)\right]}{4 + 2\imath\Omega_c}\right.$$

$$\left. + \frac{(1 + \sigma)(\rho_c + \imath\Omega_c)}{\rho_c(\rho_c - 1)}(3\rho_c^2 - 6\rho_c + 2)\right\}. \tag{5.40}$$

If $\sigma = 0$, the real parts of both coefficients vanish. For small $\sigma > 0$ we find to leading order,

$$a = \frac{\pi^2\sigma}{16\Omega_c^2}(14\rho_c^4 - 69\rho_c^3 + 70\rho_c^2 + 4\rho_c - 16) + \frac{\imath\pi^2}{4\Omega_c}(\rho_c - 1)(4\rho_c - 3)$$

$$a + b = \frac{-\pi^2\rho_c\sigma(2\rho_c^6 + 3\rho_c^5 - 33\rho_c^4 + 50\rho_c^3 - 48\rho_c + 16)}{4(4 + \Omega_c^2)(\rho_c^2 + \Omega_c^2)} + \frac{\imath\pi^2}{2\Omega_c}(3\rho_c^2 - 6\rho_c + 2).$$

These expressions will be examined further in the next section.

6 Dynamics of confined waves

A variety of interesting dynamical phenomena may be expected to occur in both the local and global envelope equations. Whereas the effects of nonlocal coupling terms have not been much studied yet, a number of results were obtained recently for locally coupled Ginzburg Landau equations [Cr86, Cr88, DaK90, DaWK91, DaKW91, DaRG93, KnP92]. In this section we review parts of these results. The starting point is provided by the rescaled system of partial differential equations,

$$A_T = \{D\partial_x^2 + s\partial_x + a_0\Lambda + a|A|^2 + b|B|^2\}A, \qquad (6.1)$$

$$B_T = \{D\partial_x^2 - s\partial_x + a_0\Lambda + a|B|^2 + b|A|^2\}B, \qquad (6.2)$$

for the envelopes of left and right going waves, with the perturbed boundary conditions

$$X = -L: \begin{pmatrix} A \\ B \end{pmatrix} = \epsilon \begin{pmatrix} \alpha & \beta \\ \gamma & \delta \end{pmatrix} \begin{pmatrix} A_X \\ B_X \end{pmatrix}$$

$$X = L: \begin{pmatrix} A \\ B \end{pmatrix} = -\epsilon \begin{pmatrix} \delta & \gamma \\ \beta & \alpha \end{pmatrix} \begin{pmatrix} A_X \\ B_X \end{pmatrix}, \qquad (6.3)$$

where $0 \leq \epsilon \ll 1$ and all other quantities are of order 1. We note that in the references cited above the coefficients in (6.3) are restricted to $\alpha = \bar{\gamma}$, $\beta = \bar{\delta}$, although this is not enforced by the symmetry. The $O(\epsilon)$-terms in the p.d.e.'s have been omitted because they preserve all three symmetries

$$\mathbf{S}^1: \quad A \to e^{\imath\phi}A, \ B \to e^{\imath\phi}B \qquad (6.4)$$

$$\mathbf{SO}(2): \quad A \to e^{\imath\psi}A, \ B \to e^{-\imath\psi}B \qquad (6.5)$$

$$\mathbf{Z}_2: \quad (A, B, X) \to (B, A, -X). \qquad (6.6)$$

under which (6.1) and (6.2) stay invariant. The symmetry breaking that destroys the translation invariance $\mathbf{SO}(2)$ occurs at $O(\epsilon)$ in the boundary conditions (6.3), hence these terms are the most relevant perturbations. We refer to (6.3) as *weak symmetry breaking boundary conditions*.

It is worth noting that a completely different situation arises when (6.3) is replaced by

$$X = -L: A = \mu B, A_X = \mu B_X; \quad X = L: B = \mu A, B_X = \mu A_X, \qquad (6.7)$$

with $\mu = e^{2\imath k_c l}$. These boundary conditions appeared in the convection system, studied in the previous section, in the stress free case. The important feature of (6.7) is that the breaking of $\mathbf{SO}(2)$ occurs already at $O(1)$. This should have strong implications on the dynamics of the coupled Ginzburg Landau equations. We refer to (6.7) as *strong symmetry breaking boundary conditions.*

The analysis of (6.1)–(6.3) is facilitated by the fact that the linearized part has a discrete spectrum, because the spatial domain is bounded, and that the eigenfunctions can be easily determined in the limit $\epsilon = 0$. When Λ is increased from below and $\epsilon = 0$, the trivial solution $A = B = 0$ encounters the first instability at a certain critical value Λ_c. This instability is characterized by two complex unstable modes. Near Λ_c one can perform a center manifold reduction and describe the dynamics by a system of normal form differential equations in $\mathbb{C}^2$, in which the effects of the symmetry breaking $O(\epsilon)$-terms are incorporated [DaK90, DaWK91]. The most prominent feature of the normal form is the presence of a two-torus that corresponds to a modulated wave in the continuous system, and causes a travelling wave to reverse periodically its direction of propagation. States of this kind have been observed in convection experiments with binary fluids in boxes with large aspect ratio [FMS88, KoS88, KoSW89, SFMR89]. We review the normal form approach in some detail in Subsections 6.1–6.4, and comment in Subsection 6.5 on the modifications which might be expected if (6.3) is replaced by (6.7).

When Λ is increased further, additional modes become unstable. The normal form description is then no longer valid and one has to resort to a numerical analysis. In Subsection 6.6 we report on some of the results obtained in [DaRG93] from Galerkin simulations of (6.1)–(6.3). The simulations indicate that the local analytical results are globally valid for real nonlinear coefficients. For complex coefficients bifurcation sequences to quasiperiodic and chaotic dynamics have been observed as successive eigenmodes become unstable.

6.1 Modal equations and center manifold reduction

We begin by deriving a system of modal equations from (6.1)–(6.3), which is useful for analytical considerations, in particular for the center manifold reduction, as well as for Galerkin simulations. As usual, the eigenfunctions of the linearized p.d.e. are used as basic modes. Consider the linearized operator $\mathcal{L}_{GL}$ of the Ginzburg Landau equations (6.1), (6.2),

$$\mathcal{L}_{GL} = \begin{pmatrix} D\partial_X^2 + s\partial_X + a_0\Lambda & 0 \\ 0 & D\partial_X^2 - s\partial_X + a_0\Lambda \end{pmatrix}, \tag{6.8}$$

acting on $\mathbb{C}^2$-valued functions $(A(X), B(X))^T$ which satisfy the boundary conditions (6.3). For $\epsilon = 0$ there is no coupling of A and B in (6.3), i.e., the envelopes satisfy simple Dirichlet boundary conditions. In this case the eigenvalues of $\mathcal{L}_{GL}$ are given by $a_0\Lambda - \Gamma_n$ for $n = 1, 2, \ldots$, where

$$\Gamma_n = D\left(\frac{n\pi}{2L}\right)^2 + \frac{s^2}{4D}. \tag{6.9}$$

Moreover, the presence of two continuous symmetries enforces two dimensional eigenspaces E_n over $\mathbb{C}$. As bases for E_n one can choose pure traveling wave envelopes,

$$E_n = \text{span}\left\{ \left(e^{-sX/2D}S_n(X), 0\right)^T, \left(0, e^{sX/2D}S_n(X)\right)^T \right\},$$

with

$$S_n(X) = \sin\left[\frac{n\pi}{2L}(X - L)\right]. \tag{6.10}$$

For $\epsilon \neq 0$ the nontrivial Robin boundary conditions induce a coupling between A and B. As a consequence, the former degenerate eigenvalues split into two eigenvalues $a_0\Lambda - \Gamma_{n\pm}$, where

$$\Gamma_{n\pm} = \Gamma_n - \epsilon Q_n \mp \epsilon R_n + O(\epsilon^2), \tag{6.11}$$

with

$$Q_n = D\left(\frac{n\pi}{2L}\right)^2 (\alpha + \delta), \quad R_n = D\left(\frac{n\pi}{2L}\right)^2 (\beta e^{-sL/D} + \gamma e^{sL/D}). \tag{6.12}$$

Let $U_{n\pm}(X)$ be the eigenfunctions of $\mathcal{L}_{GL}$ corresponding to $a_0\Lambda - \Gamma_{n\pm}$. Because the **SO**(2)-symmetry is broken, $U_{n\pm}$ cannot be represented by pure travelling wave envelopes. Instead, setting

$$S_{n\pm}(x) = \sin\left[\frac{n\pi}{2L}(X - L) - \frac{\epsilon L}{Dn\pi}(Q_n \mp R_n)X + O(\epsilon^2)X\right],$$

and defining $C_{n\pm}(X)$ analogously by replacing the sine by a cosine, we find the following representation,

$$U_{n\pm}(X) = \begin{pmatrix} e^{-sX/2D}[S_{n\pm}(X) + \epsilon P_{n\pm}C_{n\pm}(X)] \\ \pm e^{sX/2D}[S_{n\pm}(X) - \epsilon P_{n\pm}C_{n\pm}(X)] \end{pmatrix} + O(\epsilon^2), \tag{6.13}$$

53

where

$$P_{n\pm} = \frac{n\pi}{4L}(\alpha - \delta) \pm \frac{n\pi}{4L}(\beta e^{-sL/D} - \gamma e^{sL/D}).$$

These eigenfunctions induce standing waves when $\epsilon A(\epsilon x)$ and $\epsilon B(\epsilon x)$ are substituted into the representation (4.1).

For deriving modal equations we also need to consider the adjoint differential operator $\mathcal{L}_{GL}^*$. This operator acts on $\mathbb{C}^2$-valued functions $(A(X), B(X))^T$ which satisfy the boundary conditions (6.3) with γ and β interchanged. The adjoint eigenfunctions $U_{n\pm}^*(X)$ can be written in a similar form as the $U_{n\pm}(X)$,

$$\bar{U}_{n\pm}^*(X) = C_{n\pm}\text{diag}(e^{sX/D}, e^{-sX/D})U_{n\pm}(X) + O(\epsilon^2),$$

where the constant $C_{n\pm} = 1/2L + (\epsilon/2Dn\pi)(Q_n \pm R_n)$ serves for the correct normalization. We now represent (A, B) in terms of complex amplitudes $(v_n, w_n) \in \mathbb{C}^2$ as

$$\begin{pmatrix} A(T,X) \\ B(T,X) \end{pmatrix} = \frac{1}{2}\sum_{n\geq 1}\left\{[v_n(T) + w_n(T)]\,U_{n+}(X) + [v_n(T) - w_n(T)]\,U_{n-}(X)\right\},$$

(6.14)

substitute this expansion into the Ginzburg Landau equations (6.1), (6.2) and project with the adjoint eigenfunctions $U_{n\pm}^*$. In the resulting (infinite) system of differential equations for the mode amplitudes (v_n, w_n) we retain the cubic terms to order $O(\epsilon^0)$ and the linear terms to order $O(\epsilon)$. Terms of higher order are disregarded because these are also influenced by further terms which have already been omitted in (6.1)–(6.3). The equations for the v_n become

$$\begin{aligned} \dot{v}_n = \ & (a_0\Lambda - \Gamma_n + \epsilon Q_n)v_n + \epsilon R_n w_n \\ & + \sum_{m,k,l} K_{nmkl}\left(av_m\bar{v}_k + (-1)^{n+m+k+l}bw_m\bar{w}_k\right)v_l, \end{aligned}$$

(6.15)

where the dot denotes the time derivative d/dT. The equations for the w_n follow from (6.15) by the replacements $v_n \to (-1)^n w_n$. The constants appearing in the sum are

$$K_{nmkl} = \frac{1}{2L}\int_{-L}^{L} e^{-sX/D}S_n(X)S_m(X)S_k(X)S_l(X)dX,$$

with $S_n(X)$ defined in (6.10). In [DaRG93] a twenty-mode truncation of (6.15) was used in the Galerkin simulations of the initial-boundary value problem (6.1)–(6.3).

Assuming $\Re D > 0$, $\Re a_0 > 0$, we infer from (6.15) that for $\epsilon = 0$ successive modes become unstable when Λ increases through the sequence of bifurcation values $\Lambda_n =$

$\mathfrak{R}(\Gamma_n)/\mathfrak{R}(a_0)$. For $\epsilon \neq 0$ each of these bifurcation values is split into two such values, $\Lambda_{n\pm} = \mathfrak{R}(\Gamma_{n\pm})/\mathfrak{R}(a_0)$, separated by an amount of order ϵ. If (Λ, ϵ) varies in a neighborhood of $(\Lambda_n, 0)$, one can perform a center manifold reduction and describe the dynamics on the center manifold by an o.d.e.-system for the mode amplitudes (v_n, w_n) alone. This reduction is here trivial: the dominant terms of the reduced system follow simply by setting all amplitudes in (6.15) equal to zero except (v_n, w_n). For $n > 1$ there is also an unstable eigenspace present, whence the reduced system does not tell very much about the attractors of the full system, all solutions derived from the former being unstable. Nevertheless, knowledge of these unstable solutions may be useful for a global numerical investigation. For $n = 1$ the center manifold is attracting and the dynamics of the reduced system reflects that of the full p.d.e. The first step towards understanding the phenomena, which may occur in (6.1)–(6.3), is therefore provided by a study of the flow on the center manifold for Λ close to Λ_1. Retaining only the first mode amplitudes in (6.15), we write this system as

$$\dot{v} = (\lambda + \imath\omega + a'|v|^2 + b'|w|^2)v + \epsilon dw \tag{6.16}$$

$$\dot{w} = (\lambda + \imath\omega + a'|w|^2 + b'|v|^2)w + \epsilon dv, \tag{6.17}$$

where

$$\lambda + \imath\omega = a_0\Lambda - \Gamma_1 + \epsilon Q_1, \quad d = R_1, \quad (v, w) = (v_1, w_1), \tag{6.18}$$

and $\lambda, \omega \in \mathbb{R}$. Since Λ vanishes in a neighborhood of $\Lambda_1 = \mathfrak{R}(\Gamma_1)/\mathfrak{R}(a_0)$, λ is small and plays the role of a local bifurcation parameter, whereas ϵ is viewed as an imperfection parameter. The rescaled coefficients (a', b') are

$$(a', b') = K(a, b), \quad K = K_{1111} = \frac{3\pi^4 \sinh(q)}{q(q^2 + \pi^2)(q^2 + 4\pi^2)},$$

with $q = (sL/|D|^2)\mathfrak{R}(D)$ [DaKW91].

For $\epsilon = 0$, the system (6.16), (6.17) is invariant under the symmetries

$$\mathbf{S}^1: \quad (v, w) \to e^{\imath\phi}(v, w), \tag{6.19}$$

$$\mathbf{Z}_2: \quad (v, w) \to (w, v), \tag{6.20}$$

$$\mathbf{SO}(2): \quad (v, w) \to (e^{\imath\psi}v, e^{-\imath\psi}w), \tag{6.21}$$

which are a direct consequence of the symmetries (6.4)–(6.6) of the Ginzburg Landau equations (6.1) (6.2) with Dirichlet boundary conditions imposed. Again, for $\epsilon \neq 0$ the translation symmetry $\mathbf{SO}(2)$ is broken and only the two symmetries (6.19), (6.20)

55

remain. We refer to (6.16), (6.17) with $\epsilon = 0$ and $\epsilon \neq 0$ as the *unperturbed* or *perfect system* and the *perturbed system*, respectively. The limit $\epsilon \to 0$ is interpreted as the transition to an infinitely extended domain of the original evolution equation (2.1), although in this limit the asymptotic solution $u(t, x)$ vanishes. However, the rescaled Ginzburg Landau equations possess nontrivial solutions for $\epsilon = 0$, and these characterize the limiting behavior when the length of the original system tends towards infinity. In the next two Subsections we describe the basic features of the solution structure of (6.16), (6.17).

6.2   Infinite system: Hopf bifurcation with O(2)-symmetry

The system (6.16), (6.17) with $\epsilon = 0$ is well known as the normal form for a Hopf bifurcation with $\mathbf{O}(2)$-symmetry [GSS88]. The presence of both translation symmetries $\mathbf{S}^1$ and $\mathbf{SO}(2)$ is responsible for the decoupling of the amplitude and phase equations. In terms of the real variables y_j, Θ_j, $j = 1, 2$, defined by

$$v = y_1 e^{\imath \Theta_1}, \quad w = y_2 e^{\imath \Theta_2} \tag{6.22}$$

equations (6.16) (6.17) become

$$\dot{y}_1 = (\lambda + a'_r y_1^2 + b'_r y_2^2), \quad \dot{y}_2 = (\lambda + a'_r y_2^2 + b'_r y_1^2), \tag{6.23}$$

$$\dot{\Theta}_1 = \omega + a'_i y_1^2 + b'_i y_2^2, \quad \dot{\Theta}_2 = \omega + a'_i y_2^2 + b'_i y_1^2, \tag{6.24}$$

where the subscripts r and i refer to the real and imaginary parts. Consequently (6.23) suffices to determine the dynamics. These equations have the following three types of solutions (y_1, y_2): the trivial state $(0, 0)$, the travelling wave TW, $(y, 0)$ or $(0, y)$, and the standing wave SW, (y, y), with

$$\lambda + a'_r y^2 = 0 \text{ for TW}, \quad \lambda + (a'_r + b'_r) y^2 = 0 \text{ for SW}.$$

All other solutions are transient, provided the real parts a'_r and b'_r satisfy the non-degeneracy conditions

$$a'_r \neq 0, \quad a'_r + b'_r \neq 0, \quad a'_r - b'_r \neq 0. \tag{6.25}$$

In Figure 6.1 the results of analyzing (6.23) under the conditions (6.25) are summarized. This figure shows the (a'_r, b'_r)-plane together with the bifurcation diagrams characteristic for each of the six regions I-VI. As a consequence of the $\mathbf{O}(2)$-symmetry

56

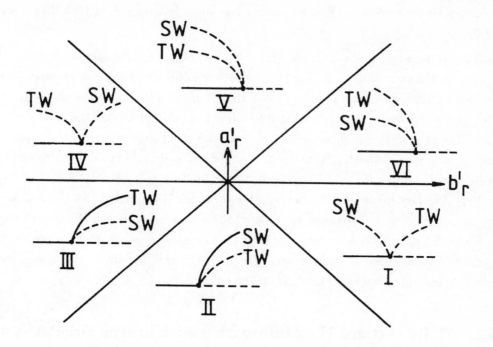

Figure 6.1: Division of the (a'_r, b'_r)-plane into 6 regions I–VI by the lines $a'_r = 0$, $a'_r \pm b'_r = 0$. In each region a characteristic bifurcation diagram occurs for the unperturbed system. Full (dashed) branches are stable (unstable).

both TW and SW bifurcate simultaneously, with stable persistent solutions present only when both are supercritical. If one of the non-degeneracy conditions is violated, i.e., (a'_r, b'_r) is on a boundary separating the regions in Figure 6.1, terms of higher order have to be taken into account. The bifurcation scenarios for these cases are classified in [Kn86a] and [CK88].

It is worth noting that (6.23), (6.24) admit a whole circle of standing wave solutions since the relative phase $\Theta_1 - \Theta_2$ is arbitrary. The standing waves are therefore neutrally stable with respect to translations according to the $\mathbf{SO}(2)$-action (6.21). This action in turn corresponds to spatial translations $e^{ik_c x} \rightarrow e^{ik_c(x+x_0)}$ $(x_0 = \Psi/k_c)$ of a wave pattern in the original system (2.1). On the other hand, the travelling waves are left unchanged by translations under $\mathbf{SO}(2)$ followed by appropriate phase shifts in $\mathbf{S}^1$. Since the action of $\mathbf{S}^1$ corresponds to temporal translations $e^{i\omega_c t} \rightarrow e^{i\omega_c(t+t_0)}$ $(t_0 = \phi/\omega_c)$ in the original system, this invariance of the travelling wave solutions is a spatio-temporal symmetry. The reflection $\mathbf{Z}_2$ also acts differently on the two types of solutions: it leaves SW unchanged, but interchanges left- and right-travelling waves $(y, 0)$ and $(0, y)$.

Let us interpret the classification of Figure 6.1 for the convection system discussed in Section 5. Validity of the local Ginzburg Landau equations requires ξ to be small, thus we can use the expressions of a, b given in (5.39) and (5.40). Numerical inspection of their real parts shows that $a_r + b_r < 0$ for all $\sigma > 0$, $\rho_c > 0$ and $a_r < 0$, $a_r - b_r < 0$ for sufficiently small ρ_c. The signs of $a_r - b_r$ and a_r change on curves $\rho_c = \rho_{c1}(\sigma)$ and $\rho_c = \rho_{c2}(\sigma)$, respectively, where $0 < \rho_{c1}(\sigma) < \rho_{c2}(\sigma)$. The limiting values on these curves are $(\rho_{c1}(0), \rho_{c2}(0)) = (1.94, 1.98)$ and $(\rho_{c1}(\infty), \rho_{c2}(\infty)) = (1.97, 2.09)$, and $\rho_{c1,2}(\sigma)$ vary monotonically between these values. We conclude that for $0 < \rho_c < \rho_{c1}(\sigma)$, $\rho_{c1}(\sigma) < \rho_c < \rho_{c2}(\sigma)$ and $\rho_c > \rho_{c2}(\sigma)$ the diagrams II, III and IV, respectively, of Figure 6.1 occur. Consequently, for large ρ_c no stable wave patterns are present in the infinitely extended system, whereas for $\rho_c < \rho_{c2}(\sigma)$ either stable standing waves or stable travelling waves bifurcate from the basic state.

6.3 Finite system: Hopf bifurcation with broken circular symmetry

The full system (6.16) (6.17) may be viewed as a perturbation of the Hopf bifurcation with $\mathbf{O}(2)$-symmetry: switching to $\epsilon \neq 0$ has the effect of adding small terms breaking the $\mathbf{SO}(2)$ translation invariance while preserving the reflection and phase shift symmetry of the unperturbed system. A detailed classification of the generic properties induced by the symmetry breaking is given in [DaK91]. In this Subsection we describe the main qualitative features of the perturbed dynamics and exemplify the results of [DaK91] in terms of some selected bifurcation diagrams.

The o.d.e.'s are most conveniently studied in terms of the variables

$$v = r \cos\left(\frac{\phi}{2}\right) e^{i(\varphi+\Theta)/2}, \quad w = r \sin\left(\frac{\phi}{2}\right) e^{i(\varphi-\Theta)/2}, \quad 0 \leq \phi \leq \pi, \qquad (6.26)$$

which leads to a closed system for (r, ϕ, Θ),

$$\dot{r} = \lambda r + \left[a'_r + \frac{1}{2}(b'_r - a'_r)\sin^2\phi\right] r^3 + \epsilon' r \sin\phi \cos\Theta \cos\alpha \qquad (6.27)$$

$$\dot{\phi} = \frac{1}{2}(b'_r - a'_r)r^2 \sin 2\phi + 2\epsilon'(\cos\phi\cos\Theta\cos\alpha - \sin\Theta\sin\alpha) \qquad (6.28)$$

$$\dot{\Theta} = (a'_i - b'_i)r^2 \cos\phi - (2\epsilon'/\sin\phi)(\sin\phi\cos\Theta\cos\alpha - \sin\Theta\sin\alpha), \qquad (6.29)$$

58

and a decoupled equation for the evolution of the total phase $\varphi = \Theta_1 + \Theta_2$,

$$\dot{\varphi} = \omega + \frac{1}{2}(a_i' + b_i')r^2 + (\epsilon'/\sin\phi)(\sin\alpha\cos\Theta - \cos\phi\cos\alpha\sin\Theta). \tag{6.30}$$

In the equations above we have set

$$d = |d|e^{\imath\alpha}, \quad \epsilon' = |d|\epsilon. \tag{6.31}$$

Steady states of the closed (r, ϕ, Θ)-system correspond to periodic solutions of the full system, with the frequency given by $\dot{\varphi}(r, \Theta)$. The coupling of the relative phase $\Theta = \Theta_1 - \Theta_2$ to ϕ and r is here a consequence of the broken $\mathbf{SO}(2)$-invariance. In contrast, the $\mathbf{S}^1$-symmetry is preserved also for $\epsilon \neq 0$, so that φ is still decoupled from the remaining variables. The meaning of the variable ϕ is, loosely speaking, that it interpolates between standing waves and pure travelling waves: for $\phi = \pi/2$ both amplitudes $y_1 = r\cos(\phi/2)$ and $y_2 = r\sin(\phi/2)$ coincide, whereas for $\phi \to 0$ or $\phi \to \pi$ one of them tends to zero. The main properties of the perturbed system are the following:

(1) *Standing waves.* Solutions in the form of standing waves (y, y) continue to exist, but there is no longer a whole circle of them. Instead, the sidewalls select two values of Θ, namely $\Theta = 0$ and $\Theta = \pi$, by a process that may be called phase pinning. We refer to these solutions as SW_0 and SW_π. Their amplitudes and frequencies $\dot{\varphi} = \Omega_{0,\pi}$ are given by

$$\lambda + \frac{1}{2}(a_r' + b_r')r^2 \pm \epsilon'\cos\alpha = 0 \tag{6.32}$$

$$\Omega_{0,\pi} = \omega + \frac{1}{2}(a_i' + b_i')r^2 \pm \epsilon\sin\alpha. \tag{6.33}$$

Thus $SW_{0,\pi}$ bifurcate no longer simultaneously from the trivial solution. The degenerate bifurcation point, $\lambda = 0$, of the unperturbed system splits into two non-degenerate primary bifurcation points $\lambda = \pm\epsilon'\cos\alpha$ at which SW_0 and SW_π are created in succession. This splitting reflects the splitting of the eigenvalues of the Ginzburg–Landau operator $\mathcal{L}_{GL}$ when the Dirichlet boundary conditions evolve into Robin boundary conditions for $\epsilon \neq 0$. Consequently, the effect at small amplitude of the presence of sidewalls is indeed to set up standing waves with a fixed overall phase, as might intuitively be expected.

(2) *Confined travelling waves.* Besides the standing waves $SW_{0,\pi}$, for which $\Theta = 0$ or π and $\phi = \pi/2$, the system (6.27)–(6.29) possesses further, asymmetric stationary solutions in which all three variables (r, ϕ, Θ) vary with λ. Since $\phi \neq 0, \pi$ along

59

this branch, the solutions may be interpreted as a superposition of left- and right-travelling waves of unequal amplitudes ($\phi \neq \pi/2$). In contrast to the unperturbed problem in which such a superposition results in a modulated (i.e. quasiperiodic) travelling wave [Kn86a], in the present case the superposition is characterized by a single frequency $\dot{\Theta}_1 = \dot{\Theta}_2 = \omega + O(r^2)$, a fact that follows immediately from the fixed point condition $\dot{\Theta} = 0$. Moreover, as r or equivalently $|\lambda|$ increases, ϕ approaches 0 or π, and this solution looks more and more like a pure travelling wave at large amplitudes. Consequently we refer to it as CTW and interpret it as a confined travelling wave. The CTW is created in a secondary bifurcation from one of the $SW_{0,\pi}$-branches, which one depending on α. A parametric representation of the CTW-branch with ϕ as parameter can be written in the form

$$(r^2, \lambda) = \frac{\epsilon'}{\delta \sin \phi} \left(\frac{2}{|b' - a'|}, \frac{-2a_r}{|b' - a'|} - \sin(\alpha - \gamma) \sin \alpha \sin^2 \phi \right),$$

$$(\cos \Theta, \sin \Theta) = \frac{1}{\delta} \left(-\cos(\alpha - \gamma), \sin(\alpha - \gamma) \cos \phi \right),$$

where

$$b' - a' = |b' - a'|e^{i\gamma}, \quad \delta = 1 - \sin^2(\alpha - \gamma) \sin^2 \phi.$$

The secondary bifurcation point on the $SW_{0,\pi}$-branch is determined by (6.32), together with the condition

$$|b' - a'|r^2 \cos(\alpha - \gamma) \pm 2\epsilon' = 0$$

for a zero eigenvalue. If $\cos(\alpha - \gamma) < 0$ and > 0, respectively, this bifurcation point is on SW_0 (positive sign) and on SW_π (negative sign). Since the bifurcation of CTW is of the pitchfork-type, there are in fact two such solutions, distinguished by their values of Θ and ϕ, as indicated in Figure 6.2. When $\epsilon \to 0$, CTW evolves towards the pure travelling branch TW of the unperturbed problem.

(3) *Modulated waves.* The $SW_{0,\pi}$-branches may also lose stability in a secondary Hopf bifurcation. This bifurcation leads to a periodic orbit in (6.27)–(6.29), and so is an oscillation in the amplitudes. We refer to this solution as MW. In the (v, w)-space the MW-solution produces motion on a two-torus and therefore corresponds to a modulated wave solution of the original system (2.1). Of the two frequencies that are present, the first is associated with the standing wave component, while the new frequency results in a slow oscillation in (r, ϕ, Θ) and hence in (y_1, y_2). The Hopf bifurcation point on the $SW_{0,\pi}$-branch is determined by (6.32) together with

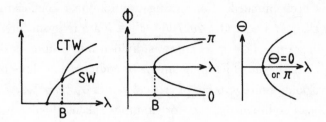

Figure 6.2: Pitchfork bifurcation of CTW from a SW-branch

the condition

$$r^2 \pm \frac{4\epsilon' \cos\alpha}{b'_r - a'_r} = 0$$

for an imaginary eigenvalue. On which of the two branches SW_0 or SW_π this point occurs depends on the sign of $(b'_r - a'_r)\cos\alpha$. The frequency Ω of the oscillation at the bifurcation point is given by

$$\Omega^2 = -4\epsilon'^2 \frac{\cos(2\alpha - \gamma)}{\cos\gamma},$$

showing that the Hopf bifurcation exists only in certain α-ranges. Near the bifurcation point (6.27)–(6.29) can be put into the normal form $\dot\rho = \lambda'\rho + \sigma\rho^3$, where λ' is the local bifurcation parameter, ρ is the amplitude of the oscillation and σ is the cubic Hopf coefficient. These quantities have been computed in [DaK91]. Because it is possible for σ to vanish, the Hopf bifurcation can be sub- or supercritical, depending on α. Of particular interest is the fact that the MW-branch does not persist to large amplitudes. As $|\lambda|$ increases, it terminates either in a global bifurcation or in a tertiary Hopf bifurcation on the CTW-branch. A key to understanding the full behavior of the MW branch is provided by the appearance of a codimension two bifurcation of the Takens-Bogdanov type. The Hopf bifurcation degenerates to such a codimension two bifurcation when $\Omega^2 = 0$, i.e., $2\alpha = \gamma \mod 2\pi$. Near this degeneracy one can describe (6.27)–(6.29) by the unfolded normal form of the Takens-Bogdanov bifurcation with $\mathbf{Z}_2$-symmetry [DaG87, GH90, KnP81]. This normal form depends on two nonlinear coefficients which have been calculated and analyzed in dependence of a', b' [DaK91]. They determine in particular the nature of the termination of MW on CTW and further global features of the MW branch.

(4) *Classification of bifurcation diagrams.* In [DaK91] a relatively complete classification of the possible bifurcation scenarios associated with the perturbed system

(6.27)–(6.29) has been pursued. Concerning the nonlinear coefficients, it turns out that only the angle γ and the ratio $2a'_r/|b' - a'| \equiv k$ are relevant for the structure of the bifurcation diagrams. The (γ, k)-space is divided into a number of open regions, bounded by certain lines on which phenomena of codimension three occur. For fixed (γ, k) in such a region a finite number of critical values α_ν ($\nu = 1, 2, \ldots$) of α exist, at which bifurcations of codimension two, for example Takens-Bogdanov bifurcations or degenerate CTW-bifurcations, occur. The ordering and types of the codimension two points do not change as long as (γ, k) does not cross a boundary. Then, for α passing through a critical value α_ν, the qualitative form of the bifurcation diagram in the (r, λ)-plane changes. This induces a finite sequence of distinct bifurcation diagrams when α traverses the circle. The classification scheme of [DaK91] classifies just these sequences. For example, perturbing the diagram III of Figure 6.1 leads to five distinct bifurcation sequences. One of these is shown in Figure 6.3. The diagrams 1 through 6 occur in succession when α increases from 0 to π. Between π and 2π the sequence is the same, but the roles of SW_0 and SW_π are interchanged. Observe that the MW branch appears and disappears as α traverses the circle. Since typically α depends on the length l of the basic spatial domain, one may expect modulated waves to come and go when l varies.

(5) *Breaking of* $\mathbf{S}^1$. In [DaK87a] it was pointed out that the $\mathbf{S}^1$-translation-symmetry of (6.16), (6.17) is also expected to be broken at some level in the perturbation analysis, because $\mathbf{S}^1$ is a pure normal form symmetry and does not correspond to any geometrical symmetry of the original evolution equation. Although it is difficult to incorporate the breaking of $\mathbf{S}^1$ into the Ginzburg–Landau description, it is clear that the main effects of this further symmetry breaking should be to disturb the quasiperiodic flow on the MW-torus, specifically near heteroclinic or homoclinic connections. In [DaK87a] the linear part of (6.16), (6.17) has been supplemented by additional small terms which break the $\mathbf{S}^1$-symmetry. A Melnikov-type analysis showed then the possibility of transversal intersections of stable and unstable manifolds of the CTW-orbits. Further numerical analysis, presented in [DaKW91], revealed that the two frequencies of the torus may also lock, in a manner as described by the perturbed circle map. Figure 6.4 shows a quasiperiodic orbit on the MW-torus and a frequency-locked state as well as a chaotic state which result from breaking $\mathbf{S}^1$. The parameters are chosen such that diagram 3 of Figure 6.3 occurs and λ increases from (a) to (c) in the range where MW exists. The numerical integrations were performed for (6.16), (6.17) with additional linear terms $\delta(\bar{w}, \bar{v})$, where $\delta/\epsilon' = 0.001 \exp(0.8\imath)$ for (a) and

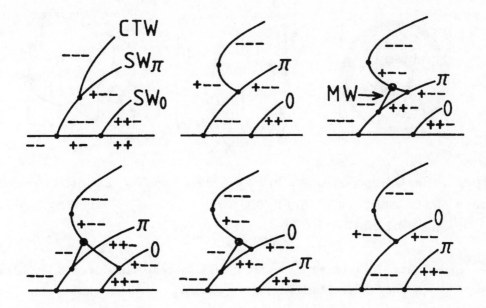

Figure 6.3: Sequence of bifurcation diagrams induced by perturbing diagram III of Figure 6.1. The signs along the branches refer to the stability symbols (real parts of eigenvalues). Small circles indicate global (here heteroclinic) bifurcations of MW.

$\delta/\epsilon' = 0.29 \exp(0.8\imath)$ for (b), (c).

6.4 Small amplitude wave patterns

Noting that (6.16), (6.17) is derived from (6.1)–(6.3) near the first instability Λ_1, we may associate approximate envelopes

$$\begin{pmatrix} \tilde{A}(T,X) \\ \tilde{B}(T,X) \end{pmatrix} = \epsilon^{-1/2} \begin{pmatrix} v(T)e^{-sX/2D} \\ w(T)e^{sX/2D} \end{pmatrix} \cos\frac{\pi X}{2L}, \qquad (6.34)$$

to any solution $(v(T), w(T))$ of the center manifold reduced system. Here, $(A, B) = \sqrt{\epsilon}(\tilde{A}, \tilde{B})$, where the factor $\epsilon^{-1/2}$ has been introduced to obtain quantities of order 1. In particular, the stationary solutions of (6.27)–(6.29) are determined analytically, thus we get analytical expressions for the envelopes describing standing and confined travelling wave solutions. Similarly, the MW solution may be calculated by numerical integrations of (6.16), (6.17) and leads again via (6.34) to temporally oscillating

63

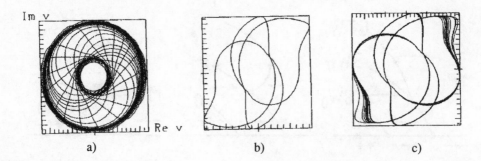

Figure 6.4: (a) Quasiperiodic flow, (b) 2:3-locked state and (c) chaotic state obtained from numerical integration of (6.16), (6.17) with additional, $\mathbf{S}^1$-symmetry breaking terms. (From [DaKW91]).

envelope patterns. All these envelopes are small amplitude solutions of (6.1)–(6.3) of order $O(\sqrt{\epsilon})$. They induce, therefore, $O(\epsilon^{3/2})$-solutions of the original system (2.1) and describe variations of R near R_c of order $O(\epsilon^3)$.

Following [DaKW91] one may proceed further and associate to (6.34) a wave pattern

$$
\begin{aligned}
u_P(t,x) \;=\; & \epsilon^{-1/2} e^{\iota\omega_c t} \left\{ e^{\iota k_c x} v(\epsilon^2 t) e^{-s\epsilon x/2D} + \right. \\
& \left. e^{-\iota k_c x} w(\epsilon^2 t) e^{s\epsilon x/2D} \right\} \cos \frac{\pi\epsilon x}{2L} + cc,
\end{aligned} \tag{6.35}
$$

that corresponds to an appropriate one-dimensional projection of the solution of the original equation (2.1), enlarged by a factor of the order $O(\epsilon^{-3/2})$. Although being small amplitude solutions, these patterns reflect the main qualitative features observed in convection experiments with binary mixtures [FMS88, KoS88]. Figure 6.5 shows the patterns (6.34) and (6.35) for a standing wave SW_0 and a right going travelling wave CTW. Note the "chevron" pattern of the standing wave envelope, resulting from the fact that it is a superposition of right and left going confined waves which peak in the left and right half of the domain, as visible in Figure 6.5(b).

Patterns corresponding to MW are shown in Figure 6.6 for (a) λ near the Hopf bifurcation point and (b) λ close to a heteroclinic connection. One observes here a periodic reversal of confined travelling waves going to the right and to the left, with the period of the reversal diverging when the heteroclinic connection is approached. These patterns are in qualitative agreement with the so called blinking state observed in convection experiments [FMS88, KoS88]. In [DaKW91] the same kind of patterns are also shown for the case in which the $\mathbf{S}^1$ normal form symmetry is broken in

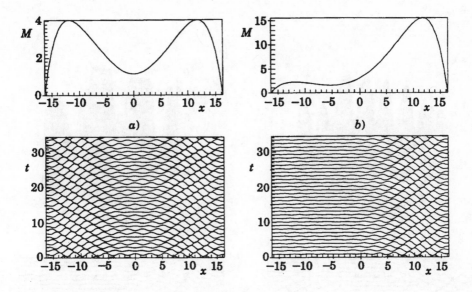

Figure 6.5: Total amplitude $M = (|\tilde{A}|^2 + |\tilde{B}|^2)^{1/2}$ (Eq. (6.34) and the pattern $u_P(t, x)$ (Eq. (6.35)) corresponding to (a): stable SW_0 for $\lambda = 0$, and (b): stable right CTW for $\lambda/\epsilon' = 2.5$. The parameters are $a = -1$, $b = -4$, $D = 1$, $s = 1.2$, $k_c = \pi/\sqrt{2}$, $\omega_c = 2.0$, $\epsilon = 0.35$, $L = 1$, $l = 15$ and $\alpha = 0.6545$, giving rise to diagram 2 of Figure 6.3. (From [DaKW91].)

addition to $\mathbf{SO}(2)$. Chaotic states lead here to aperiodic reversals.

Although the results following from a local bifurcation analysis of the Ginzburg–Landau equations describe only envelopes of very small amplitude and are valid only in a limited range of Λ, they are in excellent qualitative agreement with existing experiments as well as providing a simple analytical explanation for some of the larger-amplitude numerical solutions of (6.1)–(6.3) obtained in [Cr86, Cr88]. In particular, the larger s is the more confined are the eigenfunctions of the Ginzburg–Landau operator and hence the resulting wave patterns. This fact has also been observed by Cross [Cr86, Cr88] in his numerical solutions. We note that Cross scales (6.1) (6.2) so that $\Lambda = 1$, while allowing the length L of the domain to vary. In contrast, we have kept Λ as a bifurcation parameter, fixing the length L. These procedures are related by the scaling $\Lambda = L^2/4$, $s' = 2s/L$, where s' denotes Cross' group velocity. Furthermore, his amplitudes A_L, A_R are related to A, B according to $A_L = (2/L)A$

65

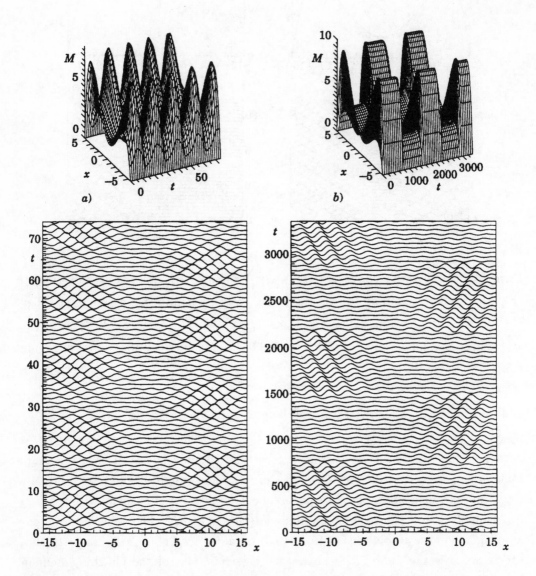

Figure 6.6: Total amplitude M and the pattern $u_P(t, x)$ following from the MW-torus. Here $\alpha = 1.1781$ and the remaining parameters are as in Figure 6.5, giving rise to diagram 3 of Figure 6.3. (a): $\lambda/\epsilon' = 1.5$ and (b): $\lambda/\epsilon' = 1.9476$. (From [DaKW91]).

66

and $A_R = (2/L)B$. Since the analysis described here is valid when Λ is close to Λ_1, we require Cross' parameters (s', L) to be near the curve $s' = 2[1 - (\pi/L)^2]^{1/2}$. Note that, since $0 \leq s' < 2$ for all s, the local analysis cannot describe Cross' regime $s' > 2$ in which he observed different shapes for the envelopes than for $s' < 2$.

The local results described in this and the preceding subsection have also been applied to the Taylor-Couette system with counter-rotating cylinders. With periodic boundary conditions in the axial direction the Hopf bifurcation from Couette flow typically gives rise to stable vortex flow (TW) and unstable ribbons (SW). In [KnP92] it is shown that in the finite system counter-propagating spirals will be the first state observed, followed by a distinct transition to either modified unidirectional spirals (CTW) or, depending on the aspect ratio, to alternating spiral vortex flow which is described by the MW-torus. Similar behavior may be expected for the oscillatory instability of convection rolls [CW89]. The effects of imperfections on flow induced oscillations in tubes can also be described by the o.d.e. normal form (6.16), (6.17) [BS90], although not in the Ginzburg–Landau context as in our case.

Strong symmetry breaking boundary conditions:

It is worth noting that the approach discussed above can be applied only if the boundary conditions may be considered as small symmetry breaking imperfections. In contrast, when (6.3) is replaced by the "strong symmetry breaking boundary conditions" (6.7), the eigenfunctions and eigenvalues of the linearized Ginzburg–Landau operator (6.8) can no longer be determined perturbatively. The small parameter ϵ appears here in a non-perturbative way through the coefficient $\mu = e^{2\imath k_c l} = e^{2\imath k_c L/\epsilon}$. The eigenfunctions have the form

$$
\begin{aligned}
A(K_\pm, X) &= e^{-sX/2D}(P_\pm e^{\imath K_\pm X} + Q_\pm e^{-\imath K_\pm X}) \\
B(K_\pm, X) &= \pm e^{sX/2D}(Q_\pm e^{\imath K_\pm X} + P_\pm e^{-\imath K_\pm X}),
\end{aligned}
$$

where

$$
P_\pm = \mu e^{-sL/D} \pm e^{2\imath K_\pm L}, \quad Q_\pm = -\mu e^{-sL/D} e^{2\imath K_\pm L} \pm 1,
$$

and the wave number $K_\pm$ runs through the infinite set of complex solutions of the transcendental equation

$$
\left(\frac{s}{2DK_\pm}\right) \sinh(sL/D - 2\imath k_c l) \pm \sin(2K_\pm L) = 0.
$$

Note that this equation is even in $K_\pm$, hence the solutions may be restricted to, for example, $\Re K_\pm > 0$. To each $K_\pm$ there corresponds an eigenvalue

$$\Gamma(K_\pm) = -D\left(K_\pm^2 + \frac{s^2}{4D^2}\right)$$

of the Ginzburg–Landau operator $\mathcal{L}_{GL}$, subject to (6.7).

In contrast to the weak symmetry breaking boundary conditions (6.3), the splitting of the eigenvalues according to the $\pm$ signs cannot be expanded in a power series in ϵ. The limit $\epsilon \to 0$ is here not well defined because both L and l appear in the wave number equation. Thus, when Λ is increased from below, the first instability will be generically to a standing wave and will have a finite distance from the next instability. This situation may be described by the standard Hopf normal form, with the cubic Hopf coefficient depending on a, b and D. The sign of this coefficient then determines the stability of the bifurcating standing wave. More information about the nontrivial solution branches of (6.1), (6.2) with (6.7) is generically not available from a local bifurcation analysis.

6.5 Numerical results

To gain some insight into the dynamics of (6.1)–(6.3) for values of Λ beyond the primary instability Λ_1, the modal equations (6.15) have been analyzed numerically in [DaRG93] for $D = 1$, $L = 1$ and $s = 0.2$. The nonlinear coefficients were chosen as $(a, b) = (-1, -4)$ and $(a, b) = (-1, -4) + 4i(1, 1)$ which will be referred to as the real and the complex cases, respectively. Both, the unperturbed problem $\epsilon = 0$ and the perturbed problem have been considered. In the latter case the numerical value of ϵ was fixed to $\epsilon = 0.05$ and the coefficients α, β, γ, δ appearing in (6.3) were selected as $\delta = \bar{\alpha} = -3.052e^{2.725i}$ and $\gamma = \bar{\beta} = 4.254e^{0.637i}$. For these values the local analysis near $\Lambda = \Lambda_1$ predicts the diagram 3 of Figure 6.3.

The validity of this diagram could be confirmed numerically up to $\Lambda \approx 7$, with the global bifurcation of MW on the CTW-branch appearing at $\Lambda \approx 2.3$. In the range $2.3 < \Lambda < 7$ the CTW solution discussed in Subsection 6.3 is the global attractor of the system. In the real case this branch persists as global attractor for all values of Λ which have been considered ($\Lambda \leq 104$). The spatial structure of the envelopes A, B for a left-going CTW-solution is shown in Figure 6.7 for three increasing values of Λ. For $\Lambda = 104$ the envelope B is confined to small ranges near the two ends, whereas A

is nearly constant in the interior of the domain, i.e., the pattern is close to an ideal travelling wave.

In the complex case already the unperturbed system ($\epsilon = 0$) displays a rich variety of dynamical behavior. Figure 6.8 shows the global bifurcation diagram for $\Lambda \leq 20$, i.e., in a range where two primary modes become unstable. The primary travelling wave branch, (b), loses stability at $\Lambda = 11.4$ in a subcritical Hopf bifurcation giving rise to a stable two-torus. In addition to the primary bifurcation of travelling and standing waves at $\Lambda = \Lambda_1$ and $\Lambda = \Lambda_2$, we also find two-mode travelling waves, (c), so called because the first two eigenmodes are of comparable magnitude. This state appears from a saddle node bifurcation at $\Lambda = 9.95$ and eventually loses stability through a Hopf bifurcation to an invariant torus at $\Lambda = 11.2$. The second primary travelling wave state, (e), bifurcates at $\Lambda = \Lambda_2$ from the trivial state and has two saddle node points. Bistability of tori can occur when a second stable torus is generated in a bifurcation from the second travelling wave at the upper saddle node point at $\Lambda = 10.98$. In addition, the first primary wave, (a), is also stable for values of Λ which overlap the range where the stable tori exist. There are then up to three attractors at ranges of Λ between the instability of the second and third eigenvalues. Beyond $\Lambda = 11.2$ no stable pure wave state can exist and the dynamics becomes more complex.

For $\epsilon = 0.05$ and complex coefficients the global bifurcation of the standing and confined travelling waves are similar to these occurring for $\epsilon = 0$, besides a shift of the bifurcation points and the splitting of the primary bifurcations. Here the CTW-state created near Λ_1 loses stability in a subcritical Hopf bifurcation at $\Lambda = 8.7$. Two-mode CTW-solutions are also present for this case appearing at $\Lambda = 7.98$. Additional tori are generated via bifurcations from confined travelling branches. These tori differ from the MW-torus determined from (6.16), (6.17) in that ($|A|$, $|B|$) oscillates about a single confined wave state. There are then again ranges of Λ in which up to three attractors exist simultaneously. Figure 6.9(a) shows the primary CTW-state, (a), and two tori, (b) and (c), which are generated from the two-mode CTW and from the second single-mode CTW, respectively. Here, $p_n = \frac{1}{2}(v_n + w_n)$ ($n = 1, 2$) are standing wave amplitudes. In the projection onto the ($|p_1|$, $|p_2|$)-plane, the CTW-orbits appear as fixed points and the tori appear as periodic orbits. At this value of Λ the basins of attraction are fractalized, resulting in chaotic transients. A chaotic trajectory which eventually approaches the state (a) is shown in Figure 6.9(b). When Λ is increased further, the additional tori are destroyed in period doubling cascades and create chaotic attractors. For details we refer to [DaRG93].

69

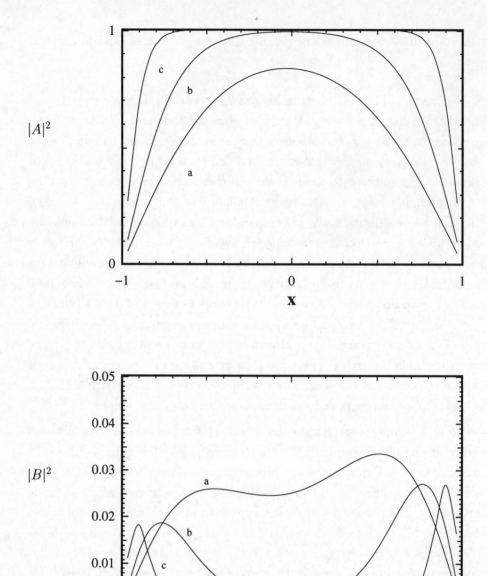

Figure 6.7: Spatial structure of the stable confined travelling wave envelopes in the case of real nonlinear coefficients. a: $\Lambda = 4$, b: $\Lambda = 14$, c: $\Lambda = 104$. (From [DaRG93].)

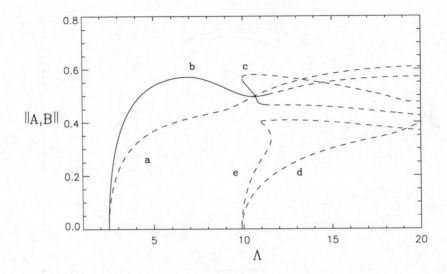

Figure 6.8: Bifurcation diagram of wave states $\|(A, B)\|$ vs. Λ for $\epsilon = 0$ in the case of complex coefficients. a: primary standing wave, b: primary travelling wave, c: two-mode travelling wave, d: second primary standing wave, e: second primary travelling wave. (From [DaRG93].)

7 Discussion and Conclusion

In this article we have studied the dynamics of waves, which are created through oscillatory instabilities in evolution systems of large spatial extent in one distinguished direction. The full solution of the basic evolution equation is described in terms of two envelopes representing slow modulations of left and right going traveling waves. In the idealized, infinitely extended case the interaction of these waves is governed by coupled envelope equations of the Ginzburg–Landau type. Of particular importance is the observation, first recognized by Knobloch and DeLuca [KD90], that the waves are coupled in rather different manners for the cases of small and large group velocities. While small group velocities lead to local couplings, the case of large group velocity induces global coupling terms in which either wave is coupled to the spatial energy

71

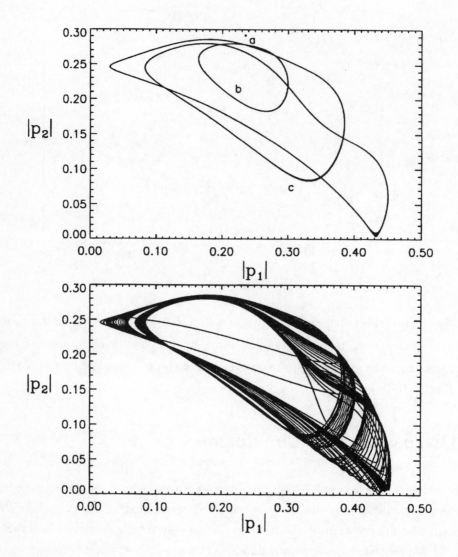

Figure 6.9: (a) Multiple stable attractors at $\Lambda = 8.65$. a: primary CTW, b and c: stable tori (see text). (b) chaotic trajectory due to fractalization of basin boundaries. (From [DaRG93].)

mean of the other wave. This means that energy is transported so rapidly that a wave cannot react to local variations of the counter propagating wave. A first step towards a rigorous justification of the nonlocal envelope equations in an infinitely extended domain has been achieved recently [PiW95] for the case of a conservative system. Also in the conservative context these equations have recently been shown to govern the stability of warer waves with bounded spectra [PiK95]. It is interesting to note that the local equations do not admit spatially quasiperiodic patterns while such patterns may occur in the globally coupled situation [KD90, MV92]. Stability calculations for this kind of states have been pursued in [MV92], where the globally coupled system is derived for the equations describing a gasless combustion front. In contrast, the local equations with unbounded X show phase turbulence in certain parameter regimes [JPBCRK92].

The two kinds of envelope equations behave also differently when distant side walls are present. In the local case the equations are only distorted in the boundary conditions which are modified from Dirichlet to Robin type, combined with a breaking of the spatial $\mathbf{SO}(2)$-translation invariance. Again, this may be interpreted as a consequence of the slow energy transport in the system, i.e., the sidewalls cause only a local interaction between incoming and reflected waves. In contrast, large group velocities enforce a global balance between the two counter propagating waves. Mathematically this is reflected by the fact that the coupled system reduces to a single integro-differential equation for one of the two envelopes, whose solution completely determines the shape of the other one. This reduction was, however, only possible if no energy loss takes place at the boundaries, i.e., if the reflection coefficient r has modulus one, $|r| = 1$. For $|r| < 1$ energy is transported rapidly through the bulk and lost at the boundaries, so that disturbances of the basic state cannot persist slightly above the stability threshold of the infinitely extended system. For a physical discussion of the meaning of the reflection coefficient in the case of the local equations we refer to [CrK92].

Owing to the decomposition of $\mathcal{H}$ into finite dimensional subspaces, invariant under $\mathcal{L}(\mu, \imath k, r)$, the linear stability analysis of Section 2 reduced to an investigation of a set of determinantal equations $g_n(\mu, k^2, R) = 0$. Moreover, this decomposition allowed us to compute the boundary conditions (4.17), (4.18) and (4.21) entirely from the restriction of $\mathcal{L}$ to the critical subspace $\mathcal{H}_1$. Such decompositions are typical for convection systems with stress free boundary conditions at the top and bottom and for certain classes of reaction diffusion equations. For such problems it is often possible to calculate the leading linear and nonlinear coefficients of the Ginzburg–

Landau equations as well as the coefficients occurring in the boundary conditions analytically. As an example for such an ideal situation we discussed in Section 5 two-dimensional Boussinesq convection in a vertical magnetic field. For convection systems with realistic (rigid) boundary conditions or for reaction diffusion equations in cylindrical domains with boundary conditions of the mixed (Robin) type the situation is more complicated. The linearized equation (2.7) is then no longer separable into a sequence of linear p.d.e.'s posed in finite dimensional spaces, and often already the linear stability analysis must be pursued numerically.

Formally, the linear stability analysis requires in such cases the solution of an eigen-value problem. When $\mathcal{L}(\mu, \imath k, R)$ ($\mu \in \mathbb{C}$, $k \in \mathbb{R}_+$) is an elliptic differential operator over a bounded domain, its spectrum is discrete and one can replace the determinantal equations $g_n(\mu, k^2, R) = 0$ by $\rho_n(\mu, k^2, R) = 0$ ($n = 1, 2, \ldots$), where the ρ_n are the real eigenvalues of $\mathcal{L}$. An oscillatory instability occurs then if for some fixed m a set of par-ameters $(\imath\omega_c, k_c^2, R_c)$ exists such that $\rho_m(\imath\omega_c, k_c^2, R_c) = 0$ and $\Re\mu < 0$ holds whenever (μ, k^2) is a solution of one of the equations $\rho_n(\mu, k^2, R) = 0$ for $R < R_c$. Generically, $\rho_m(\mu, k, R_c)$ depends smoothly on the parameters for (μ, k^2, R) sufficiently close to $(\imath\omega_c, k_c^2, R_c)$, although the global variation of the ρ_n is continuous, but generally non-smooth. The leading order coefficients that depend entirely on the linearized system, like the group velocity c_g or the diffusion constant D, may then be expressed in terms of derivatives of $\rho_m(\mu, k^2, R)$, evaluated at $(\imath\omega_c, k_c^2, R_c)$, in the same way as described in Sections 2 and 3 in terms of derivatives of $g_1(\mu, k^2, R)$.

A more general approach to the derivation of boundary conditions for the envelopes, including the case in which $\mathcal{L}$ and the boundary operators do not admit separation into finite dimensional problems, has not yet been developed. Nevertheless, in prac-tice one has to proceed numerically anyway and this is most conveniently done by means of a suitable Galerkin projection. The projected space is finite dimensional, so that the procedure discussed in Section 4 can be applied, although some care is necessary when numerical computations are combined with asymptotic expansions. The expressions for the coefficients appearing in the envelope equations (3.45), which are summarized in Section 3 and the appendix, do not depend on the separability of the linearized problem. They require only knowledge of the critical quantities fol-lowing from the linear stability analysis. For these computations the Fourier space formulation employed in Section 3 turned out to be very useful.

An advantage of the local Ginzburg–Landau equations is that they provide a well posed parabolic system of partial differential equations over a finite domain. In particular, the transition from the infinitely extended system to a system with sidewalls is simply described by small symmetry breaking terms in the boundary conditions of the envelopes. This allows small amplitude solutions to be analyzed to a large extent analytically, by means of a rigorous center manifold reduction near the first instability. It is worth mentioning that the o.d.e.-normal form of Subsection 6.3 captures not only oscillatory instabilities in extended systems. It may also be applied to problems with a slightly distorted *geometrical* O(2)-symmetry, like the study of the effects of rotational symmetry breaking imperfection on flow-induced oscillations in tubes [BS90]. In the o.d.e.-normal form context, each wave pattern in a large but finite system has its counterpart in an oscillatory pattern of a system with broken geometrical O(2)-symmetry. For example, in the tube problem the two standing waves are identified with planar motion in one of the two planes selected by the perturbation, the confined travelling waves correspond to motion in ellipses (as opposed to circles) and the modulated waves represent here modulated (i.e. precessing) oscillations. Other applications in that direction include reaction diffusion equations in slightly distorted annuli [DaK89].

An interesting observation in the numerical analysis [DaRG93] of the local Ginzburg–Landau equations (Subsection 6.6) is that the solution structure predicted by the o.d.e-normal form persists over the full range of the bifurcation parameter when the coefficients are real. In contrast to that, in the complex case the dynamics becomes extremely complicated beyond the validity regime of the o.d.e.-normal form. Both chaotic attractors and fractalized basins of attraction have been observed already beyond the second instability. Moreover, we find that the qualitative global dynamics are similar in both the perfect and perturbed systems, except for a shift in bifurcation values. Thus the global behavior of the system is not essentially changed by the symmetry breaking.

When spatial modulations are neglected and no sidewalls are present, both the local and global Ginzburg Landau equations reduce to the o.d.e.-normal from for a Hopf bifurcation with O(2)-symmetry. This normal form and its generic solutions provided the first and basic step towards the understanding of wave propagation in dissipative systems [GSS88] with one allowed direction of propagation. Analogously, attempts to generalize the theory to two unbounded directions should be based on the Hopf bifurcation with O(2) × O(2)-symmetry. If the system is isotropic, the imposition of doubly periodic boundary conditions leads to an extension of $\mathbf{O}(2) \times \mathbf{O}(2)$ by $\mathbf{Z}_6$, $\mathbf{Z}_4$ or $\mathbf{1}$, depending on the chosen lattice. The Hopf bifurcation normal forms for these symmetries have been analyzed in [RSW86], [SK91] and [SRK92], [We93], respectively, and from a unified point of view in [DGSS95]. In contrast to the simple $\mathbf{O}(2)$-case, these normal forms give rise to a rich variety of dynamical behaviors. In particular, there is the possibility of structurally stable heteroclinic cycles. Exploring the effect of spatial modulations on these states is a challenge for future research.

Acknowledgements First of all I would like to express my sincere gratitude to M. Wegelin for his help in typesetting this manuscript. I also appreciate fruitful and stimulating discussions with him and the other colleagues at the Institut für Informationsverarbeitung, particularly with C. Geiger, J. Hettel, J. Oppenländer and J. D. Rodriguez. Much of the results reviewed in Section 6 originated from the collaboration with E. Knobloch, J. D. Rodriguez and M. Wegelin. I also benefitted from helpful discussions with my colleagues at the Forschungsschwerpunkt *Nonlinear Dynamics of Continuous Systems* of the state Baden-Württemberg: from B. Fiedler I learned much about homoclinic bifurcations, W. Güttinger stimulated the analysis of magnetoconvective instabilities and K. Kirchgässner and A. Mielke explained to me the functional analytic aspects of pattern formation in large systems. Finally I would like to thank E. Knobloch, A. Mielke and G. Schneider for carefully reading the manuscript and drawing my attention on some related papers, specifically on recent work concerning the rigorous justification of the Ginzburg-Landau formalism.

Appendix

In this appendix we summarize explicit expressions for the vectors occurring in (3.43) and (3.45). We first introduce some notation. Besides the operators

$$\mathcal{L}_+ = \mathcal{L}(\imath\omega_c, \imath k_c; R_c), \quad \hat{\mathcal{L}}_+ = Q_+\mathcal{L}|_{\mathcal{H}_c^+},$$

where $\mathcal{H}_c^+ = Q_+\mathcal{H}_c$ and Q_+ is the complementary projection $Q_+ = id - \langle \cdot, v_0^* \rangle v_0$ in $\mathcal{H}_c$, we need some of the Taylor terms of $\mathcal{L}$ at criticality. For these we use the notation (cf. Eqs. (3.13),(3.28)),

$$\mathcal{L}_{\alpha\beta}^+ \equiv \mathcal{L}_{\alpha\beta0}(1,1) = \frac{1}{\alpha!\beta!} \frac{\partial^{\alpha+\beta}}{\partial(\imath\omega)^\alpha\partial(\imath k)^\beta}\mathcal{L}(\imath\omega, \imath k; R_c)|_{\omega=\omega_c, k=k_c}$$

$$\mathcal{L}_1^+ \equiv \mathcal{L}_{001}(1,1) = \frac{\partial}{\partial\lambda}\mathcal{L}(\imath\omega_c, \imath k_c; R_c(1+\lambda))|_{\lambda=0}\,.$$

In terms of these operators, the vectors w_{10}, w_{01} and w_1 that appear at linear order in (3.45), are given by

$$w_{10} = \mathcal{L}_{10}^+ v_0\,, \quad w_{01} = \mathcal{L}_{01}^+ v_0\,, \quad w_1 = -\mathcal{L}_1^+ v_0\,, \tag{A.1}$$

and from the first two of these the leading vectors in (3.43) are uniquely defined by the linear equations

$$\hat{\mathcal{L}}_+ v_\sigma = -\imath Q_+ w_{10}\,, \quad \hat{\mathcal{L}}_+ v_\kappa = -\imath Q_+ w_{01}\,. \tag{A.2}$$

The second order linear terms in (3.45) then follow from v_σ, v_κ and v_0 as

$$\begin{aligned} w_{20} &= \mathcal{L}_{20}^+ v_0 - \imath \mathcal{L}_{10}^+ v_\sigma \\ w_{02} &= \mathcal{L}_{02}^+ v_0 - \imath \mathcal{L}_{01}^+ v_\kappa \\ w_{11} &= \mathcal{L}_{11}^+ v_0 - \imath \mathcal{L}_{10}^+ v_\kappa - \imath \mathcal{L}_{01}^+ v_\sigma\,. \end{aligned} \tag{A.3}$$

This completes the derivation of the linear coefficient vectors in (3.43) and (3.45). Note, however, that the relevant linear coefficients that appear in the nonlocal and the local rescaled envelope equations (4.32) and (4.42),(4.43) (D and D_{02} for $c_g = O(1)$ and $c_g \approx 0$, respectively, as well as a_0) can be directly inferred from the determinant $g_1(\mu, k^2, R)$ according to (3.47) and (3.48).

The expressions for the nonlinear coefficient vectors involve also the "non-critical" operators

$$\mathcal{L}_0(n,m) \equiv \mathcal{L}(\imath n\omega_c, \imath m k_c; R_c)\,.$$

In addition we need the following symmetrized version of the kernel G_2, Eq. (3.8),

$$\begin{aligned} G_2^s(n_1, m_1 | n_2, m_2)[u_1, u_2] &\equiv G_2(\imath n_1\omega_c, \imath m_1 k_c | \imath n_2\omega_c, \imath m_2 k_c)[u_1, u_2] \\ & \quad G_2(\imath n_2\omega_c, \imath m_2 k_c | \imath n_1\omega_c, \imath m_1 k_c)[u_2, u_1]\,, \end{aligned}$$

but fortunately local derivatives of the kernels with respect to the Fourier variables are not needed up to the order we present. The nonlinear coefficient vectors v_{nm}^0 for $n, m \in \{0, 2\}$ in (3.43) are then uniquely defined by

$$\begin{aligned} \mathcal{L}_0(0,0)v_{00}^{(0)} &= -G_2^s(1,1|-1,-1)[v_0, \bar{v}_0] \\ \mathcal{L}_0(2,0)v_{20}^{(0)} &= -G_2^s(1,1|1,-1)[v_0, Sv_0] \\ \mathcal{L}_0(0,2)v_{02}^{(0)} &= -G_2^s(1,1|-1,1)[v_0, S\bar{v}_0] \\ \mathcal{L}_0(2,2)v_{22}^{(0)} &= -G_{20}^s(1,1|1,1)[v_0, v_0]\,, \end{aligned} \tag{A.4}$$

since the operators on the l.h.s. are all invertible. From these vectors one finally obtains the cubic coefficient vectors in (3.45) as

$$
\begin{aligned}
w_+ &= -G_2^s(1,1|0,0)[v_0, v_{00}^{(0)}] - G_2^s(-1,-1|2,2)[\bar{v}_0, v_{22}^{(0)}] - \tilde{w}_+ \\
w_- &= -G_2^s(1,1|0,0)[v_0, Sv_{00}^{(0)}] - 2G_2^s(1,-1|0,2)[Sv_0, v_{02}^{(0)}] \\
&\quad -G_2^s(-1,1|2,0)[S\bar{v}_0, v_{20}^{(0)}] - \tilde{w}_-
\end{aligned}
\tag{A.5}
$$

Here, $\tilde{w}_+$ and $\tilde{w}_-$ are defined in terms of the cubic kernel G_3,

$$
\begin{aligned}
\tilde{w}_+ &= G_3(+|+|-)[v_0, v_0, \bar{v}_0] + G_3(+|-|+)[v_0, \bar{v}_0, v_0] + G_3(-|+|+)[\bar{v}_0, v_0, v_0] \\
\tilde{w}_- &= \sum_{\pi \in S_3} G_3(\pi(1), \pi(2), \pi(3))[v_{\pi(1)}, v_{\pi(2)}, v_{\pi(3)}],
\end{aligned}
$$

where in the first equation the "+" and the "-" signs stand for $(\imath\omega_c, \imath k_c)$ and $(-\imath\omega_c, -\imath k_c)$, respectively. In the second equation we use the notation

$$
(1|2|3|) = ((\imath\omega_c, \imath k_c), (\imath\omega_c, -\imath k_c), (-\imath\omega_c, \imath k_c), \quad (v_1, v_2, v_3) = (v_0, Sv_0, S\bar{v}_0),
$$

and the sum runs over all six permutations $(1,2,3) \rightarrow (\pi(1), \pi(2), \pi(3))$ of S_3.

Chapter 2
Global Pathfollowing of Homoclinic Orbits in Twoparameter Flows

by **B FIEDLER**

Contents

1 Introduction

A *heteroclinic orbit* of a flow is by definition a solution $x = h(t)$, $t \in \mathbb{R}$, of a differential equation

$$(1.1) \qquad \dot{x} = f(\lambda, x), \quad x \in X = \mathbb{R}^N$$

with or without parameters λ, such that $h(t)$ converges for both $t \to +\infty$ and $t \to -\infty$,

$$(1.2) \qquad A_\pm := \lim_{t \to \pm\infty} h(t), \quad A_+ \neq A_-.$$

The distinct limits are equilibria of (1.1), by invariance of ω-limit sets. If $A_+ = A_-$, then $h(t)$ is called a *homoclinic orbit*. In either case, we call $A_\pm$ the *associated equilibria* of the (homoclinic or heteroclinic) orbit $h(t)$. See figure 1.1. For abbreviation we occasionally write hom/het for a homoclinic/heteroclinic orbit. Technically speaking, if $A_\pm$ are hyperbolic equilibria, then homs or hets arise as nontrivial intersections of the respective unstable and stable manifolds

$$W^u(A_-) \cap W^s(A_+).$$

See for example [CH82], [Ar93], [GH90], [Wi88], [Ku95] for a general background.

From an applied point of view, homs serve as one possible cause of complicated dynamics. We just mention the homoclinic explosion generating the Lorenz attractor, see e.g. [Sp82], and the celebrated example due to Shilnikov generating shift dynamics for flows. See examples 2.2, 5.4, 5.8, and definition 2.3 below.

Partial differential equations are another important source of applications. We explicitly mention travelling waves, solitons, boundary layers in singular perturbation

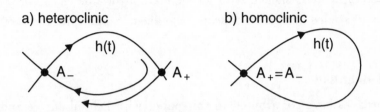

Figure 1.1: A heteroclinic orbit (a) and a homoclinic orbit (b).

problems, and viscous profiles for systems of conservation laws. Already ten years ago, for example, Shilnikov type shift dynamics have been discovered for travelling waves of the FitzHugh-Nagumo model of nerve conduction, see [EFF82], [Fe82], [Ha82]. For further applications see the contributions by Dangelmayr, Kirchgässner, Mielke in this volume.

An embracing systematic theory of homoclinic bifurcations in two parameter systems, $\lambda \in \mathbb{R}^2$, is not in sight. Rather, there appear to be scores of different cases which need to be analyzed separately. Some of the cases which have been analyzed successfully, so far, will be discussed here. In fact we will not add to that list. Instead we will introduce some new concepts, based on a few relevant examples, which allow for a global *pathfollowing approach* to homoclinic bifurcation. Essentially we try to follow paths of homoclinic orbits up to regions in parameter and phase space with complicated recurrent dynamics. Our main result in this direction, theorem 3.4, is formulated in terms of the new concepts: *tame* versus *chaotic* homoclinic orbits, *stratified* versus *non-stratified* codimension 2 *loops*, and *bounded* versus *unbounded* ϵ-*length*. The concept of stratification is open in the following sense: given a particular codimension 2 homoclinic or heteroclinic loop, it may not be obvious at all, whether this loop is stratified or not. We discuss some examples below, for which this question has been settled. In general, the question is open. Our theorem, as it is formulated here, can only gain scope in the future, when more and more particular cases of homoclinic bifurcation will be understood.

We conclude this introduction with an outline. The new concepts mentioned above are introduced in section 2, together with a select few examples. Section 3 explains our global continuation idea, in those terms, and presents the main result, theorem 3.4. Proofs are relegated to section 4. In section 5, we demonstrate how our continuation process works for some, mostly local, examples from the literature. Two more globally-minded examples are designed, with an inclination towards chaos, in section 6. We finish with a detailed discussion of scope, significance, and limitations of our results, in section 7.

Acknowledgement.

Many colleagues have contributed to this paper with helpful advice and generously shared insight. In particular I would like to mention Valja Afraimovich, Jay Alexander, Don Aronson, Shui-Nee Chow, Gerhard Dangelmayr, Bo Deng, Kazimierz Gęba, Jörg Härterich, Ale Jan Homburg, Klaus Kirchgässner, John Mallet-Paret, Alexander Mielke, Björn Sandstede, Arnd Scheel, Misha Shashkov, Floris Takens, Dima Turaev, André Vanderbauwhede, and J. Yorke. For careful typesetting I am much indebted to E. Schlumberger and R. Löhr.

2 Basic examples and definitions

In this section, we collect and adapt some known results and examples related to homoclinic bifurcation. Based on five examples, we introduce the notions of *tame* versus *chaotic* homoclinic orbits, of *ε-length*, and of *stratified* homoclinic or heteroclinic loops. These concepts are central to our subsequent analysis.

2.1 Example: B-point

Consider the planar two-parameter system

$$(2.1) \quad \begin{aligned} \dot{x}_1 &= x_2 \\ \dot{x}_2 &= \lambda_1 + \lambda_2 x_1 + x_1^2 + x_1 x_2. \end{aligned}$$

As noticed by [Ar72], [Ta74], system (2.1) describes a generic local normal form of an equilibrium with algebraically double eigenvalue zero, truncated at second order, and unfolded on the linear level.

In parameter space $\lambda = (\lambda_1, \lambda_2) \in \Lambda = \mathbb{R}^2$, the bifurcation diagram of figure 2.1 has been obtained by [Bo81]. See also the accounts in [Ar83] and [GH90]. Crossing the fold

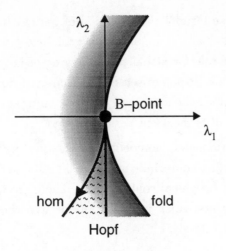

Figure 2.1: Bifurcation diagram of a B-point, example 2.1.

curve in figure 2.1 from right to left, two equilibria appear by saddle-node bifurcation. Crossing the Hopf line, one of these equilibria undergoes a Hopf bifurcation. Crossing the hom line, the generated periodic orbit becomes homoclinic to the associated saddle equilibrium, and disappears. Moving the parameter λ along the hom curve towards the B-point at $\lambda = (0,0)$, the homoclinic orbits shrink to the equilibrium $x = 0$ in phase space.

System (2.1) is a universal unfolding, as was proved by [Bo81]. In fact, consider a general smooth two-parameter system

$$(2.2) \qquad \dot{x} = f(\lambda, x), \quad \lambda \in \Lambda = \mathbb{R}^2, \quad x \in X = \mathbb{R}^N,$$

with an equilibrium at $(\lambda_0, x_0) = (0,0)$. Assume the linearization $D_x f(\lambda_0, x_0)$ possesses, aside from a hyperbolic part, an algebraically double eigenvalue zero with precisely one 2×2 Jordan block. Then, locally in a center manifold, system (2.2) can be transformed to form (2.1), provided a certain nondegeneracy condition holds. The transformation is diffeomorphic in λ, homeomorphic in x, and possibly requires a reversal of time. In particular, the bifurcation diagram of figure 2.1 still holds locally up to a diffeomorphism.

For brevity, we call the Arnol'd-Bogdanov-Takens singularity (λ_0, x_0) a B-point henceforth. More precisely, we require in addition to the above nondegeneracy assumptions that the B-point itself is not associated to any (large) homoclinic orbit. We note here that these assumptions are all generic in the sense of definition 2.4 below.

B-points play a central role for global bifurcation of periodic orbits in systems with one or two parameters [Fi86b]. It is our main goal below to show how B-points play an equally important role for global pathfollowing of homoclinic orbits in two-parameter flows. Intuitively, B-points serve as a source of homoclinic branches, in two parameters, in very much the same way as Hopf bifurcations, in one parameter families, serve as a source of branches of periodic orbits.

The similarity between periodic and homoclinic bifurcations will in fact be more than a superficial analogy. Viewing homoclinic orbits as limits of periodic orbits will serve as our basic technical tool for global continuation. Therefore we now consider some examples of homoclinic orbits which generate unique hyperbolic periodic orbits or shift dynamics nearby, respectively.

2.2   Example: one-parameter homoclinics

Consider a smooth one-parameter family

(2.3) $\dot{x} = f(\lambda, x)$,

$\lambda \in \mathbb{R}$, $x \in X = \mathbb{R}^N$, with a homoclinic orbit $x = h(t)$ and associated equilibrium $x = A_0$ at $\lambda = \lambda_0 = 0$. Let spec denote the spectrum of the linearization $D_x f(\lambda_0, A_0)$. The *principal eigenvalues* $\mu_\pm$ are those elements of spec with $\pm Re \geq 0$ closest to the imaginary axis. We assume $\mu_\pm$ to be simple and, up to complex conjugacy, unique. We distinguish five cases:

(2.4.a)   spec is hyperbolic, and the principal eigenvalues $\mu_\pm$ are both real with $\mu_+ + \mu_- \neq 0$;

(2.4.b)   spec is hyperbolic, and $\{\mu_+, \overline{\mu}_+\}$, $\{\mu_-, \overline{\mu}_-\}$ are conjugate complex pairs with $Re\mu_+ + Re\mu_- \neq 0$;

(2.5.a)   spec is hyperbolic, and one of $\mu_\pm$ is real while the other forms a complex conjugate pair such that the real part of the complex pair is *larger* than the real eigenvalue, in absolute value;

(2.5.b)   spec is as in (2.5.a), but the real part of the complex pair is *less* than the real eigenvalue, in absolute value;

84

(2.6) $\mu_+ = \mu_- = 0$ is just one simple eigenvalue, and (λ_0, A_0) lies on a nondegenerate quadratic, stationary fold $\lambda = \lambda(\sigma)$, $A = A(\sigma)$ such that $A(0) = A_0$, $A'(0) \neq 0$; $\lambda(0) = \lambda_0$, $\lambda'(0) = 0$, $\lambda''(0) \neq 0$.

In the hyperbolic cases (2.4) and (2.5), the equilibria near A_0 are parametrized over λ as $A = A(\lambda)$, locally near $\lambda = \lambda_0$, due to the implicit function theorem. We obtain associated unstable and stable manifolds $W^u = \bigcup_\lambda W_\lambda^u$, $W^s = \bigcup_\lambda W_\lambda^s$, which we may consider as immersed submanifolds of $\mathbb{R} \times X$, smoothly fibered over the parameter $\lambda \in \mathbb{R}$. We assume

(2.7) $W^u \pitchfork W^s$ along (λ_0, h) in $\mathbb{R} \times X$.

As usual, $\pitchfork$ indicates transverse intersection: at any intersection point the two tangent spaces of W^u, W^s together span the whole space $\mathbb{R} \times X$. In terms of the fibers W_λ^u, W_λ^s, we can express this assumption by saying that W_λ^u, W_λ^s cross each other, along h, with nonzero speed at $\lambda = \lambda_0$.

In the fold case, by assumption (2.6), the stable set

$$W_{\lambda_0}^s(A_0) := \{x_0 \mid x(t) \to A_0 \text{ for } t \to +\infty\}$$

is also an immersed submanifold of $X = \mathbb{R}^N$ with boundary given by the strong stable manifold $W_{\lambda_0}^{ss}(A_0)$. Similarly, $W_{\lambda_0}^u(A_0)$, the unstable set, will be an immersed submanifold with boundary $W_{\lambda_0}^{uu}(A_0)$. We assume transversality again:

(2.8) $W_{\lambda_0}^s(A_0) \pitchfork W_{\lambda_0}^u(A_0)$ along h in X.

Note that we have fixed the λ_0-fiber, this time, requiring that the tangent spaces of the *fibers* together span X.

Finally we assume that

(2.9)   the homoclinic orbit $h(t)$ approaches A_0 along the principal directions, that is, along the eigenspaces of $\mu_\pm$, for $t \to \mp\infty$.

Under the conditions (2.7) – (2.9), the celebrated work of [Shi62, Shi65, Shi66, Shi67, Shi68, Shi69, Shi70] essentially leads to the following results. Fix a small neighborhood $\mathcal{N}$ of h in $X = \mathbb{R}^N$. In cases (2.4.a), (2.5.a), (2.6), the neighborhood $\mathcal{N}$ then contains at most one non-stationary periodic orbit, for λ near λ_0. In fact, one periodic orbit exists precisely for λ on one side of $\lambda_0 \in \mathbb{R}$. The periodic orbit is hyperbolic.

In contrast, consider cases (2.4.b), (2.5.b). In these cases, $\mathcal{N}$ contains Smale horseshoes. In particular, a shift on finitely many symbols can be embedded into the given flow at, and near, $\lambda = \lambda_0$.

eigenvalues	μ_+ real	μ_+ complex
μ_- real	tame	tame
μ_- complex	chaotic	chaotic

Table 2.1: Homoclinic orbits with principal eigenvalues $Re\mu_+ > |Re\mu_-| > 0$.

We condense these results into the following definition.

2.3 Definition: tame, chaotic homoclinics

Consider a homoclinic orbit h as in example 2.2. In particular, assume that nonde-generacy assumptions (2.7) – (2.9) hold. We call h tame in cases (2.4.a), (2.5.a), (2.6). We call h chaotic in cases (2.4.b), (2.5.b).

In short, tame homs are accompanied by a unique branch of hyperbolic periodic orbits, while chaotic homs are accompanied by shift dynamics. The hyperbolic cases are listed in table 2.1. Note that our definition of tame and chaotic is invariant under time reversal.

We emphasize the local character of our definition of tame versus chaotic with an example due to [Shi69]. Consider the fold case (2.6) of example 2.2 again. Conceivably, the stable and unstable sets $W_{\lambda_0}^s(A_0)$ and $W_{\lambda_0}^u(A_0)$ could intersect, transversely as in (2.8), along several distinct homs $h_1, \ldots, h_m$ simultaneously. In that case, the shift on m symbols embeds into the flow for those values of λ near λ_0, for which the equilibria have disappeared. Nevertheless, we would call each individual orbit h_j tame, because the shift dynamics is encoded by excursions along the respective orbits $h_1, \ldots, h_m$, and thus leaves a fixed small neighborhood $\mathcal{N}_j$ of h_j, in general.

The nondegeneracy assumptions entering into examples 2.1, 2.2 and definition 2.3 are somewhat awkward to explicate. To handle nondegeneracy conditions more summarily, we use the notion of genericity.

86

2.4 Definition: generic

Let $\mathcal{G}$ be a subset of a topological Baire space $\mathcal{F}$. We call $\mathcal{G}$ generic if $\mathcal{G}$ contains a countable intersection of open, dense subsets of $\mathcal{F}$. In particular, generic sets are still dense. We likewise call elements of $\mathcal{G}$ generic, as well as the properties defining $\mathcal{G}$.

In our setting, $\mathcal{F} = C^k(\Lambda \times X, X)$, $\dim \Lambda < k \leq \infty$, with the (compact-open) Whitney topology. For example, we claim that the zero set of $f \in \mathcal{F}$ is an embedded submanifold of $\Lambda \times X$ of dimension $\dim \Lambda$, generically. Indeed, the set $\mathcal{G}$ of such f contains the set $\mathcal{G}_0$ of f for which zero is a regular value. Clearly $\mathcal{G}_0$ is open, if $\Lambda \times X$ is any closed ball in $\mathbb{R}^2 \times \mathbb{R}^N$. By Sard's theorem, $\mathcal{G}_0$ is also dense. This proves our claim. Note that $\mathcal{F}$ is a Baire space and therefore generic subsets are always dense, see e.g. [Hi76]. A property defining $\mathcal{G}$ is said to be generic in one parameter, if $\mathcal{G}$ is generic for $\Lambda = \mathbb{R}$. Similarly, $\Lambda = \mathbb{R}^2$ defines two-parameter genericity.

2.5 Proposition

Generically, for smooth one-parameter families of vector fields, homoclinic orbits are either tame or chaotic.

Proof:

The proof proceeds very much along the lines of the Kupka-Smale theorem, see for example [AR67]. For a list of generic one parameter bifurcations of stationary and periodic solutions see proposition 1.1 in [Fi86b]. Genericity of the transversality assumptions (2.7), (2.8) follows as for the Kupka-Smale theorem. Note, however, that

$$\dot{h} \in T_h W^s \cap T_h W^u$$

requires the additional parameter $\lambda \in \mathbb{R}$ in order that transversality holds. Once we restrict the parameter value $\lambda = \lambda_0$ to belong to a homoclinic orbit h we may perturb the eigenvalue structure at the equilibrium, such that indeed one of the cases (2.4) - (2.6) occurs for the principal eigenvalues. Assumption (2.9) is generic, by small perturbations along the homoclinic orbit h. This completes a sketch of proof of proposition 2.5. For more detailed genericity arguments, we also refer to [Fi85]. $\square$

We now return to two parameters $\lambda = (\lambda_1, \lambda_2) \in \Lambda = \mathbb{R}^2$. We will restrict our attention to generic phenomena. Generically, we distinguish the following types of equilibria

(2.10)   (i) hyperbolic

      (ii) nondegenerate folds

     (iii) Hopf points

     (iv) cusps

      (v) fold-Hopf points

     (vi) B-points

    (vii) Hopf-Hopf points

   (viii) Hopf side switching

See for example [Fi86b] as well as [Ar72], [GH90], [Ku95] and the references there. Case (i) occurs for open regions in parameter space Λ, cases (ii), (iii) along differentiable curves. Cases (iv) – (viii) occur at isolated parameter values, if the set of equilibria is bounded in X. We will return to some of these cases later. Unfortunately, cases (vii),(viii) were ignored in [Fi86b]; fortunately, this does not affect any of the results proved there. Case (vii) describes equilibria with two distinct non-resonant pairs of purely imaginary eigenvalues, all simple. Case (viii), analyzed by Bautin in 1949, refers to a change from subcritical to supercritical Hopf bifurcation of periodic orbits along one-parameter sections transverse to a path of Hopf points. In particular a path of periodic folds is generated there, and the side of the Hopf path to which periodic orbits bifurcate, locally, switches.

As we have seen for the B-point, example 2.1, homoclinic orbits occur along curves in two-parameter problems. In particular, a tame hom in a one-parameter problem gives rise to a local path γ of tame homs, in two parameters. Our notion of tame then refers to restricting λ to one-dimensional curves which intersect the hom path transversely. This way, we may talk about *tame paths*.

Taking a more global point of view, we ask for limits of homoclinic orbits, for example along a tame path. This question leads to our definition of the ϵ-length of a homoclinic orbit.

2.6   Definition: ϵ-length

Fix $\epsilon > 0$. Let $\mathcal{E} = \mathcal{E}(\lambda) \subseteq X$ denote the set of equilibria and $U_\epsilon(\mathcal{E})$ its open ϵ-neighborhood. Let h denote a homoclinic orbit of $f(\lambda, \cdot)$. We call

$\ell_\epsilon(h) :=$ *length of h outside* $U_\epsilon(\mathcal{E})$

the ϵ-length of h. Note that

$$\ell(h) := \lim_{\epsilon \downarrow 0} \ell_\epsilon(h) = \int\limits_{-\infty}^{+\infty} |\dot{h}(t)|_2 dt \leq \infty$$

is the length of the homoclinic orbit h. Similarly, we can define the ϵ-length of any, not necessarily homoclinic orbit.

We say that a set (or a path) of orbits has unbounded *ϵ-length, if $\ell_\epsilon(\cdot)$ is unbounded on this set for some, and hence for all, small ϵ. In contrast, we say that the set has* bounded *ϵ-length, if $\ell_\epsilon(\cdot)$ is bounded for each fixed $\epsilon > 0$, but not necessarily uniformly in ϵ.*

2.7 Example: fold-Hopf

In principle, a set of homoclinic orbits with bounded ϵ-length $\ell_\epsilon(\cdot)$ can have unbounded length $\ell(\cdot)$.

Fold-Hopf points in $X = \mathbb{R}^3$ with eigenvalues $0, \pm i$ provide specific examples. Unlike B-points, a universal unfolding has not been established for the fold-Hopf (and probably does not exist). We recall some results for the truncated normal form here. See [GH90], p. 74 and [Ku95], chapter 8.5 for more details on the local analysis. In cylindrical coordinates coordinates (r, φ, z) for $x \in \mathbb{R}^3$, the linearly unfolded truncated normal form reads

(2.11)

$$\begin{aligned} \dot{r} &= \lambda_1 r + \alpha r z, \\ \dot{z} &= \lambda_2 + \beta r^2 - z^2 + \gamma z^3 \end{aligned}$$

with fixed parameters $\alpha \in \mathbb{R}$ and $\beta = \pm 1, \gamma \neq 0$. Note that the angle coordinate φ associated to the radius r does not appear because of S^1-equivariance of the normal form. Moreover $|\dot{\varphi}| \approx 1$ is uniformly bounded below. We only mention two particular cases of several inequivalent bifurcation diagrams here:

(a) $-1 < \alpha < 0, \beta = +1,$ and

(b) $\alpha < -1, \beta = -1$.

Local phase portraits, at $\lambda = 0, r = z = 0$, obtained by a blow-up technique, are given in figures 2.2.a,b. Note the shaded sectors, solid cones in $\mathbb{R}^3$. In either case, these sectors allow for two-parameter families of large homoclinic orbits towards $x = 0$ in $\mathbb{R}^3$, with λ fixed at $\lambda = 0$.

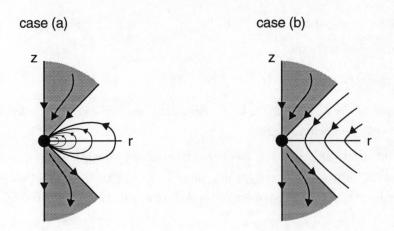

case (a)   case (b)

Figure 2.2: Local phase portraits of (2.11) for cases (a), (b), $\lambda = 0$.

In case (b), the $r > 0$ boundaries of the shaded sector could belong to the same trajectory, globally. Breaking the S^1-equivariance of the normal form, by higher order or flat terms, the two-dimensional boundaries in $\mathbb{R}^3$ can be assumed to intersect transversely. This would imply recurrent behavior of shift type, even at $\lambda = 0$.

Consider case (a) next. Analyzing the local unfolding, as in [GH90], [Ku95], large homoclinic orbits cannot exist for $\lambda_2 \neq 0$. Indeed, there are no equilibria for $\lambda_2 < 0$. For $\lambda_2 \geq \lambda_1^2/\alpha^2$, one of the two equilibria is a local attractor while the other is a repeller. For $0 < \lambda_2 < \lambda_1^2/\alpha^2$, the only saddle equilibrium connects to a unique periodic orbit. In either region, homoclinic orbits cannot occur. Note, however, the possibility of a persistent large homoclinic orbit along the fold $\lambda_2 = 0$. See figure 2.3.

Following such a hom branch, which locally coincides with the stationary fold in Λ, the homoclinic orbits typically develop unbounded length while their ϵ-length remains bounded. Indeed, for $|\lambda_1|$ small, consider the intersection points $h_\pm = h_\pm(\lambda_1)$ of the homoclinic orbit h with a fixed small neighborhood of the fold-Hopf point. From that (large positive or negative) time t on, the normal form (2.11) describes $h(t)$ well. Passing to the fold-Hopf limit $\lambda_1 = 0$, we may assume that $h_\pm(\lambda_1 = 0)$ does not lie on the z-axis, generically. By blow-up arguments, the quotient

$$\lim_{t \to \pm\infty} z(t)/r(t) = q_\pm$$

then exists and is finite, at $\lambda_1 = 0$; see figure 2.2(a). In particular, (2.11) at $\lambda_1 = \lambda_2 = 0$ implies

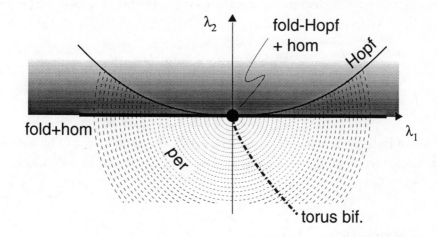

Figure 2.3: A possible bifurcation diagram for case (a).

$$|\dot{r}(t)| \leq c_2 r(t)^2$$

for some $c_2 > |\alpha q_{\pm}|$ and for $|t|$ large enough. Therefore

$$r(t) \geq c_3/|t|,$$

for some $c_3 > 0$ and all $|t|$ large. On the other hand, since $|\dot{\varphi}| \approx 1$ is bounded below,

$$(2.12) \qquad \int |\dot{h}(t)| \mathrm{d}t \geq \int r(t)|\dot{\varphi}(t)| \mathrm{d}t \geq c_1 \int r(t) \mathrm{d}t,$$

for some positive c_1. By our lower estimate on $r(t)$, the integral (2.12) diverges and the length of h becomes unbounded for $\lambda_1 \searrow 0$. Clearly the ϵ-length of h will remain bounded, if the time which h spends outside our fixed neighborhood of the fold-Hopf point remains uniformly bounded. We return to this example in section 5.5.

Tame homoclinic orbits, by transversality, can be continued locally to form a tame path as we have seen above. We now consider limits of homoclinic orbits of bounded ϵ-length.

Consider a bounded sequence of homoclinic or heteroclinic orbits $h_n(\cdot)$ of uniformly locally Lipschitz vector fields

$$(2.13) \qquad \dot{x} = f_n(x)$$

in X. In other words, the α- and ω-limit sets $\alpha(h_n), \omega(h_n)$ are contained in the respective sets $\mathcal{E}_n$ of equilibria of f_n. Note that we do not require convergence to a single equilibrium $A_\pm$ here, thus extending our original definition (1.2) slightly. Generically the set $\mathcal{E}$ is discrete by our list (2.10), and therefore $\omega(x_0)$, $\alpha(x_0)$ will be single points. Assume

(2.14) $\qquad f = \lim_{n \to \infty} f_n$

exists, locally uniformly. Let

(2.15) $\quad H := \{x \in X \mid x = \lim_{m \to \infty} x_m$ holds for some sequences

$$x_m \in h_{n_m}, n_m \nearrow \infty\}$$

be the set of limiting points of the sequence $h_n(\cdot)$. We define a *loop* to be a nonempty, compact, connected subset of X which consists of equilibria, of homoclinic, and of heteroclinic orbits, only.

2.8 Proposition: loops

In the above setting, assume the sequence $h_n(\cdot)$ and the ϵ-lengths of the sequence $h_n(\cdot)$ are bounded, uniformly in n for each fixed $\epsilon > 0$. In addition, assume that there exists a converging sequence $A_n \in clos(h_n)$.

Then H is a loop.

Proof.

Compactness and connectedness of H follow from [Wh68] as in [Fi88], lemma 7.1. We will show that

(2.16) $\qquad \omega(x_0) \subseteq \mathcal{E},$

for all $x_0 \in H \setminus \mathcal{E}$. Here $\mathcal{E}$ denotes the set of equilibria of f. Reversing time, $\alpha(x_0) \subseteq \mathcal{E}$ follows analogously. Therefore, all non-stationary orbits in H are then homoclinic or heteroclinic, as claimed. Clearly, the ϵ-length of any trajectory in H is bounded by $\sup \ell_\epsilon(h_n)$.

To show (2.16), we proceed indirectly. Suppose there exists $\xi \in \omega(x_0) \setminus \mathcal{E}$ for some $x_0 \in H \setminus \mathcal{E}$. Let $x(\cdot)$ denote the solution of $\dot{x} = f(x)$ with $x(0) = x_0$. Suppose $x(\cdot)$ is not periodic. Then existence of ξ implies

(2.17) $\qquad \ell_\epsilon(x(\cdot)) = \infty$, for all $\epsilon < dist(\xi, \mathcal{E})$.

Indeed, note that $x(\cdot)$ must reenter a flow box around ξ infinitely often, at distinct points. Now consider a convergent (sub-)sequence

(2.18) $\qquad x_m = h_m(0) \underset{m \to \infty}{\longrightarrow} x_0 \in H,$

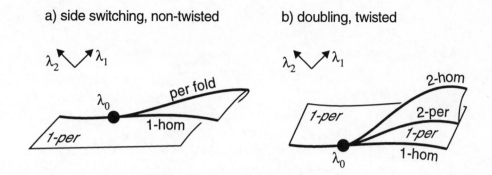

Figure 2.4: Resonant homoclinic bifurcations.

according to (2.15). By continuous dependence on initial conditions, (2.17) implies

(2.19) $\ell_\epsilon(h_m(\cdot)) \xrightarrow[m\to\infty]{} \infty$

is unbounded, contrary to our boundedness assumption on $\ell_\epsilon(h_m(\cdot))$. This contradiction completes the proof for non-periodic $x(\cdot)$.

If $x(\cdot)$ is periodic, still the $h_m(\cdot)$ cannot be periodic because they were assumed to be homoclinic or heteroclinic. Therefore, (2.18) still implies (2.19).

This completes the proof.   □

It is worthwhile to reinterpret the proposition in terms of generic two parameter vector fields for which $f_n = f(\lambda_n, \cdot)$, $\lambda_n \to \lambda$. The set $\mathcal{E}$ of equilibria of $f(\lambda, \cdot)$ is then discrete by genericity, recalling our list (2.10). In principle, the limiting loop H can contain infinitely many homoclinic or heteroclinic orbits. In our generic setting, however, this is clearly impossible by the local flow structure near equilibrium, except perhaps at a fold-Hopf point, cf. figure 2.2.a. Even in that case, however, at most finitely many orbits of H will leave a fixed ϵ-neighborhood of $\mathcal{E}$, the remaining ones being locally homoclinic to the fold-Hopf point.

We consider two specific generic examples, next, where tame paths of homoclinic orbits limit onto more degenerate codimension two homoclinic orbits.

2.9   Example: resonant homoclinic

One way to produce degenerate, codimension two homoclinic bifurcations is by violating nondegeneracy conditions of our generic one-parameter examples 2.2. Here

93

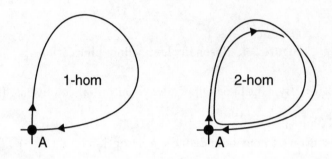

Figure 2.5: Phase portraits of 1-hom and 2-hom orbits.

we violate the nonresonance condition (2.4.a), $\mu_+ + \mu_- \neq 0$, for the real principal eigenvalues $\mu_\pm$ of the hyperbolic equilibrium A_0 associated to the homoclinic orbit. Let us assume an eigenvalue resonance

$$(2.20) \qquad \mu_+ + \mu_- = 0,$$

at some parameter $\lambda = \lambda_0 \in \Lambda = \mathbb{R}^2$. Under additional nondegeneracy assumptions which are generic in two-parameter vector fields, this situation was analyzed in detail in [Chow et al.] (1990 a). Depending on a global twist condition, one of the two alternatives described in figures 2.4(a,b) is selected. In both cases there is a local path of 1-hom(oclinic) orbits, tame except for the bifurcation point $\lambda = \lambda_0$. In analogy to period doubling, we distinguish here between 1-hom and 2-hom orbits near bifurcation: where 1-hom orbits traverse a fixed small Poincaré section to the flow only once, 2-hom orbits do so twice. In case b), there is a tame branch of 2-hom orbits limiting, as a set, onto the 1-hom orbit at $\lambda = \lambda_0$. Note that the corresponding ϵ-lengths limit onto twice the ϵ-length of the limiting 1-hom orbit, for any small $\epsilon > 0$.

All tame hom paths are accompanied by per(iodic) sheets. In case a), the per sheet develops a fold and limits onto the hom path from opposite sides, on opposite sides

94

of the bifurcation point. Hence the name *resonant homoclinic side switching*.

In case b), the 1-per sheet limits onto the 1-hom path from just one side. The bifurcating 2-hom path generates a 2-per sheet which limits onto the 1-per sheet, at its other boundary, along a path of periodic orbits with Floquet multiplier -1: classical period doubling. We have termed this case *resonant homoclinic doubling*. Notably the bifurcations in both cases are exponentially flat, not of any finite polynomial order.

Resonant homoclinic side switching was discovered, in the plane, by [Le51]. Resonant homoclinic doubling, in contrast, occurs in its simplest form at eigenvalue resonance in a Möbius strip. This properly accounts for the global twist condition mentioned above. For a first investigation of several non-chaotic homoclinic doubling mechanisms in $\mathbb{R}^3$ see [Ya87]. Notably, [KKO93] prove, in the general generic case considered here, that no further homoclinic orbits bifurcate: there are no bifurcating n-hom orbits with $n \geq 2$ (case a)), $n \geq 3$ (case b)), respectively. By [Sa92], the same statement also holds for n-per orbits. Moreover, the 2-per sheet in case b) is confined in between the per doubling and the 2-hom path in parameter space Λ. In fact, [Sa92] has success-fully constructed a two-dimensional homoclinic bifurcations (and many other cases), in a full neighborhood of $\gamma = \gamma_0$. Topologically, this manifold is an annulus in case a), and a Möbius strip in case b). This effectively excludes any more complicated recurrences than the ones described above.

2.10   Example: Tresser's tame 8, saddle focus

Tame homoclinic paths may cross each other, in two parameters and associated to the same equilibrium, see figures 2.6, 2.7. Such a crossing can produce very intricate bifurcation patterns; Tresser [Tr84] has provided a very interesting example analyzed in more detail in [Tu88], [Ga87]. At the transverse intersection point $\lambda = \lambda_0$ of the tame paths, it is assumed that (2.5.a) holds for the principal eigenvalues $\mu_\pm$; specifically suppose

(2.21) $0 < \mu_+ < -Re\mu_-.$

Moreover, let the unstable dimension be one. We assume the geometric situation of figure 2.6, that is,

$$\dot{x}(t)/|\dot{x}(t)|$$

limits onto opposite eigendirections of μ_+ along the two homoclinic orbits, for $t \to -\infty$. By these assumptions, the "figure 8" formed by the two homoclinic orbits is

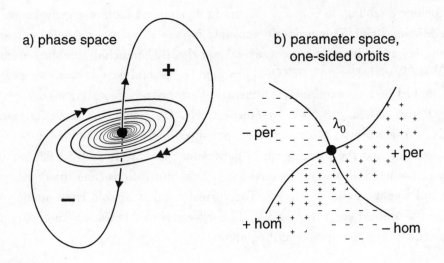

Figure 2.6: Tresser's tame 8.

an attractor. As in the previous example, the two-parameter bifurcation diagram depends on a certain global twist quantity along the homoclinic orbits. Here, we assume both homoclinic orbits to be non-twisted.

A partial bifurcation diagram of only some homoclinic orbits is sketched in figure 2.7. The hom paths are labeled by finite sequences of $\pm$ signs, according to their excursions along the original $\pm$homoclinic orbits. Descriptively, there is a $-+$hom path, oscillating into $\lambda = \lambda_0$ around the $+$hom path. At any intersection point of these two paths, the original bifurcation pattern at $\lambda = \lambda_0$ is re-ignited with $-$ replaced by $-+$. By this "self-similarity", arbitrarily long (but not all) finite sequences of the symbols $+, -$ occur in the bifurcation diagram of figure 2.7, under generic assumptions. For a more complete description using Farey sequences, see [Ga87]. Note that homoclinic orbits of all these types accumulate at the bifurcation point $\lambda = \lambda_0$. A similar accumulation occurs at all the other intersection points.

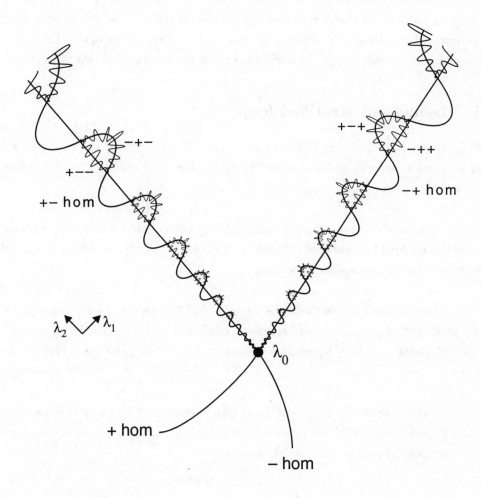

Figure 2.7: Two-parameter bifurcation diagram for Tresser's tame 8.

Examples 2.1, 2.2, 2.7, 2.9, 2.10 give a first impression of the complexity of homoclinic bifurcations in two-parameter space $\lambda \in \Lambda = \mathbb{R}^2$. We now distinguish a class of "simple" bifurcations, which we call *stratified*. Recall from proposition 2.8, that bounded sequences of homoclinic orbits with bounded ϵ-lengths limit onto a loop H: a compact connected set in which every trajectory tends to equilibrium for $t \to \pm\infty$. In the following we use loop in that sense; to avoid trivial repetitions, we also assume that a loop does not consist of just a single tame or chaotic homoclinic orbit. Excluding generic codimension one "loops", we will thus focus on codimension two phenomena.

2.11 Definition: stratified loop

Let $H \subseteq X$ be a loop at $\lambda = \lambda_0 \in \Lambda$. We call $\{\lambda_0\} \times H$ stratified, if there exist closed bounded neighborhoods V_λ of λ_0 in Λ and V_x of H in X such that (i), (ii) below both hold.

(i) *$V_\lambda \setminus \{\lambda_0\}$ contains only finitely many paths of homoclinic orbits in V_x; all these paths are tame, of bounded ϵ-length, and do not touch the boundary of V_x, but leave $V_\lambda \times V_x$ through the boundary of V_λ.*

For some smooth closed Jordan curve S in $V_\lambda \setminus \{\lambda_0\}$, transverse to the tame paths, and for any continuum (i.e. locally compact, connected set) $C \subseteq S \times V_x$ of periodic orbits which contains a Hopf bifurcation point or a homoclinic orbit in its closure, we also require

(ii) *the periodic orbits in C do not touch the boundary of V_x, and any sequence of periodic orbits in C with unbounded virtual periods contains a subsequence converging to a homoclinic orbit, as a set.*

Above, τ is called a *virtual period* of a periodic orbit $p(\cdot)$ with minimal period T, if τ is the minimal period of a pair $(p(\cdot), q(\cdot))$, where $q(\cdot)$ solves the linearized equation along $p(\cdot)$. In particular $\tau = mT$, for some integer m. If $m \neq 1$, then in our generic two-parameter setting $p(\cdot)$ possesses nontrivial Floquet multipliers which are m-th roots of unity. The notion of virtual period was introduced in [MY82], [CMY83]; see also [Fi88]. Effectively, virtual periods provide a necessary criterion for bifurcation: limits of minimal periods of $p_n(\cdot)$ are virtual periods of the limit $p(\cdot)$.

Note that the B-point, example 2.1, is stratified. The two resonant homoclinic bifurcations of example 2.9 are also stratified because the center manifold reduction by [Sa92] shows that these bifurcations in fact occur on an annulus or a Möbius strip. According to our discussion, the fold-Hopf example 2.7(a) may be stratified but 2.10, Tresser's tame 8, is not. More precisely, let H_1^* indicate the homoclinic orbit $+$, whereas H_2^* indicates $-$, and H_3^* is the union of both. Then H_1^*, H_2^* will be "stratified", separately, since they are tame. But their union, H_3^*, is not stratified.

3 Global continuation

We develop our main result in this section. In section 2 we have set up the distinction between tame and chaotic homoclinic orbits, we have introduced the concept of ϵ-length, and we have discussed stratified loops of codimension 2; see definitions 2.3, 2.6, 2.11. From now on, we consider vector fields

(3.1) $\qquad \dot{x} = f(\lambda, x)$

with two-dimensional real parameter $\lambda \in \Lambda$ and real N-dimensional $x \in X$.

In two parameters, tame homoclinic orbits typically occur along one-dimensional paths. Below, we define an orientation for such paths (definition 3.1). The orientation matches, locally near the B-points of example 2.1, an index $B = \pm 1$ of these B-points. The B-index was originally designed to describe global bifurcation of periodic orbits in two-parameter flows, see [Fi86b]. It turns out that $B = +1$ provides a source of a tame path, oriented away from the B-point as in figure 2.1, whereas $B = -1$ provides a sink; see definition 3.2 and lemma 3.3. To procure some confidence in our set-up, we indicate the consistency of the orientation of tame paths through a codimension 2 loop: the resonant homoclinic bifurcations of example 2.9. Up to this point, the present theory has already been sketched by the author in [CDF90], pp. 233–236.

Our main continuation result is codified in theorem 3.4 and its variants, remarks 3.5 (i) – (iii). For bounded, oriented tame hom paths with bounded ϵ-length, these results basically leave the following alternatives:

(i) connecting B-points in pairs, from sources to sinks, or

(ii) hitting non-stratified loops, or

(iii) becoming chaotic.

Theorem 3.4 will be proved in section 4.

We define an *orientation* of (some) tame paths of homoclinic orbits as follows. Recall that a tame hom path γ, in parameter space Λ, is locally accompanied by hyperbolic periodic orbits $p(\cdot)$, for λ on one side of γ. The orientation of γ will be defined via an orbit index Φ of the periodic orbits $p(\cdot)$. This index has been introduced by [Mallet-Paret & Yorke] (1982) for the purpose of global continuation of periodic orbits. It is given by

$$(3.2) \qquad \Phi(p(\cdot)) := \tfrac{1}{2}((-1)^{\sigma_+} + (-1)^{\sigma_+ + \sigma_-}),$$

where σ_+, σ_-, respectively, count the real Floquet multipliers of $p(\cdot)$ in $(1, \infty)$, $(-\infty, -1)$, with their algebraic multiplicity. In other words, Φ is the average of the local Brouwer fixed point indices of the first and second iterate of the Poincaré map associated to $p(\cdot)$. In particular, Φ turns out to be homotopy invariant through generic one-parameter bifurcations of periodic orbits: saddle-node bifurcation and period doubling. Note that

$$(3.3) \qquad \Phi \in \{-1, 0, +1\}$$

at any hyperbolic periodic orbit.

3.1 Definition: orientation of tame paths

Let γ be a tame path, locally accompanied by periodic orbits $(\lambda, p(\cdot))$. If $\Phi(p(\cdot)) = 0$, we do not define an orientation of γ. If $\Phi(p(\cdot)) = \pm 1$, we define an orientation of the tame path γ such that

> *λ is on the right of γ, if $\Phi = +1$;*
> *λ is on the left of γ, if $\Phi = -1$.*

In other words, the orientation of γ is the induced boundary orientation in Λ, if the parameter region of the accompanying periodic orbits is given the orientation $-\Phi$.

Note that the (local) orientation of γ is well-defined, because σ_+, σ_- do not change as long as $p(\cdot)$ remains hyperbolic. Therefore, in fact, the orientation of a path is defined consistently, as long as the path remains tame.

The orientation defined above matches the definition of a *B-index* given in [Fi86b], see definition 2.1 and lemma 2.3 there. In terms of the *B*-point example 2.1, figure 2.1, the *B*-index is defined geometrically as follows.

3.2 Definition: B-index

Let $(\lambda, x) \in \Lambda \times X$ denote a B-point. Orient the stationary fold through λ, locally in parameter space Λ, such that the parameter values of the bifurcating pair of stationary solutions are to the left of the fold. In other words, the orientation of the fold is the induced boundary orientation in Λ, if the adjoining parameter region of the bifurcating stationary solutions is given a positive orientation. Pick another point (λ', x') on the fold, near (λ, x), such that the given local orientation of the fold is from λ to λ'. Define the B-index

$(3.4)\qquad B(\lambda, x) := (-1)^{E(\lambda', x')},$

where $E(\lambda', x')$ counts the strictly positive real eigenvalues of the linearization $D_x f(\lambda', x')$ with their algebraic multiplicity.

3.3 Lemma

Any B-point generates an oriented tame path of homoclinic orbits, locally. This path is oriented

   away from the B-point, if $B = +1$,
   towards the B-point, if $B = -1$.
In other words, $B = +1$ indicates a source, and $B = -1$ a sink of an oriented tame path.

Proof.

We first show that the homoclinic path generated by a B-point (λ, x) is tame, locally. We then prove the claims about orientation. We distinguish two geometrically different bifurcation diagrams, see figure 3.1, a and b. In parameter space, the two cases are related by a reflection.

We will show below that

$(3.5.\text{a})\qquad \Phi = B$ in case a), and
$(3.5.\text{b})\qquad \Phi = -B$ in case b),

for the periodic orbits generated at the B-point. This will turn out to be sufficient, by our definition of an orientation of tame homoclinic paths.

Along the local homoclinic path, the associated equilibria are hyperbolic: within the two-dimensional center manifold describing the universal unfolding of the B-point,

101

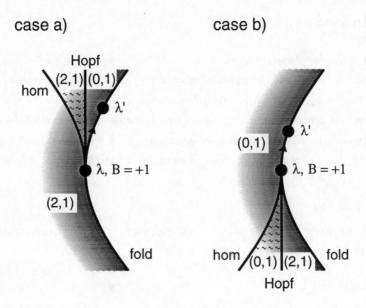

Figure 3.1: Unstable dimensions (u_1, u_2) of equilibria near B-points.

as well as in full phase space X. The principal eigenvalues are real. They appear within the center manifold and describe the asymptotic decay of the homoclinic orbits. Their sum is nonzero. The tangent spaces of the stable and unstable manifold of the associated equilibria intersect one-dimensionally, along $\dot{x}$, tangentially to the center manifold. Indeed, the remaining strong stable and unstable directions are inherited from the B-point. Crossing the homoclinic path in Λ transversely, the stable and unstable manifolds therefore also cross transversely, since they do so within the center manifold. Therefore, the homoclinic orbits along our path are tame, of type (2.4.a), locally near the B-point.

To prove (3.5), we may again restrict our attention to the center manifold. Indeed, a (strong) unstable dimension u, at the B-point, contributes as u to both $E(\lambda', x')$ and σ_+, in the notation of (3.2) and (3.4). This accounts for a common factor $(-1)^u$ in both

(3.6) $\quad \Phi = \frac{1}{2}((-1^{\sigma+} + (-1)^{\sigma+ + \sigma-}),$ and

(3.7) $\quad B = (-1)^{E(\lambda', x')}.$

Note here that the Floquet exponents of the periodic orbits are, up to a small perturbation, the eigenvalues of the linearization at the B-point. In particular, σ_- is zero (or even) in (3.6) and hence can be omitted. Therefore it is sufficient to prove

(3.5.a,b) or, more specifically,

(3.8.a) $\quad \sigma_+ \equiv E(\lambda', x')$ (mod 2),   for case a),

(3.8.b) $\quad \sigma_+ \equiv E(\lambda', x') + 1$ (mod 2),   for case b),

within the center manifold of the B-point.

To prove (3.8), we may assume $E(\lambda', x') = 0$ without loss of generality. Indeed, reversing time (within the two-dimensional center manifold) changes both σ_+ and $E(\lambda', x')$ by 1. We then have to show

(3.9.a) $\quad \sigma_+ \equiv 0$ (mod 2), and

(3.9.b) $\quad \sigma_+ \equiv 1$ (mod 2),

for the respective cases.

Since $E(\lambda', x') = 0$, we can fill in the unstable dimensions of the two equilibria in the sectors of the bifurcation diagram, figure 3.1, in the form (u_1, u_2). Note that, at Hopf bifurcation, the affected equilibrium has to change its unstable dimension from 0 to 2. By elementary exchange of stability, the unstable dimensions of the equilibria at Hopf bifurcation imply

(3.10.a) $\quad \sigma_+ = 0,$

(3.10.b) $\quad \sigma_+ = 1,$

in the respective cases. This proves (3.9), and the lemma. $\square$

The above proof is straightforward and, in some sense, elementary. Using the center index ⌗ ("zhong"), introduced in [MY82], and the properties of the B-index from [Fi86b], the proof can be recast as follows. Consider a small circle S in Λ around the B-point parameter λ. Let S be oriented positively. Then ⌗ $= B$ at the Hopf bifurcation point on S. By exchange of stability,

(3.11.a) $\quad \Phi = $ ⌗,

(3.11.b) $\quad \Phi = -$⌗

in cases a) and b), respectively. Again this proves (3.5), and the lemma.

In section 4 we will show how the orientations of tame homoclinic paths fit together in a neighborhood V_x of a stratified codimension 2 loop at $\lambda = \lambda^*$. In fact, let η_+ denote the number of oriented paths in $\Lambda \times V_x$ with orientation towards λ^*, and let η_- count the paths directed away from λ^*. Then

(3.12) $\quad \eta_+ = \eta_-,$

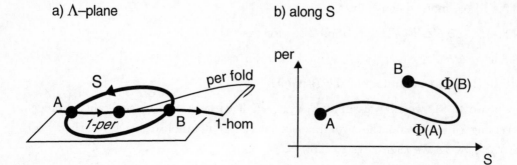

a) Λ–plane

b) along S

Figure 3.2: Orientations and resonant homoclinic side switching.

as we will see in (4.4), lemma 4.1.

This crucial fact allows us to match up tame hom paths in pairs at any stratified loop: to any path oriented towards the stratified loop in $\Lambda \times V_x$ we may associate, bijectively, one which leaves. In this manner, *orientations of tame paths can be extended consistently through stratified codimension 2 loops.*

We illustrate this important fact with our resonant homoclinic bifurcation examples 2.9, a) and b). Consider a), resonant homoclinic side switching, first. For definiteness assume the geometric situation of figure 3.2.a. In figure 3.2.b, the parameters are restricted to a small, positively oriented circle S around the bifurcation point. Let $\Phi(A), \Phi(B)$ denote the periodic orbit indices near the homoclinic orbits A, B. By homotopy invariance of Φ, see [MY82], we obtain

(3.13) $\Phi(A) + \Phi(B) = 0.$

By definition 3.1 of orientation, this implies that either both tame half-branches of homoclinics are non-oriented ($\Phi(A) = 0 = \Phi(B)$), or are oriented in the same direction. Indeed, $\Phi(A) = -\Phi(B)$ and the periodic orbits appear on opposite sides of the tame path, near A and B. In particular, $\eta_+ = \eta_- = +1$. The orientation in figure 3.2.a corresponds to the choice $\Phi(A) = +1$.

Consider resonant homoclinic doubling next for definiteness, in the geometric situation of figure 3.3.a. Again, figure 3.3.b shows the periodic orbits along the small, positively oriented circle S. Suppose $\Phi(C) = +1$, for example, near C. By homotopy invariance of Φ,

(3.14) $\Phi(B) + \Phi(C) = \Phi(A).$

104

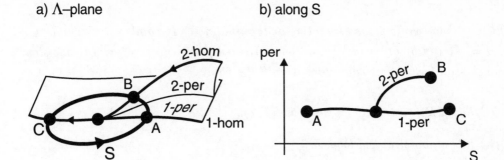

a) Λ–plane   b) along S

Figure 3.3: Orientations and resonant homoclinic doubling.

Since σ_- changes by 1, along the primary periodic branch at period doubling, we have

$$\sigma_-(A) \equiv \sigma_-(C) + 1 \quad (\text{mod } 2).$$

But $\Phi(C) \neq 0$ implies, by (3.2), that $\sigma_-(C)$ is even and hence $\sigma_-(A)$ must be odd. In particular, $\Phi(A) = 0$, by (3.2). Therefore (3.14) implies

(3.15) $\Phi(B) + \Phi(C) = 0,$

similarly to the previous case. Since the periodic orbits appear on opposite sides of the tame paths, near B and C, we again obtain a consistent orientation through the bifurcation point as in figure 3.3.a. Note that $\eta_+ = \eta_- = +1$ as in (3.12) above. All remaining combinations of orbit indices, and all stratified homoclinic bifurcations, will be dealt with in lemma 4.1.

We are now ready to state our main result. Our formulation is in terms of smooth compact manifolds Λ, X of real dimension $2, N$, respectively. We assume Λ is oriented, so that our orientation of tame paths is defined consistently, and that X is Riemannian, so that ϵ-length is defined. All other concepts used are local, and therefore carry over to manifolds easily.

3.4   Theorem

Let Λ, X be smooth compact manifolds, X Riemannian, Λ oriented and two-dimensional. Consider smooth vector fields

$$\dot{x} = f(\lambda, x), \quad \lambda \in \Lambda, x \in X,$$

in the topology of C^k-uniform convergence, for any k. Then for generic f the following *holds.*

The tame homoclinic paths emanating or terminating at B-points with index $B = +1$ or $B = -1$ can be oriented. These tame oriented paths can be extended consistently through stratified codimension 2 loops until the path

(i) hits a B-point of the opposite index, or

(ii) hits a non-stratified loop, or

(iii) reaches the boundary of the set of chaotic homoclinic orbits, or

(iv) develops unbounded ϵ-length.

Here, a tame path hitting a B-point or a non-stratified loop H is understood in the sense of proposition 2.8: the limiting set is a B-point or is contained in a non-stratified, connected loop H.

3.5 Remarks

(i) Stated loosely, we may rephrase the conclusion of the theorem as follows. Either, our oriented tame hom paths join up B-points in pairs of opposite index, (i), or else they lead into regions with complicated recurrent dynamics, (ii) – (iv).

(ii) The theorem remains valid, but is weakened, if ϵ-length in (iv) is replaced by ordinary length. Indeed, unbounded ϵ-length implies unbounded length, but not conversely.

(iii) The theorem generalizes directly to the case where the total space $\Lambda \times X$ is replaced by a compact (Riemannian) fiber bundle with oriented base Λ and flow invariant fiber X. Indeed, all concepts used in the theorem and its proof are local in the parameter, although the result is global.

(iv) In applications, $\Lambda = \mathbb{R}^2$ and $X = \mathbb{R}^N$ are typically non-compact. But suppose, for example, that f is (locally uniformly) dissipative: for each λ, there exists a ball in X which attracts all bounded sets; assume this ball can be chosen uniformly for all parameters in a neighborhood of λ. Then the theorem remains valid, if we add the possibility that λ might become unbounded along the homoclinic path. In particular, the theorem then remains valid as it stands, if homoclinic orbits do not exist for large $|\lambda|$. Similarly, suppose f is not dissipative but the set of (associated) equilibria remains bounded. Then homoclinic orbits $(\lambda, x(\cdot))$ may also become unbounded in $x(\cdot)$; but they will have to develop unbounded length along the way.

All these variants follow easily from the theorem. Just assume boundedness of the relevant homoclinic paths. Then modify f, for large λ, x, to make f dissipative and to guarantee

$$\dot{x} = f(\lambda, x) = -x, \quad \text{for } |\lambda| \text{ large}.$$

This defines a flow on the compactifications $\Lambda^c = S^2$, $X^c = S^N$, and the theorem applies.

For a more detailed discussion we refer to sections 5 – 7.

4 Proof

In this section, we prove our main result, theorem 3.4. The proof basically reduces to showing that, at a stratified loop, the same numbers of oriented tame homoclinic paths enter and leave. This step, also of independent interest, is accomplished in lemma 4.1.

We will proceed as follows. By propositions 2.5 and 2.8, homoclinic orbits arise as tame or chaotic homs, and limit onto (stratified or non-stratified) loops, as long as ϵ-length remains bounded. To prove the theorem, we suppose that none of the cases (ii) – (iv) occurs, that is, ϵ-length stays bounded, our path stays away from the chaotic region, and only hits stratified loops. Under these assumptions we then have to show that such a path joins pairs of B-points of opposite B-index.

We comment on our notion of a path, first. Let

(4.1) $\mathcal{E}_\lambda := \{x \in X \mid f(\lambda, x) = 0\}$

denote the set of equilibria of $f(\lambda, \cdot)$, and

(4.2) $\mathfrak{H} := \text{clos}\{(\lambda, x) \in \Lambda \times X \mid x \notin \mathcal{E}_\lambda, \alpha(x) = \omega(x) \subseteq \mathcal{E}_\lambda\}$

the closure of the set of homoclinic orbits. Note that all B-points lie in $\mathfrak{H}$, as do all loops H in the setting of proposition 2.8 with $f_n = f(\lambda_n, \cdot)$. Any maximal connected component $\mathfrak{P}$ of $\mathfrak{H}$ is called a *(global) path*.

107

As an interlude, consider $\mathfrak{H}$ near a B-point (λ_0, x_0). Let $\mathfrak{P}$ denote the path in $\mathfrak{H}$ containing (λ_0, x_0). For generic f, we claim that there exists a neighborhood $V \subseteq \Lambda \times X$ of (λ_0, x_0) such that $\mathfrak{P} \cap V$ consists of only the B-point and the local tame homoclinic path originating from it, provided that $\mathfrak{P}$ has bounded ϵ-length. Indeed, suppose the claim is not true. Then (λ_0, x_0) is part of a nontrivial loop H by proposition 2.8. (Note that H need not be contained in V.) By genericity of the vector field $f(\lambda_0, \cdot)$ we may assume that all equilibria in $\mathcal{E}_{\lambda_0}$ are hyperbolic, except for the B-point x_0. Moreover, the respective stable and unstable manifolds (respectively sets, for x_0) are mutually transverse. The dimensions of the stable and unstable sets of (λ_0, x_0), immersed manifolds with boundary, add up to $N = \dim X$ by local normal form analysis. Therefore, x_0 cannot be associated with a homoclinic orbit h, having $\dot{h}$ as an intersection of the tangent spaces of these sets. For the same reason, x_0 cannot belong to a circular chain of heteroclinic orbits involving other, hyperbolic equilibria. This contradiction proves our claim.

By this claim, the path emanating or terminating at a B-point is now uniquely identified, locally in Λ, as the oriented tame path from the normal form analysis of example 2.1.

Let $\mathfrak{P}$ denote a global path. Let $\mathfrak{P}'$ denote a maximal connected component of the set $\mathfrak{P}$ after all non-oriented tame paths in $\mathfrak{P}$ have been removed. Following our outline of proof, we may assume that the ϵ-length of $\mathfrak{P}'$ is bounded, $\mathfrak{P}'$ is bounded away from the chaotic region, and $\mathfrak{P}'$ contains only stratified loops. Let (λ_j, x_j) enumerate all B-points in $\mathfrak{P}'$; by compactness and genericity their number is finite. We claim

(4.3) $\sum_j B(\lambda_j, x_j) = 0.$

To prove (4.3) we note that $\mathfrak{P}'$ has the structure of a finite oriented graph. The vertices are the B-points and the stratified loops in $\mathfrak{P}'$, and the edges are the oriented tame paths of homoclinics. Finiteness of the graph follows from bounded ϵ-length, compactness, convergence proposition 2.8, and local finiteness at each stratified vertex.

It is a somewhat subtle point here that, at the same parameter value λ^*, several loops H_j^* might belong to $\mathfrak{P}'$. For a (non-stratified) example, recall Tresser's 8 in 2.10 where again H_1^* indicates the tame homoclinic orbit $+$, H_2^* indicates $-$, and H_3^* is the non-stratified union of both. In our general setting, we have by now required the union of the loops H_j^* at λ^* to be stratified. This allows us to ignore the individual loops H_j^*, replacing them by their union throughout.

Clearly, any B-point of $\mathfrak{P}$ is still contained in some component $\mathfrak{P}'$, by lemma 3.3. Note that a B-point (λ_j, x_j) with $B = +1$ provides a *source* of $\mathfrak{P}'$, while $B = -1$ provides a *sink*. Moreover, (4.3) follows if we prove "Kirchhoff's law" at each stratified vertex $\{\lambda^*\} \times H^*$. More precisely, let η_+ and η_- denote the number of oriented tame paths in $\mathfrak{P}'$ oriented towards (away from) λ^*, respectively, as in (3.12). As for balancing currents in an electrical network, it is then equivalent to (4.3) to prove $\eta_+ = \eta_-$ at each stratified vertex $\{\lambda^*\} \times H^*$ of the oriented graph $\mathfrak{P}'$.

This fact, $\eta_+ = \eta_-$, is of independent practical significance in the analysis of homoclinic and heteroclinic bifurcations. Even in purely numerical investigations, it can provide a valuable clue to oriented tame homoclinic branches which might easily be overlooked otherwise. We single out this fact as lemma 4.1; see also figure 4.1. With the proof of this lemma, our proof of theorem 3.4 will also be complete, since the additional assumption below about absence of B-points at λ^* is generic in two parameters.

4.1   Lemma

Let $\{\lambda^*\} \times H^*$ be a stratified loop with η_+ tame homoclinic paths oriented towards it and with η_- tame homoclinic paths oriented away from it. Assume that none of the equilibria associated to H^* is a B-point. Then

$$(4.4) \qquad \eta_+ = \eta_- .$$

Proof.

For the stratified loop $\{\lambda^*\} \times H^*$ we choose neighborhoods V_λ of λ^*, V_x of H^* and a smooth closed Jordan curve S in $V_\lambda \setminus \{\lambda^*\}$, as in definition 2.11. Let S be oriented positively, and restrict $f(\lambda, \cdot)$ to the one-parameter family $\lambda \in S$. By assumption, we may require that the interior of S does not contain any B-point.

The proof now consists of a careful analysis of global continua of periodic solutions along the curve S, in the spirit of [MY82].

First we approximate f, along S, by a sequence

$$(4.5) \qquad g_n \to f$$

of generic one-parameter vector fields g_n in the sense of [MY82], [Fi85]. Because S intersects the tame paths transversely, the tame homoclinic orbits on S persist under these small generic perturbations and remain tame.

For the generic vector fields g_n, periodic orbits are hyperbolic except at saddle-node

109

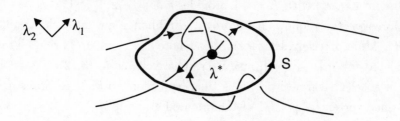

Figure 4.1: A stratified loop in parameter space.

and period doubling bifurcations along S. The orbit index

$$\Phi = \tfrac{1}{2}((-1)^{\sigma_+} + (-1)^{\sigma_+ + \sigma_-}),$$

as defined in (3.2) has homotopy invariance properties at these bifurcations. Following [MY82], we may orient paths of hyperbolic periodic orbits

along S, if $\Phi = +1$,

opposite to S, if $\Phi = -1$.

If $\Phi = 0$, we do not define an orientation. Note that hyperbolic periodic orbits are parametrized over S, locally. Homotopy invariance of Φ implies that the orientation extends, consistently, through saddle-node and period doubling bifurcations. At period doubling, note that precisely two of the three half-branches are oriented: the secondary, period doubled branch and one half of the primary branch. See also the resonant homoclinic doubling example 2.9.b, figure 3.3, and (3.14), (3.15). Maximal connected sets of oriented periodic orbit paths are called *snakes*.

In [MY82], creation of snakes at Hopf bifurcation points (alias "centers") was also considered. Their center index $\Cup$, mentioned above when proving lemma 3.3, is defined such that the originating snake is oriented locally

away from the center, if $\Cup = +1$,

towards the center, if $\Cup = -1$.

The center was called a source or sink, accordingly. By [Fi86b],

$$(4.6) \qquad \sum_{S} \Cup = 0$$

along the Jordan curve S, since the interior of S does not contain any B-points. In other words, the number of sources equals the number of sinks, as far as Hopf bifurcation is concerned.

110

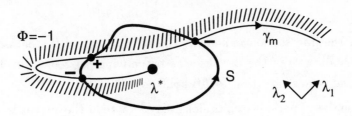

Figure 4.2: The intersection indices i_j of S with γ_m.

We establish an analogous result for bifurcation from tame homoclinic orbits on S next. Let $\tilde{\lambda}_j$ enumerate the intersection points with S of the oriented tame homoclinic paths γ_m near the stratified loop $\{\lambda^*\} \times H^*$. Let $i_j = \pm 1$ denote the local intersection index, at $\lambda = \lambda_j$, of the positively oriented smooth closed Jordan curve S with the oriented tame homoclinic path γ_m. In other words, $i_j = +1$ if the tangent vectors of S and γ_m, in this order and along the given orientations, are a positively oriented pair. Define $i_j = -1$ otherwise; see figure 4.2. Define $i(\gamma_m) = +1$, if γ_m is oriented towards λ^*, and $i(\gamma_m) = -1$ otherwise. Using $i(\gamma_m)$ we can rewrite our claim (4.4) as

$$(4.7) \qquad \sum_m i(\gamma_m) = 0.$$

Also note that for any fixed m

$$(4.8) \qquad \sum_{\lambda_j \in \gamma_m} i_j = i(\gamma_m),$$

as in figure 4.2, by homotopy invariance with respect to γ_m. Thus it is sufficient to show

$$(4.9) \qquad \sum_j i_j = 0,$$

where the sum over j now enumerates all intersection points $\tilde{\lambda}_j$ of all the homoclinic paths γ_m with S.

Our proof of (4.9) is indirect: suppose

$$(4.10) \qquad \sum_j i_j \neq 0.$$

Recall the generic approximation $g_n \to f$. Since tame homoclinic paths remain tame, under this approximation, and since the tame paths γ_m intersect S transversely, the intersection points $\tilde{\lambda}_j$ and their homoclinic orbits h_j are slightly perturbed for g_n to become

$$\tilde{\lambda}_{j,n} \to \tilde{\lambda}_j \, ,$$
$$h_{j,n} \to h_j \, .$$

In particular, the intersection indices i_j, the orientations of the paths $\gamma_{m,n} \to \gamma_m$, and the orbit indices Φ of the per sheet accompanying γ_m remain unchanged along S. Therefore we can assume that (4.10) holds for all g_n.

Consider the snakes of periodic orbits along S, generated at the intersections $(\tilde{\lambda}_{j,n}, h_{j,n})$. We claim that the snake generated at $h_{j,n}$ is oriented

$$\text{away from } (\tilde{\lambda}_{j,n}, h_{j,n}) \quad if \quad i_j = +1,$$
$$\text{towards } (\tilde{\lambda}_{j,n}, h_{j,n}) \quad if \quad i_j = -1.$$

In other words, $i_j = +1$ is a source of a snake whereas $i_j = -1$ is a sink (at the limit of infinite period of the periodic solutions in the snake).

To prove the above claim, first recall that i_j is the local intersection index of S with γ_m (we omit the additional subscript n, for a while). The snake, secondly, is oriented along S, for $\Phi = +1$, and anti S, for $\Phi = -1$. Thirdly, the orientation of γ_m is related to Φ, by definition 3.1, as follows. (For an example, see figure 4.2 again). Consider the ordered pair at $\tilde{\lambda}_j$, given by the tangent vector to S, pointing to the side of the per sheet, and the tangent vector to γ_m, along its orientation. This ordered pair has orientation $\Phi = \pm 1$. Using the tangent vector to S along its (positive) orientation, instead, the orientation of the same pair is i_j. Thus the per sheet bifurcates in the direction of the positive orientation of S if $i_j \cdot \Phi = +1$, and in the negative direction if $i_j \cdot \Phi = -1$. In particular, $i_j = +1$ is a source and $i_j = -1$ is a sink of a snake at infinite period.

In view of (4.10), the number of sources of snakes differs from the number of sinks, at infinite period and in $S \times V_x$.

Since $\sum \Phi = 0$, by (4.6), this imbalance remains in effect if centers are taken into account as sources and sinks of snakes at finite period. Therefore, for any small $\delta > 0$ there must exist a *global snake* $\mathfrak{S}_n$ of g_n along S: a snake which

(4.11.a) touches the boundary of V_x, or

(4.11.b) contains points, at a distance at least δ from the homoclinic orbits, which lie on periodic orbits of arbitrarily large minimal period.

By construction, the snake $\mathfrak{S}_n$ bifurcates at a tame homoclinic orbit or at a Hopf bifurcation point. Note that δ can be chosen independently of n, because tameness of homoclinic orbits persists.

Passing to the limit $n \to \infty$, $g_n \to f$, we will obtain a contradiction, thereby proving our original claims (4.4), (4.7), (4.9). Indeed, $\mathfrak{S}_n$ limits onto a continuum $\mathfrak{S}$ of periodic orbits in $S \times V_x$ such that (4.11.a,b) remain valid if we replace "minimal period" by *virtual period*. For a proof see [Fi85] and chapter 7 in [Fi88]. We now recall that $\{\lambda^*\} \times H^*$ is assumed to be stratified; in particular definition 2.11 (ii) holds for $\mathcal{C} := \mathfrak{S}$. Hence $\mathfrak{S}$ cannot touch the boundary of V_x as in (4.11.a). Therefore, (4.11.b) applies: $\mathfrak{S}$ must contain periodic orbits of arbitrarily large virtual periods at a distance at least δ from the homoclinic orbits. This clearly contradicts definition 2.11 (ii), where convergence of a subsequence to a homoclinic orbit is required.

This contradiction completes the indirect proof of the lemma, and the proof of theorem 3.4. □

5 More examples

In this section we illustrate scope and limitations of our pathfollowing approach by eight additional examples. Examples 5.1 – 5.3 consider stratified loops, example 5.3 showing some global interaction in local unfoldings of certain codimension three degenerate planar vector fields. The remaining examples 5.4 – 5.8 address the issue of non-stratified loops and chaotic dynamics, including the Belyakov transitions to the Shilnikov region, 5.4, and the Lorenz homoclinic explosion, 5.8.

5.1 Example: fold hitting

Locally at a nondegenerate stationary fold point (λ, x), the stable set $W^s(\lambda)$ is an immersed manifold with boundary given by $W^{ss}(\lambda)$, the strong stable manifold. Similarly, the local unstable set $W^u(\lambda)$ has the strong unstable manifold $W^{uu}(\lambda)$ as a boundary. Denoting the respective dimensions by $s, ss, \ldots$, we have

$$ss + 1 = s,$$
$$uu + 1 = u,$$
$$s + u = N + 1,$$

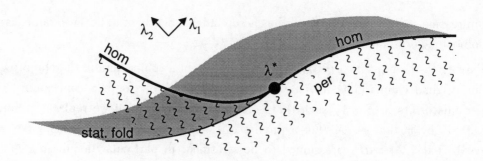

Figure 5.1: The fold hitting.

where $N = \dim X$ is the dimension of the phase space. We have already observed in example 2.7 that a homoclinic path can, even generically, coincide with a fold. This occurs when $W^s(\lambda)$ and $W^u(\lambda)$ intersect transversely (and not on a boundary). Moving (λ, x) along the fold curve in two parameters, such a transverse intersection can disappear, for example, by $W^s(\lambda)$ sliding across the boundary $W^{uu}(\lambda)$ of $W^u(\lambda)$, at $\lambda = \lambda^*$ and at an associated loop

$\quad H^* = \mathrm{clos}(W^s(\lambda^*) \cap W^{uu}(\lambda^*))$.

Under generic nondegeneracy assumptions, it turns out that a homoclinic path persists but detaches from the fold at $\lambda = \lambda^*$. Tracking the hom path in the opposite direction we call the homoclinic bifurcation at $\lambda = \lambda^*$ a *fold hitting*. This bifurcation was first investigated by [Lu82], in the plane. Schecter [Sc87a, Sc87b] gives a revised account and an application to the Josephson junction. For a detailed analysis in $\mathbb{R}^N$ see [CL90] and [De90]. For the bifurcation diagram see figure 5.1.

It turns out that the hom path is accompanied by a unique sheet of uniformly hyperbolic periodic orbits. In particular the hom path is tame. If the periodic orbits have nonzero orbit index Φ, then the tame path is oriented accordingly, consistently through the stratified homoclinic loop $\{\lambda^*\} \times H^*$.

Let us augment this example, assuming that, at the hitting value $\lambda = \lambda^*$, there exists still another homoclinic orbit to the fold via another transverse intersection $W^u(\lambda^*) \cap W^s(\lambda^*)$, not on the boundary. This requires $\dim X \geq 3$. While the original hom path γ_1 detaches from the stationary fold, the transverse intersection generates a second path γ_2 which stays on the fold. By [Shi69], shift dynamics on two symbols occurs near the homoclinic paths γ_1, γ_2 where they both stick to the fold. The two symbols 1, 2 encode excursions along the original homoclinic orbits of γ_1, γ_2, respectively, which

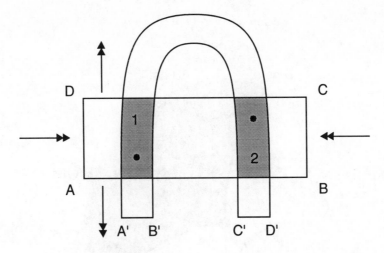

Figure 5.2: Fixed points in the horseshoe.

occur for parameter values beyond the fold curve, where the stationary points have annihilated. In spite of this chaotic dynamics, note that each of the paths γ_1, γ_2 will be tame, separately, because the shift type orbits necessarily leave a neighborhood of each. In effect, homoclinic pathfollowing can remain an option, even in the presence of chaotic dynamics.

Following γ_1, γ_2 further, along the fold and away from λ^*, we may also consider their subsequent annihilation: the respective transverse intersections $W^s(\lambda) \cap W^u(\lambda)$ approach each other and disappear at a non-transverse interior intersection $\{\tilde{\lambda}\} \times \tilde{H}$. Although pathfollowing through $\{\tilde{\lambda}\} \times \tilde{H}$ seems obvious, joining the paths γ_1 and γ_2, we do not expect $\{\tilde{\lambda}\} \times \tilde{H}$ to be stratified. In fact, note that near λ^* the periodic orbits generated from γ_1, γ_2, with corresponding trivial sequences $\mathbf{1} = \dots 111 \dots$, $\mathbf{2} = \dots 222 \dots$, respectively, are both hyperbolic with equal unstable dimensions, due to the hyperbolic structure of the Smale horseshoe in which they must occur. Due to subsequent annihilation at $\tilde{\lambda}$, the geometry of the Poincaré map near λ^* is as indicated in figure 5.2 for $\dim X = 3$. In particular, note that $\Phi = -1$ at the fixed point $\mathbf{1}$ whereas $\Phi = 0$ at $\mathbf{2}$. Therefore, only γ_1 will be oriented. (In fact, a similar reasoning applies to all periodic orbits in the horseshoe: again $\Phi \in \{-1, 0\}$). In particular, Lemma 4.1 fails at $\{\tilde{\lambda}\} \times \tilde{H}$. Therefore $\{\tilde{\lambda}\} \times \tilde{H}$ cannot be stratified. Moreover γ_1 does not continue into the non-oriented tame path γ_2, all the way back to λ^*. Instead, we expect further non-stratified loops or a cascade of (non-resonant) homoclinic doublings to occur.

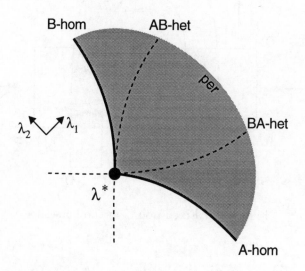

Figure 5.3: Bi-contractive het loop, no twist.

5.2 Example: bi-contractive het loop

A heteroclinic loop consists, in the simplest case, of a pair of distinct equilibria A, B and a pair AB, BA of heteroclinic orbits joining them, from A to B and vice versa, respectively. For early results see [Re79], [No85, No82]. With slight adaptations, we follow [CDT90, CDT91], [De91a, De91b] who also give applications to travelling waves of the FitzHugh-Nagumo system. For a series of excellent and more complete results on not necessarily bi-contractive het loops we refer to [Sh92, Sh94] and [ShT96]. We assume that A, B are both hyperbolic, only one-dimensionally unstable, the principal stable eigenvalue μ_- is simple real (as the (principal) unstable eigenvalue μ_+), the heteroclinic orbits limit tangentially onto the associated principal eigenvectors, and each equilibrium is locally contractive:

$$
\begin{aligned}
&\mu_-(A) + \mu_+(A) < 0, \\
&\mu_-(B) + \mu_+(B) < 0.
\end{aligned}
$$
(5.1)

Under these assumptions and canonical transversality conditions with respect to parameter dependence of the stable and unstable manifolds, the bifurcation diagrams in figures 5.3 – 5.5 were established, see [De91a], [Sh92, Sh94, ShT96]. Figures 5.3 and 5.5 can be realized by planar vector fields.

116

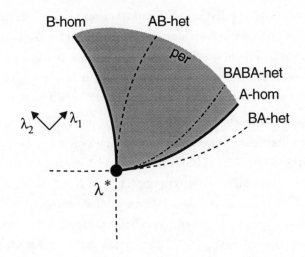

Figure 5.4: Bi-contractive het loop, one twist.

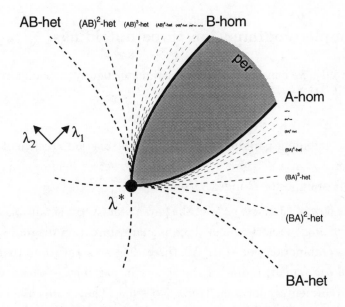

Figure 5.5: Bi-contractive het loop, double twist.

The three cases in figures 5.3 – 5.5 differ by a global twist condition along the heteroclinic orbits AB and BA. Geometrically, the twist arises by collapsing the fibers of the associated strong stable foliation with contraction rates exceeding $\mu_-(A), \mu_-(B)$, locally. Then the het loop occurs within a two-dimensional annulus, each half of which may be twisted or not. The non-twisted and the doubly twisted case, figures 5.3 and 5.5, provide an orientable annulus and thus can occur in the plane. One twist, alias the Möbius band, requires at least $X = \mathbb{R}^3$ for an embedding. See also [Sh92, Sh94].

In all three cases, we note the two het paths crossing at the bifurcation point $\lambda = \lambda^*$. Moreover, two hom paths emanate from λ^*, one for each associated equilibrium A, B. Also note the additional $ABAB$ het path in figure 5.4 and, in figure 5.5, the sequences of heteroclinic bifurcation curves $(AB)^k, (BA)^k$, $k = 1, 2, 3, \ldots$ of orbits which pass near A, B successively, $k - 1$ times, before converging to A or B. In all three cases, there is a unique sheet of hyperbolic periodic orbits extending from the A-hom path to the B-hom path. Both hom paths are therefore tame, in each case, and the corresponding het loops are stratified. The same holds true, if the unstable dimensions of the equilibria A, B exceed one but remain equal. This requires a center manifold reduction; see [Sa92], [Sh94].

5.3 Example: codimension three unfoldings

Following [DRS91], we consider the three parameter family of planar vector fields

$$
(5.2) \quad
\begin{aligned}
\dot{x}_1 &= x_2 \\
\dot{x}_2 &= \lambda_1 + \lambda_2 x_1 + a x_1^3 + x_2(\lambda_3 + b x_1 + x_1^2).
\end{aligned}
$$

with small real unfolding parameters $\lambda_1, \lambda_2, \lambda_3$ and constant coefficients $a = \pm 1, b > 0, b \neq 2\sqrt{2}$. For λ, x in a neighborhood of zero, the vector field describes coalescence of two B-points of opposite B-index. See also [Me85].

To describe the flow of (5.2), we restrict the parameters λ to a small sphere centered at the origin; removing the north pole, three planar bifurcation diagrams are obtained for respective combinations of a,b. All three diagrams turn out to be essentially independent of the (small) radius of the sphere in parameter space. We sketch the three cases in rudimentary form as figures 5.6 – 5.8. They were called saddle, focus, and elliptic to distinguish the behavior of the degenerate germ at $\lambda = 0, x = 0$. The notation (u) or (u_1, u_2, u_3) again indicates unstable dimensions of hyperbolic equilibria.

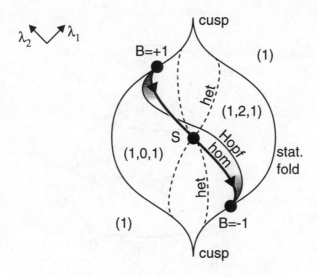

Figure 5.6: Codimension three, saddle, $a = 1$.

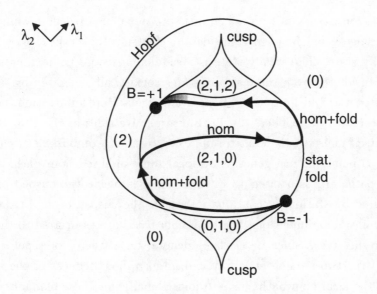

Figure 5.7: Codimension three, focus, $a = -1$, $0 < b < 2\sqrt{2}$.

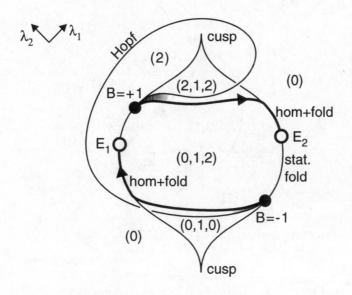

Figure 5.8: Codimension three, elliptic $a = -1$, $b > 2\sqrt{2}$.

We illustrate consistency of these diagrams with our global pathfollowing approach. We consider figure 5.6 first, in some detail. Outside the cusped oval, there is a unique steady state. The steady state is a saddle: hence no homoclinic or periodic recurrence. Inside the oval, three steady states coexist, all generated by saddle-node bifurcations on the fold line. One of them undergoes Hopf bifurcation, becoming a local attractor, (0), or repeller, (2), on the respective sides of the Hopf path. The Hopf path itself joins two B-points of opposite B-index, computable from (3.4). As usual, each B-point in turn generates a local tame oriented hom path. Note that these local paths are associated to different saddles inside the cusped oval, which are not related by stationary continuation inside the cusped oval. In the $(1, u_2, 1)$ notation for unstable dimensions, the local hom paths are associated to the first and last entry, respectively, since $u_2 \in \{0, 2\}$ denotes an attractor or repeller. Assume genericity of the two-parameter family and uniform boundedness of the hom paths next. Then the local hom paths have to join, globally, for these planar vector fields. The necessary exchange of associated equilibria can happen only via at least one stratified heteroclinic loop. In figure 5.6 this stratified bifurcation point is denoted by S. Detailed analysis by [DRS91] shows that the het loop at S is not bi-contractive, and example 5.2 does not contain the correct local bifurcation diagram at S. Besides a tame hom path and two het paths, AB and BA, through S, a per fold also terminates there. This bifurcation was also discussed by [Sh94] from a theoretical point of view. The het paths from S hit the stationary fold and persist there. Since S is stratified, the hom path can be oriented consistently through S. With the hom path oriented successfully, note that the accompanying per sheet lies on opposite sides, near the two B-points. Alternatively, note that the periodic orbits are stable near $B = -1$, but unstable near $B = +1$, in the present case. The switching is achieved via the (omitted) per fold, originating at the Hopf path at a Hopf side switching, (2.10)(viii), and terminating at S.

We briefly discuss figures 5.7, 5.8 next. In figure 5.7, the Hopf path connects the B-points via the exterior region of the cusped oval. The local hom paths are trapped inside the oval, both associated to the second entry in $(u_1, 1, u_3)$, because $u_1, u_3 \in \{0, 2\}$ indicate local attractors or repellers. The local hom paths remain bounded, thus joining up to a single hom path with consistent orientation. It turns out, due to planar phase plane geometry, that the global oriented hom path hits/leaves the stationary fold curves at four stratified fold hitting bifurcations as discussed in example 5.1. Again the per sheets lie on opposite sides of the oriented hom path, near the two B-points. Therefore, resonant homoclinic side switching as in example 2.9 (a) must also occur here. Recall that this involves an exponentially flat bifurcation of a periodic saddle-node curve from the hom path.

In figure 5.8, finally, each local hom path hits the fold curves as in example 5.1. At E_1, E_2 the hom paths escape to infinity, leaving in particular the neighborhood of validity of the unfolding. This illustrates yet another option of theorem 3.4 and remark 3.5 (iv).

5.4   Example: Belyakov transitions

Belyakov [Be80, Be84] has investigated tame hom paths transgressing into the Shilnikov chaotic region. His results are very illuminating and, at the same time, slightly discouraging from a global pathfollowing point of view. Expressed in terms of the eigenvalues μ_j of the associated equilibrium in $X = \mathbb{R}^3$ along a hom path the two transitions are

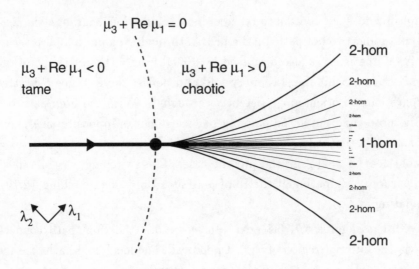

Figure 5.9: Splitting Belyakov transition.

a) originally $0 < -\mu_1 < -\mu_2 < \mu_3$, and then μ_1, μ_2 join up and split, becoming conjugate complex, $0 < -Re\mu_1 = -Re\mu_2 < \mu_3$;

b) throughout $Re\mu_1 = Re\mu_2 < 0 < \mu_3$ with a complex pair $\mu_2 = \overline{\mu}_1$, and $\mu_3 + Re\mu_1$ changes sign.

A partial bifurcation diagram of the splitting case a) showing only 1-hom and 2-hom paths is given in figure 5.9; see [Be80]. Clearly, the transition a) is not stratified, since infinitely many 2-hom branches accumulate at the bifurcation point. A similar statement holds for case b). But even if we are willing to push definitions aside, for a moment, it is not clear at all how we should continue, if we happen to approach the transition point from within the chaotic region, following one of the 2-hom paths. Therefore, we stop the pathfollowing process, in theorem 3.4, as soon as a hom path reaches the boundary of the chaotic region.

5.5 Example: the fold-Hopf, revisited

We return to examples 2.7 (a), (b) with pathfollowing in mind. Consider case (a), figures 2.2 (a) and 2.3, first. Note that the fold-Hopf at $\lambda = 0, x = 0$ may be stratified in that case. Indeed consider orbits with α- and ω-limit set given by the origin.

Assume this closure is an isolated invariant set with isolating neighborhood V_x in the sense of [Co78], at $\lambda = 0$. Then V_x contains the fold-hom path, for small $|\lambda|$. For small $|\lambda| \neq 0$ homs in V_x can only arise as fold-homs, by local normal form analysis. In particular, a small circle S around $\lambda = 0$ in parameter space will provide the two hom orbits on the fold, in V_x, but no further homoclinic orbits. Moreover, periodic orbits in V_x cannot touch the boundary of the isolating neighborhood V_x. Small periodic orbits are generated at the two Hopf points on S. Large periodic orbits are generated at the two homoclinic orbits on S. The large periodic orbits snake just inter-connects the two homoclinic orbits along S. Choosing S such that the torus bifurcation for $\lambda_2 < 0$ occurs at irrational rotation number, the small periodic orbit just inter-connects the two Hopf points on S. For stability reasons, a similar torus bifurcation has to accur along the per branch which connects the two fold-homs along S. Again, we may assume irrational rotation number at bifurcation. This then proves that the fold-Hopf loop is stratified. In particular, the orientation of the hom paths along the fold extends consistently through the fold-Hopf loop. (Note that the orbit index Φ does not change at the torus bifurcations.)

The other case, 2.7 (b), is even less innocent. We have already mentioned that, breaking S^1-equivariance of the normal form, shift dynamics can occur, even at $\lambda = 0$. Above the Hopf curve of figure 2.3, a periodic saddle with transverse intersections between stable and unstable manifolds can exist. Strictly speaking, however, shift dynamics could occur even near a stratified loop. Still, similarly to our augmented example 5.1, we do not expect the fold-Hopf 2.7 (b) to be stratified with such complicated dynamics nearby.

5.6   Example: Bykov's semi-robust het loop

We indicate a two-parameter example in $X = \mathbb{R}^3$ investigated by Bykov [By78, By80, By88]. As in the bi-contractive case of example 5.2 consider a pair A, B of distinct hyperbolic equilibria joined by a pair AB, BA of heteroclinic orbits at parameter $\lambda = 0$. Again suppose that B is one-dimensionally unstable with real eigenvalues $\mu_+(B) > 0 > \mu_-(B) > \mu_{--}(B)$ and AB approaches B along the principal eigendirection of $\mu_-(B)$. Deviating from example 5.2, however, assume that B is expansive,

(5.3) $\mu_-(B) + \mu_+(B) > 0,$

and that the unstable dimension of A is two, differing from the unstable dimension of B. Under the usual nondegeneracy assumptions, this implies that the heteroclinic

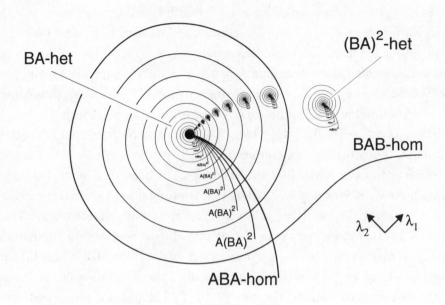

Figure 5.10: A semi-robust het loop.

orbit $AB(\lambda)$ from A to B persists under small perturbations of λ. In contrast, BA immediately disappears for $\lambda \neq 0$. We therefore call the het loop *semi-robust*.

Following Bykov, we finally assume $\mu_-(A) < 0$ is real and $\mu_+(A)$ form a complex pair with

(5.4) $\mu_-(A) + Re\mu_+(A) > 0.$

Note that an A-hom can be tame in that case; we are not in the situation of Shilnikov chaos.

A partial bifurcation diagram was given in [By78]; for some aspects see figure 5.10. For a more complete, detailed, and complicated description see [By80, By88]. Descriptively, there is a logarithmic spiral BAB towards $\lambda = 0$ where B-hom orbits exist which pass near A once. Also there is a non-spiraling curve ABA of A-hom orbits which pass near B once. Note that each of these homoclinic paths is tame in our setting, separately, except that the twist type may change due to (omitted) bifurcations.

Besides the path of ABA-hom orbits, cycling through the het loop once, there is also an infinite sequence of paths of $A(BA)^2$-hom orbits which cycle through the het loop twice. By this fact alone, Bykov's het loop is not stratified.

Similarly to Tresser's 8, a selfsimilarity is present. In fact, there is a sequence of $(BA)^2$ heteroclinic orbits at $\lambda = \lambda_m$, accumulating to the original het loop at $\lambda = 0$ for $m \to \infty$. Since the AB heteroclinic orbit persists, near $\lambda = 0$, this implies that the entire bifurcation scenario repeats near each λ_m, including sequences $\lambda_{m,m'} \to \lambda_m$ for $m' \to \infty$, replacing the sequence λ_m, and so forth. Clearly, Bykov's het loop is not stratified in any neighborhood.

The case when (5.4) is replaced by the opposite inequality

$$(5.5) \qquad \mu_-(A) + Re\mu_+(A) < 0$$

has been analyzed by Kokubu [Ko88, Ko91, Ko93] in an AB-het, BC-het, AC-het context. He obtains the ABA and BAB hom paths and points out Shilnikov chaos along the ABA hom path.

5.7 Example: tame 8, doubly twisted real saddle

Following [Ga87], [Tu85], we consider a real variant of Tresser's tame 8 from example 2.10. Again two tame hom paths cross each other transversely, as in figure 2.6(b). But this time it is assumed that all eigenvalues at the associated equilibrium A are real, and the two homoclinic orbits form a figure 8 following the simple principal eigendirections from respectively opposite sides. To utilize a foliation argument in [Ga87], we assume the unstable dimension of A to be one. Moreover, contractivity

$$(5.6) \qquad \mu_-(A) + \mu_+(A) < 0$$

is assumed for the principal eigenvalues $\mu_\pm(A)$. Standard nondegeneracy conditions are also imposed. Collapsing strong stable fibers associated to the rapidly contracting non-principal stable eigendirections, the flow near each homoclinic orbit can be viewed as a flow on a two-dimensional annulus. We assume a double twist, that is, each annulus is a Möbius strip.

Labeling excursions along the two homoclinic orbits by $+, -$, as in example 2.10, the hom bifurcation diagram of figure 5.11 emerges. The diagram follows from an analysis of one-dimensional return maps on the doubly twisted double annulus. Most periodic sheets are omitted here. Note the tame +hom and −hom paths and the additional $(+-)$hom, $(-+)$hom, $+(-+)^k$hom, and $-(+-)^k$hom half-branches.

In fact the bifurcation diagram coincides with figure 5.5 of a doubly-twisted bi-contractive het loop, replacing AB-het by +hom, BA-het by −hom, B-hom by −+hom, A-hom by +−hom, $(AB)^k$-het by $+(-+)^{k-1}$hom, and $(BA)^k$-het by

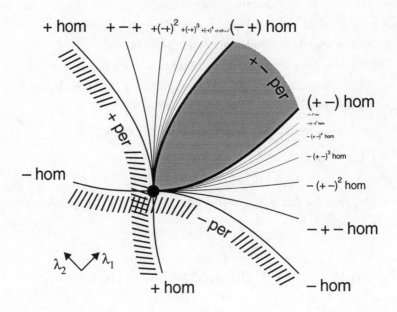

Figure 5.11: Bifurcation diagram of a doubly twisted, real saddle 8.

$-(+-)^{k-1}$hom. To understand this "coincidence", consider figure 5.12 where a figure 8 saddle is artificially split into two equilibria A, B along the dotted line. Clearly, the above correspondences can be read off from figure 5.12. Note, however, that +per and −per orbits, for example, become invisible in figure 5.12, in contrast to +−per orbits.

In particular, we conclude that the doubly twisted real contractive 8 is not stratified in a neighborhood of the figure 8 because infinitely many tame oriented hom paths of type $+(-+)^n, -(+-)^n, n = 1, 2, 3, \ldots$ bifurcate. For $n \to \infty$, the first of these families accumulates to the $-+$hom path, and the second to the $+-$hom path. Nevertheless, each of the generating +hom and −hom paths is tame and oriented, separately. Also, we see that bifurcations of periodic orbits must occur since the periodic orbits generated by the $+(-+)^n$ hom family have different winding type, containing a $++$ pair, than those of the $-(+-)^n$ family which contain a $--$ pair.

By the exact same correspondence, the bifurcation diagram of the non-twisted contractive real 8 follows from the corresponding bifurcation diagram 5.3 of the non-twisted bi-contractive het loop. In fact we again have to replace, in figure 5.3, AB-het by +hom, BA-het by −hom, B-hom by −+hom, A-hom by +−hom paths, and the per

126

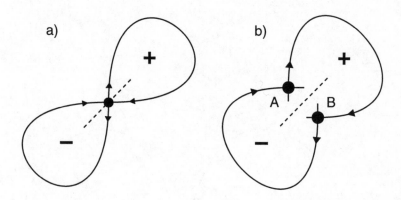

Figure 5.12: Homoclinic 8 and heteroclinic loops.

sheet by +−per. Only the +per and −per sheets are then omitted in figure 5.3; they extend to the left of the original AB-het and below the BA-het curves, respectively.

A similar correspondence finally holds for the bifurcation diagrams of a het loop and a homoclinic 8 with a single twist, respectively.

In summary, we have seen how a stratified het loop, figure 5.5, instigates a homoclinic 8 to be non-stratified, figure 5.11, under a slight change of view point.

5.8 Example: tame butterfly, real saddle

Following [Ga87] again, we consider a double homoclinic loop to a real saddle A in $\mathbb{R}^3$. All assumptions and notations are as in example 5.7, except for the asymptotic behavior of the two homoclinic loops $h_+(t), h_-(t)$. For $t \to -\infty$, $h_+(t)$ and $h_-(t)$ are still assumed to approach the equilibrium along the principal unstable eigenvector, from opposite sides. For $t \to +\infty$, however, we assume $h_\pm(t)$ to approach the equilibrium along the principal stable eigenvector *from the same side*, see figure 5.13. Such a configuration was called "butterfly" in [Ga87]. Again, standard nondegeneracy conditions are imposed.

127

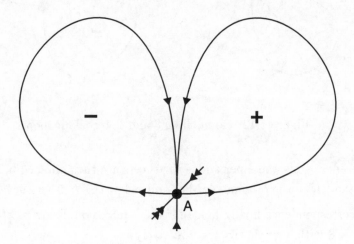

Figure 5.13: A tame butterfly.

We consider the expansive case

(5.7) $\mu_-(A) + \mu_+(A) > 0$

first. We do not give a complete two-parameter bifurcation diagram. Instead, we pass to a double cover, splitting A into two copies A, B as in figure 5.14. Next we associate to the thus created het loop a flow on a (twisted or non-twisted) annulus, identifying fibers of the strong stable foliation. We consider the case where the orientations are as indicated in figure 5.14. Such a foliation was established in the geometric Lorenz model, for example, see [Rb81]. Reversing the flow in the annulus, the results of example 5.2 apply. Note that the standard (i.e. non-twisted) Lorenz situation leads to a doubly twisted het loop, figure 5.5, in our case. In particular, the bifurcation diagram

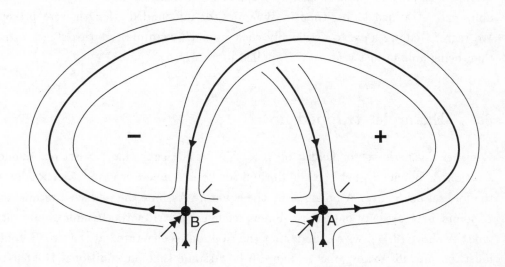

Figure 5.14: A split butterfly.

of figure 5.11 still forms a partial bifurcation diagram for the present situation. Hom branches which contain $++$ or $--$ blocks in their coding have been lost, of course, when splitting the butterfly. (So better leave the butterflies intact). But the branches found are sufficient to prove that the tame butterfly is not stratified.

The contractive case

$$(5.8) \qquad \mu_-(A) + \mu_+(A) < 0,$$

also considered in [Ga87], can be treated analogously but without reversing the flow in the annulus.

6 Homoclinic trapping

When reviewing some examples of stratified loops, we have discussed three more global homoclinic bifurcation diagrams, in example 5.3, which can be understood in terms of stratified loops alone. In the present section, in contrast, we give two examples where global homoclinic pathfollowing runs into complicated recurrent dynamics. This conclusion will be drawn from local information at the equilibria. The first example involves a "lip" on an S^2-isola. The second example, however, is dissipative. In both examples homoclinic paths would be trapped in certain regions of parameter space, unless non-stratified loops, chaotic dynamics, or unbounded ϵ-length were present. We regret that we have to artificially cook up such examples, at this stage, rather than being able to just refer to the applied literature.

6.1 Example: trapped lip

Consider a generic vector field f on $\Lambda \times X, X = \mathbb{R}^4$, such that the set of bounded solutions is bounded. Let the equilibrium set be a 2-sphere isola $S^2 \subseteq \Lambda \times X$. Let the fold set on S^2 over Λ be given by the equator. Assume the fold set contains two B-points, necessarily of opposite B-index, see figure 6.1. Relating B-index and center index as above (3.11), we conclude that the Hopf paths generated at the two B-points must join up, for example as in figure 6.1. Assume that an additional Hopf path, following approximately a longitude circle, separates the two B-points. Generically, this generates at least three codimension two bifurcation points: one Hopf bifurcation with two non-resonant purely imaginary pairs, and two fold-Hopf points. In a center manifold of each fold-Hopf point, we assume the local bifurcation diagram of figure 2.3. However, we require that none of the two fold-Hopf points is part of a loop, in the sense of proposition 2.8. For the assumed global bifurcation diagram of steady states and Hopf points see figure 6.2. There are two regions of type (2,3); by (2,3)' we denote the one adjacent to the fold. In the interior (2,3) region, away from the fold, we finally assume that the four eigenvalues of the 2-dimensionally unstable "top" equilibrium form two complex pairs.

Now consider the global tame oriented hom path generated at the B-point with $B = +1$, by theorem 3.4. Because the set of bounded solutions is assumed to be bounded, this path remains bounded. We claim that the path hits a non-stratified loop, or hits the boundary of the chaotic homoclinic set, or develops unbounded ϵ-length within a

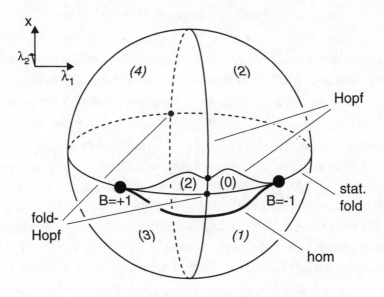

Figure 6.1: Stationary isola with B-points and separating Hopf circle.
Unstable dimensions u are denoted as (u).

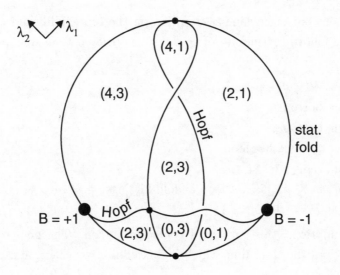

Figure 6.2: Bifurcation diagram of figure 6.1. Unstable dimensions at
equilibria are denoted as (u_1, u_2).

131

bounded set. In short, complicated recurrence must occur.

We prove our claim indirectly. If none of the above occurs, then our tame oriented hom path must extend tamely to $B = -1$, possibly through a series of stratified het loops and fold hittings involving the upper and lower hemispheres of the stationary isola S^2. But there are "forbidden regions", which render such a path impossible.

We enumerate the "forbidden regions" next. Trivially, equilibria with unstable dimension 0 or 4, i.e. local attractors or repellers, cannot be associated to a homoclinic orbit. Each fold-Hopf point possesses a neighborhood which the hom path cannot enter, by proposition 2.8. Indeed, no fold-Hopf point is part of a loop. Moreover, neighborhoods of the Hopf paths are forbidden. Indeed, outside the already forbidden neighborhoods of fold-Hopf points we find the Shilnikov conditions for chaotic dynamics satisfied (unless we have an attractor or repeller, anyway). The top (2)-equilibrium in the interior region (2,3) away from the fold is forbidden, for a similar reason.

Now follow the tame oriented hom path generated at $B = +1$. The associated equilibrium, (3), lies in the bottom hemisphere. It can jump to the top (hemisphere), or cross the fold, but only into the (2,3)' region. Even after any such a transition, the path is trapped by forbidden regions, just as it was in the bottom hemisphere. This proves our claim.

Note that without our eigenvalue assumption on the top equilibrium in the interior (2,3) region the following itinerary of equilibria associated to a tame path would be possible:

start at $B = +1$,
(4,3) region, bottom,
(2,3) region, bottom,
(2,3) region, top, via het loop,
(2,1) region, top,
(2,1) region, bottom, via het loop or fold hittings,
terminate at $B = -1$.

Under our assumptions, this path is of course forbidden. The hom path indicated in figure 6.1 crosses the Hopf line, which induces Shilnikov chaos, and completes the trapping of the lip.

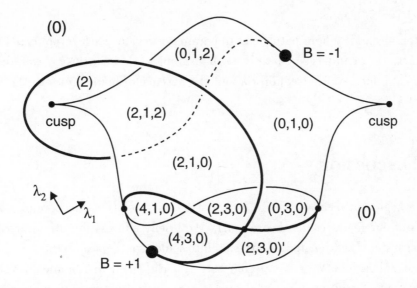

Figure 6.3: Dissipative trapping. Unstable dimensions at equilibria are
denoted (u) or, "top to bottom", (u_1, u_2, u_3).

6.2   Example: dissipative trapping

The previous example cannot be dissipative, without further equilibria besides the S^2-isola, because dissipative systems must always contain an equilibrium [Ha88]. Here now is a dissipative example.

We do not build our dissipative example from scratch. Rather, we modify the trapped lip example. Again we work in $\Lambda \times X, X = \mathbb{R}^4$. For the global bifurcation diagram of steady states and Hopf paths see figure 6.3. Consider the region $\lambda_2 \leq 0$, below the λ_1-axis first. We have a sheet of stable equilibria at the "bottom". The remaining equilibria form a set equivalent to the 2-sphere S^2 of our previous example, but with a small open neighborhood of the B-point $B = -1$ removed. In particular, we make the same assumptions for these equilibria, as in the previous example. Consequently, the tame oriented hom path emanating at $B = +1$ is trapped again. Indeed, the only additional equilibrium, being a local attractor throughout, cannot interfere with trapping.

The diagram above the λ_1-axis, $\lambda_2 > 0$, indicates a completion to a globally dissipative system. Of course, the tame oriented hom path emanating, in reverse orientation, at

$B = -1$ is also trapped.

Therefore both local hom paths lead to parameter regions with complicated recurrent dynamics. We emphasize that, in both examples, this conclusion was drawn just from local information at equilibria and the absence of homoclinics to the fold-Hopf equilibria.

7   Discussion

We will begin our discussion with a short summary, relating our examples to several aspects of our main result, theorem 3.4. We briefly address infinite-dimensional dynamical systems next. After a digression into some aspects of numerical pathfollowing of homoclinic orbits, we compare our setting with a real line boundary value problem approach, based on functional analysis, and with global continuation results of Leray-Schauder type. Since we have imposed strong genericity assumptions, we also comment on certain important classes of vector fields with additional "non-generic" structure, specifically on Hamiltonian vector fields, time reversibility, and group equivariance. We postpone a discussion of length, ϵ-length, and topologies for the real line boundary value problem up to that point. Returning to our original setting, we then briefly review our results as an attempt of approaching, and possibly crossing, the boundary of the set of Morse-Smale systems. After some final (self-)criticism of certain inadequacies of our attempt at global homoclinic pathfollowing, we hastily indicate how our present results will improve with future work by the reader.

Recall our main result, theorem 3.4. Stated loosely, as in remark 3.5(i), a bounded tame oriented hom path emanating at a B-point with $B = +1$ faces the following alternative, globally. Either the path terminates at a $B = -1$ point, or else it eventually encounters complicated recurrent dynamics. For the first possibility see example 5.3.

The second alternative further splits into three cases. First, the hom path may hit a non-stratified loop. For examples see 2.10 and 5.5 – 5.8. Second, the hom path may reach the boundary of the set of chaotic homoclinic orbits. For examples see the Belyakov transitions 5.4. Third, ϵ-length may become unbounded, even though hom orbits and parameters remain uniformly bounded in $\Lambda \times X$.

We did not give an example for the third case above. In a way, this case is analogous to minimal period becoming unbounded along a bounded continuum of periodic orbits which stays away from equilibria. For periodic orbits, this possibility was called a "blue sky" catastrophe by Palis and Pugh, 1974. See examples by [No85], [Me80], [LZ91], as well as a recent generic one-parameter example due to [Tu96]. In fact, the "flow plug" construction, [HY83], manages to interrupt orbits by trapping them within a small cross-sectional box, by a smooth flow provided $\dim X \geq 4$. For $\dim X = 3$ the problem is related to counterexamples to the so-called "Seifert conjecture": constructing vector fields on S^3 without stationary points or periodic orbits. Such counterexamples of class C^1, C^2, respectively, were constructed by [Sw74] and [Har88]. They are relevant for global continuation of periodic orbits, see [AY78], and similarly for blow-up of ϵ-length along global paths of homoclinic orbits.

Another very illuminating example related to the third case was studied by [AS74], [AS91] in the context of boundaries of Morse-Smale systems. Consider an orbit $\tilde{h}$ which is homoclinic to a saddle-node periodic orbit p with zero strong unstable dimension; see figure 7.1. This is a codimension one generic situation. Breaking p into a pair of hyperbolic periodic orbits, the vector field becomes Morse-Smale, locally near the original homoclinic loop: a finite number of periodic orbits (here: two), all hyperbolic, with transverse intersections of their stable/unstable manifolds. Perturbing into the opposite direction, a strange attractor of fractal dimension appears; for estimates and more details see [AS91]. Introducing a second parameter, we may consider $\tilde{h}$ hitting a two-dimensionally unstable hyperbolic saddle A, thus forming a heteroclinic loop between an equilibrium and a periodic saddle-node; see figure 7.1. Perturbing such that the periodic saddle-node disappears, we will also find a path γ of homoclinic orbits with associated equilibrium near A. Clearly ϵ-length blows up, as γ approaches the periodic saddle-node. Recall that typically a strange attractor of fractal dimension appears near termination of the A-hom path. Summarizing, each case of theorem 3.4 may in fact occur.

Theorem 3.4, as it stands, is formulated only for finite-dimensional flows. For infinite-dimensional dynamical systems, results are available concerning the transition from homoclinic to periodic orbits, and also for time periodic forcing of homoclinic orbits. For promising techniques, especially for delay equations, we mention [Li90] and [Wa89]. Using the method of [Li90], for example, it is fairly clear that typical homoclinic orbits will again be accompanied by a sheet of periodic orbits just as in the finite-dimensional case. For the contractive case see also [CD89]. Index theories for periodic orbits and their bifurcations are readily available, along with a notion of genericity, see for example [CM78], [Fi85, Fi86b]. Global center manifold reductions

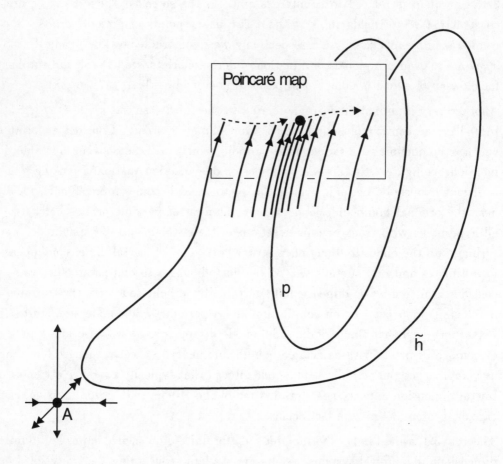

Figure 7.1: Generating a het loop between an equilibrium and a periodic saddle-node.

for compact analytic semigroups in a full neighborhood of homoclinic orbits were recently achieved by [Sa92]. These reductions also provide the first codimension two bifurcations of homoclinic orbits in an infinite-dimensional setting, if spectral gaps allow for sufficient regularity. Due to compactness of the semiflows, we expect that theorem 3.4, together with remark 3.5 (iv), carries over to the infinite-dimensional settings used in the above references.

For Volterra type integral equations, the situation is more hazy. In [Fi86a] a global Hopf bifurcation result was proved, by finite-dimensional ODE approximation. But our present results on homoclinic continuation critically depend on our genericity assumptions which, of course, need not survive an approximation process. We will return to this genericity question below.

Numerical pathfollowing of periodic orbits for one-parameter flows has become standard, by now. See e.g. [DoK85], [KKLN93], [KM83], [RB83], [Se88]. For the analogous problem of numerical pathfollowing of homoclinic orbits for two-parameter flows, there are only relatively few results available; for a line of research see [HW79], [Le80], [Be90], [DoF89], [Ku90, Ku95]. For recent progress on numerical detection of codimension 2 homoclinic bifurcations see [CK94]. To our knowledge, a systematic homoclinic pathfollowing code which would also allow for start-off from, and approach to, stratified loops along "secondary" tame hom paths is not yet available.

From a somewhat simplistic view point, time discretization might be considered as an approximation of a continuous time dynamical system by discrete time dynamics. In particular, even tame homoclinic orbits should in principle be considered as transverse, numerically. It turns out that the transversality effects, and the related Smale horseshoe shift dynamics, are exponentially small in the discretization step size; see [FiS88]. For discretizations of Runge-Kutta type, there even is a general principle of exponentially small discrepancy between discrete time and - suitably adjusted - autonomous flows; see [HL95]. Although we may therefore safely neglect the accompanying transversality effects as "spurious", in most cases of practical interest, these effects may cause problems at exponentially flat homoclinic bifurcations such as resonant homoclinic doubling, see example 2.9 and [CDF90]. It is particularly in such numerically rather delicate situations where we expect theorem 3.4 and its proof to help. Indeed, we might choose to avoid the intricate tangles of discrete time analogues of our homoclinic bifurcations. Instead, we could isolate the suspected singularity by a surrounding circle S, in parameter space. The proof in section 4 then suggests to follow paths of periodic orbits along such a circumventing circle S and match pairs of tame hom paths, according to our index theory, without ever touching the messy interior of S. There is no need to ever determine precisely what type of homoclinic bifurcation the circle S circumvents, as long as it is stratified. The only one-parameter homoclinic "bifurcation" which has to be understood well then, numerically, is that of periodic orbits limiting onto a tame homoclinic orbit. Unlike poor Laokoon, we should faithfully follow the periodic snakes along S, rather than fighting against them. This way we can avoid getting entangled too much in our pathfollowing attempts. Because our theoretical results are of a global nature, the diameter of the circle S is not a critical quantity. For example, the diameter of S need not be comparable to the size of a single, small numerical hom pathfollowing step. Of course, the interior of S will remain a white spot in the bifurcation diagram unless the homoclinic orbits are traced further inward. We realize that it is a long way from the above remarks to a fully implemented efficient algorithm. Still, we hope that such an algorithm will appear, tracing out the alternatives of theorem 3.4 in specific applied context.

One popular approach to homoclinic orbits $h(t)$ is to view them as solutions of a problem

(7.1) $\frac{d}{dt}x(\cdot) - f(\lambda, x(\cdot)) = 0$

with $x(\cdot) \in \mathfrak{X}$, a suitable Banach space of IR^N-valued functions on the real line, $t \in \mathrm{IR}$, for example $\mathfrak{X} = BC^0(\mathrm{IR}, \mathrm{IR}^N)$ or $BC^1(\mathrm{IR}, \mathrm{IR}^N)$ with sup norms. In other words, (7.1) is a boundary value problem on the whole real line, and the space $\mathfrak{X}$ imposes boundary conditions "at infinity". Under exponential dichotomy assumptions, the linearization with respect to $x(\cdot)$ becomes a Fredholm operator,[Pa84], and local continuation via implicit function theorem applies, [CH82]. Numerical discretization of $h(t)$ can be treated in the same spirit, by adding a time periodic perturbation $\epsilon^p g(\epsilon, \lambda, t/\epsilon, x(\cdot))$ with amplitude ϵ^p and time period ϵ determined by numerical order p and step size ϵ; see [FiS88]. Inverting $\frac{d}{dt}$ in $\mathfrak{X}$, and rewriting (7.1) as

(7.2) $x(\cdot) - (\frac{d}{dt})^{-1}f(\lambda, x(\cdot)) = 0,$

Leray-Schauder degree can be applied since (7.2) is a compact perturbation of identity. For the technically much more demanding problem of solitary water waves and elliptic partial differential equations in unbounded strips this idea has been pursued in [AK89].

Global bifurcation of periodic orbits has successfully been treated in a similar boundary value problem setting, with $\mathfrak{X}$ carrying periodic boundary conditions and period scaled out as an explicit parameter. See for example [Iz76]. More recent variants, [DGJM91], [DGJM91], [IMV89, Iz76], make explicit use of advanced topology and of

$SO(2)$-equivariance of (7.1), (7.2) with respect to time shifts.

We did not favor the boundary value problem approach, here, because in the above form it is intrinsically inadequate to treat most homoclinic or heteroclinic bifurcations. Consider resonant homoclinic doubling, 2.9, for example. Although the 2-hom branch converges to the 1-hom branch (traced out twice), in phase space $\Lambda \times \mathbb{R}^N$ and at bifurcation, the corresponding profiles will not converge in $\mathfrak{X}$, be it a C^k-, a Hölder-, or a Sobolev space. Following the 1-hom path, the 2-hom path remains far away in all these topologies. Approaching the bifurcation along the 2-hom path, a limiting profile need not exist. Our notion of homoclinic continuation through a stratified loop just cannot be expressed in terms of any such space $\mathfrak{X}$.

Our results crucially hinge on a genericity assumption for the vector field $f(\lambda, x)$. By genericity, we have B-points as starting points of homoclinic pathfollowing. By genericity, we obtain the sheet of periodic orbits which accompanies hom paths. Also the validity of the concept of isolated, codimension two stratified loops is critically tied to this assumption, as are all our examples of homoclinic bifurcations. When additional structure of the equations is known, there are two basic ways to incorporate this information. First, we can aim for a result which survives the limiting process of generic systems approximating non-generic ones. For example, Brouwer degree $\deg(f, \Omega, y)$ may first be defined for regular values y of f on Ω; in a second step arbitrary y are approximated by nearby regular values. For examples of global Hopf bifurcation involving approximations $p_n(\cdot) \to p(\cdot)$ of periodic orbits p see [Fi85, Fi86b, Fi88] and the references there. This approximation can be understood in terms of "virtual periods", see definition 2.11 above. Recall that virtual periods provide a necessary criterion for bifurcations: limits of minimal periods of $p_n(\cdot)$ are virtual periods of the limit $p(\cdot)$. When the ultimate goal is a universal result which survives the approximation process, only density of generic sets is used. In particular, genericity is then as appropriate a concept as "prevalence", introduced in [HSY93], or any other approximation process. Unfortunately, however, a tool analogous to virtual periods is still missing, for homoclinic orbits.

The second way to incorporate additional structure of the underlying equations is by restricting the notion of genericity, accordingly: consider open dense subsets within the particular class of vector fields under consideration. The two basic ingredients which we then need for our pathfollowing approach are the following:

(7.3.a)   transition from homoclinic to periodic orbits, and

(7.3.b)   an index theory for periodic orbits,

each in the appropriate structural context. Specifically we comment on equivariance, reversibility, and Hamiltonian structure.

Consider equivariant vector fields f first. For example $X = \mathbb{R}^N$ and f commutes with a compact Lie group G acting linearly on $X = \mathbb{R}^N$,

(7.4) $f(\lambda, gx) = gf(\lambda, x)$,

for all $g \in G, (\lambda, x) \in \Lambda \times X$. In particular, $t \mapsto gx(t)$ is a time trajectory if and only if $x(t)$ is; for example $x(t)$ could be stationary, periodic, homoclinic, or heteroclinic. An index theory in the spirit of definition 3.1 is available for periodic orbits in the presence of equivariance, see [Fi88]. The relevant groups here are cyclic factors in G. A two-parameter extension including B-points has not yet been developed.

Accounting for the transition (7.3.a) from homoclinic to periodic orbits, in a systematic equivariant way, is a somewhat involved problem; see for example [AGH88], [KM91], [Ch93], [SC92], and the references there. To sketch the basic problem, consider a homoclinic trajectory $x = h(t)$ with associated equilibrium A. Let $G_A \supseteq G_h$ denote the respective subgroups of elements in G fixing $A, h(t)$ pointwise. We may work in the linear, flow invariant subspace $Fix(G_h)$ of vectors in X which are fixed under G_h. If $G_h = G_A$, then the standard transition from homoclinic to periodic trajectories works well within $Fix(G_h)$. Next suppose G_h is strictly contained in G_A and, for simplicity, G_h is a normal subgroup of G_A. Then G_A acts on $Fix(G_h)$ and maps $h(t)$ to several copies, all homoclinic to A. A "trivial" caveat is the Lorenz explosion where $G = G_A = \mathbb{Z}_2$ and $G_h = id$.

Widening the notion of homoclinicity somewhat, we may require $h(t)$ to just be heteroclinic from A to an equilibrium $B = gA$ in the same group orbit as A. Assume, again for simplicity, that G_A is a normal subgroup of G and the factor group G/G_A of order $|G/G_A|$ is finite cyclic, generated by gG_A. Acting with G on A and h then gives rise to a heteroclinic loop composed of $|G/G_A|$ copies of h. In a non-equivariant generic setting it requires $|G/G_A|$ parameters to unfold such a loop. In contrast, such a loop can be codimension one, or even structurally stable, within the equivariant category. It can be a challenging problem in itself to analyze the nearby periodic trajectories and their symmetries. Specific examples are discussed in the above references. Although the combined complexity of forced symmetry breaking and homoclinic bifurcation may seem overwhelming, we still believe that our approach can be helpful at least in some particular cases, say with cyclic symmetries.

Time reversibility is a somewhat more subtle "symmetry" constraint. Let R be a linear involution on X; for example suppose R fixes the first N components on $X = \mathbb{R}^{2N}$ and reverses the signs of the remaining N components. Time reversibility means that

(7.5) $f(\lambda, Rx) = -Rf(\lambda, x),$

for all $(\lambda, x) \in \Lambda \times X$. Second order systems

(7.6) $\ddot{q} + v(\lambda, q) = 0, \qquad q \in \mathbb{R}^N$

provide an example with $R(q, \dot{q}) = (q, -\dot{q})$. In general $x(t)$ is a solution if, and only if, $Rx(-t)$ is. Thus reversibility imposes a constraint on those single trajectories which are invariant under R, as a set. Equivalently, these are the trajectories intersecting $Fix(R)$. Call such trajectories *symmetric*. It was observed by [De77] that symmetric homoclinic orbits are typically persistent under reversible perturbations, and are accompanied by a one-parameter family of likewise symmetric periodic orbits; see also the account in [VaF92]. Of course, the associated equilibrium A must lie in $Fix(R)$, and the linearization at A therefore defines a linear reversible vector field. In particular, the spectrum is point symmetric with respect to 0 in $\mathbb{C}$, and hence also with respect to reflection at the imaginary axis. Therefore, reversibility induces resonance relations for the principal eigenvalues $\mu_\pm$:

(7.7) $Re\mu_+ + Re\mu_- = 0,$

in contrast to nonresonance conditions (2.4.a,b) for tame homoclinic orbits; see also our (non-reversible) resonance example 2.9. For real $\mu_\pm$, some homoclinic bifurcations involving symmetric A have been discussed in [Pe92]. One interesting example involves a family of symmetric periodic orbits limiting onto a figure 8 symmetric pair of non-symmetric homoclinic orbits with the same, symmetric associated equilibrium A. In the case of complex, not purely imaginary $\mu_\pm$ the presence of n- homs, for any n, was established by Härterich. Shift dynamics is known to occur in the related Hamiltonian case [De76], but remains an open question in the reversible case.

An additional complication arises since simple, purely imaginary eigenvalues at A cannot be perturbed to have nonzero real parts, due to symmetry of the spectrum. Again, associated homoclinic orbits with shift dynamics have been analyzed in the Hamiltonian but not in the reversible case; see [HMR92]. Global pathfollowing faces a serious obstacle, even in one parameter and with real eigenvalues. The simplest bifurcation case is the coalescence of two such "tame" homs. It was shown in [FiT96], that this leads to an ellipticity region in the associated per sheet. The associated cascades of reversible subharmonics effectively obstruct hom pathfollowing.

The Hamiltonian case, for example

$$(7.8) \qquad \dot{x} = J \operatorname{grad}_x H, \qquad J = \begin{pmatrix} 0 & id \\ -id & 0 \end{pmatrix},$$

$x = (q, p) \in X = \mathbb{R}^{2N}$, has been studied quite extensively, at least as far as time periodic solutions are concerned. The reversible second order system (7.6) is a specific example, if we assume $v = \operatorname{grad}_q V$ is a gradient, $H(p, q) = p^2/2 + V(q)$. For surveys see, for example, [Ek90], [Ra86], [MW89], [St90], and the references there. Recently homoclinic orbits are receiving increased attention, mostly in the autonomous and periodically forced case; see, e.g., [An92], [CES90], [CoR91, CoR92], [HW90]. All these results are based on variational methods, alias least action principles. Notably, the possibility of homoclinic bifurcation is related to failure of the Palais-Smale compactness condition. Continuation of solutions is traditionally not considered, perhaps because minimizers need not vary continuously with parameters (but critical points do). Avoiding the variational setting, [AY78] use a one-parameter dissipative embedding

$$(7.9) \qquad \dot{x} = J \operatorname{grad}_x H + \nu \operatorname{grad}_x H, \quad \nu \in \mathbb{R},$$

to utilize one-parameter results on global Hopf bifurcation for Hamiltonian systems. Indeed, periodic solutions can occur only for $\nu = 0$ because H serves as a Lyapunov functional for $\nu \neq 0$. In this way, two-parameter results on homoclinic continuation should provide one-parameter results for Hamiltonian systems or, more generally, for systems with a first integral H. As we have mentioned above, we are unfortunately lacking a concept like virtual periods which would survive an approximation of the dissipative embedding (7.9) by generic vector fields. We note that the ingredients (7.3 a,b) to our alternative approach, genericity within the class of Hamiltonian systems, are available: see [VaF92] for the transition homoclinic to periodic, and [MY82] for a Hamiltonian orbit index. Still a pathfollowing approach in the present spirit will probably be limited to real principal eigenvalues, due to the shift dynamics results by [De76] and [HMR92] mentioned above.

Typical spaces $\mathfrak{X}$ for homoclinic orbits $h(\cdot)$ in a variational Hamiltonian context are L^p- and $W^{1/2}$-spaces, similarly to the boundary value problems sketched above. Measuring the length of $h(\cdot)$ corresponds, of course, to a $W^{1,1}$-norm. In contrast, ϵ-length is somewhat like a $W^{1,1}_{loc}$-topology, as long as only one associated equilibrium A is involved and an appropriate parametrization is chosen for $h(\cdot)$. For heteroclinic loops,

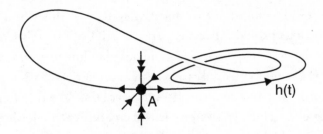

Figure 7.2: An orbit which is heteroclinic from A to an A-hom orbit;
see [Ho95].

this analogy fails. A very interesting non-local matching construction, which over-
comes the boundary value problem topology obstacle and resembles shadowing ideas
in dynamical systems, is presented in [CoR91]. In fact, shift type dynamics near
transverse homoclinic points is imitated in a variational context. This allows generalization to elliptic equations in several unbounded space dimensions which originally
lack a dynamical systems interpretation; see [CoR92].

In figure 7.1, we have indicated one way how ϵ-length might become unbounded along
a bounded hom path. That mechanism was related to the hom path approaching the
boundary $\mathcal{B}$ of the closure of the set of Morse-Smale systems. Also, Belyakov transitions 5.4 lie on $\mathcal{B}$, at least locally, and chaotic homoclinic orbits lie beyond. Here
"beyond" means "outside the closure of the set of Morse-Smale systems". Generic
folds with at least two associated homoclinic orbits also lie beyond $\mathcal{B}$, recalling example 5.1. Consider a non-stratified loop H in the sense of proposition 2.8 and definition
2.11 next. Suppose H cannot be perturbed to become stratified. Then H (or, in fact,
the vector field belonging to H) lies on $\mathcal{B}$ or beyond. On the other hand, stratified
loops may also lie on $\mathcal{B}$, or beyond. Even tame homoclinic orbits may. An example considered by [Ho95] is sketched in figure 7.2. It involves a dissipative saddle A
on an invariant 2-manifold with a homoclinic orbit h, attractive from "inside". The
remaining "left" half of the unstable manifold of A is then fed back to the "inside"
of h, accumulating onto clos(h) for $t \to +\infty$. This situation corresponds to an A-to-
periodic heteroclinic orbit, when the periodic orbit has just become homoclinic to A.
For simplicity we assume A to be only one-dimensionally unstable. We only consider
the case where the limiting A-to-h heteroclinic orbit is non-twisted; the bifurcation

143

can then be realized by vector fields on a 2-torus embedded e.g. in $\mathbb{R}^3$. Again, such a system is on the Morse-Smale boundary $\mathcal{B}$. As long as the periodic orbit associated to h exists, that orbit is an attractor and the system is Morse-Smale with two A-to-periodic heteroclinic orbits. After the periodic orbit has dissolved via h, however, the dynamics can be described by a discontinuous one-dimensional map. Phenomena like intermittency, Cantor type bifurcation sets, and renormalization structures arise. The homoclinic orbit h itself, however, is tame according to our definitions. It is not the pathfollowing of h, which causes problems here, but pathfollowing of the infinity of homoclinic orbits which start out opposite to h, for $t \to -\infty$, and re-settle into A only after n revolutions along h for any large integer $n \geq n_0$.

It is these intricacies which make our pathfollowing attempt inadequate, unsatisfactory, and incomplete. On the one hand, it seems quite easy to probe into the labyrinth of chaotic complexity along a homoclinic path, at least for a while. On the other hand, once inside, there is no visible Ariadne thread to help us move around. The above example shows that, while following h, we may in fact have already entered the labyrinth without being much aware of it. Tresser's non-stratified figure 8, composed of two separately tame homoclinic orbits, is another such case; see 2.10 and figure 2.7. It is even quite feasible, in that case, to choose arbitrarily small circles S in Λ around the figure 8 parameter value $\lambda = \lambda_0$, such that S "sees" only finitely many homs and attached per branches. In particular, pathfollowing would be possible, matching hom paths along S as in section 4. The interior of S then remains a "white spot" in the global bifurcation diagram. Unfortunately, adopting such a strategy, we may inadvertently have to remove countably many discs – effectively wiping out major parts of the bifurcation diagram. Before we will be able to do better than that, we will need a deeper understanding of that tremendously rich zoo of homoclinic and heteroclinic bifurcations.

Quoting Poincaré [Po1892] in the translation by Ekeland [Ek90], periodic solutions appear " ... to be the only opening through which we can force our way into an otherwise impenetrable citadel ... ". Extending to homoclinic orbits, we have still largely followed Poincaré's advice. Our contribution is just an attempt to detect some order in the complexities ruling homoclinic orbits. We have, in other words, investigated the homoclinic part of the large period limit of periodic orbits. Introducing terminology like "stratified loops" is a risk. It is not clear, at present, whether or not such a notion will be very appropriate. Mainly, we are still lacking a sufficient catalogue of examples. But there is some consolation, too. In the future, more and more examples will enter the homoclinic bifurcation catalogue. Among them, some ought to be stratified. This will only widen the scope of the results presented here. Hopefully, *you* will contribute ...

Chapter 3
Stability of Fronts for a KPP-System
- the noncritical case -

by **K KIRCHGÄSSNER AND G RAUGEL**

Contents

1   Introduction

In this contribution we study parameter–dependent semilinear systems of parabolic equations living on infinitely extended space–domains. In fact, we shall finally treat the case of a single unbounded space–variable, called $\eta \in I\!R$. The family of elliptic differential operators $I\!\!L(\lambda, \partial_\eta)$ acts on vector–valued functions $\underline{u}(\eta, \tau) \in I\!R^{n+1}$, $\underline{u} = (u_0, \underline{u}_1)$ and λ is a real parameter varying near zero. We assume that $I\!\!L(\lambda, \partial_\eta)$ has constant coefficients and that its principal part $I\!\!L_\omega(\lambda, \partial_\eta)$ does not depend on λ [Bu72]. Furthermore, $I\!\!L(\lambda, \partial_\eta)$ is supposed to be smooth in λ. For the sake of simplicity, we shall finally require that $I\!\!L(\lambda, \partial_\eta)$ is a symmetric operator.

We study the parabolic system

$$\partial_\tau \underline{u} + \mathbb{L}(\lambda, \partial_\eta)\underline{u} - f(\lambda, \underline{u}) = \underline{0} \tag{1.1}$$

where $f : \mathbb{R} \times \mathbb{R}^{n+1} \to \mathbb{R}^{n+1}$ is a smooth non–linear function, satisfying $f(\lambda, \underline{0}) = \underline{0}$ for all λ. We always have the trivial solution $\underline{u} = \underline{0}$.

We seek travelling waves solutions of (1.1) and study the long–time behaviour of local perturbations of these waves. When considered in an appropriate moving frame $\xi = \eta + \nu\tau$, say, travelling waves are steady solutions of a modified system (1.1), where $\mathbb{L}(\lambda, \partial_\eta)$ is replaced by $\widetilde{\mathbb{L}}(\lambda, \nu, \partial_\xi) = \mathbb{L}(\lambda, \partial_\xi) + \nu I \partial_\xi$. If $\underline{u}^*(\cdot, \lambda) \neq 0$ is such a steady solution which emanates from $\underline{u} = \underline{0}$ at $\lambda = 0$, and if the hyperbolic part ($Re\rho \neq 0$) and the central part ($Re\rho = 0$) of the set

$$C(\nu, \lambda) := \{\rho \in \mathbb{C} / \det(\widetilde{\mathbb{L}}(\lambda, \nu, \rho) - D_u f(\lambda, \underline{0})) = 0\} \cap i\mathbb{R}$$

can be "separated", and if the inverse of $\widetilde{\mathbb{L}}(\lambda, \nu, ik)$ has "minimal" growth for $|k| \geq k_0$, $k \in \mathbb{R}$ and suitable $k_0 \geq 0$, then $\underline{u}^*$ lies in a center–manifold, when the amplitude of $\underline{u}^*$ is sufficiently small ([Ki82], [Mi88]). For some concrete application see [IoK92].

Now, a natural question to ask is, whether the long–time dynamics of "local" perturbations of travelling waves–solutions of (1.1) can be determined by a reduced system living on the same subspace as the center–manifold in the steady case.

In this contribution we show the validity of this conjecture for the case of a front–wave under the condition that, in some adequate weighted functional space, the essential spectrum of $\widetilde{\mathbb{L}}(\lambda, \nu, \partial_\xi) - D_u f(\lambda, \underline{u}^*(\cdot, \lambda))$ is bounded away from the imaginary axis. In a forthcoming paper [RK95], we treat the critical case when the essential spectrum and $i\mathbb{R}$ are connected. The results will be relatively robust, so that generalisations to infinite dimensions and thus to several space dimensions are possible. Special as the case treated may seem to be, it will nevertheless turn out to have a relatively wide range of applications.

Remark: The set $C(\nu, \lambda) \cap i\mathbb{R}$ is finite, if (ν, λ) lies in a sufficiently small neighbourhod of $(0, 0)$. This follows from the fact that $\det(\widetilde{\mathbb{L}}_\omega(\lambda, \nu, ik)) \neq 0$ for all $k \in \mathbb{R}$ and thus $\det(\widetilde{\mathbb{L}}(\lambda, \nu, ik)) \neq 0$ for $|k|$ sufficiently large.

We assume for the rest of this paper that $\mathbb{L}(0, \partial_\eta)$ is the diagonal operator

$$\mathbb{L}(0, \partial_\eta) = \mathrm{diag}\,(-\partial_{\eta\eta}^2 + \alpha_j)$$

148

where $\alpha_0 = 0$, $\alpha_j > 0$ for $j = 1, ..., n$. Without loss of generality, we also assume that, for $|\lambda|$ small, $\mathbb{L}(\lambda, \partial_\eta)$ has the form

$$\mathbb{L}(\lambda, \partial_\eta) = \operatorname{diag}(-\partial_{\eta\eta}^2) + \operatorname{diag}(-\lambda + a_0(\lambda), \alpha_j + a_j(\lambda)) \tag{1.2}$$

where $a_0(\lambda) = O(\lambda^2)$, $a_j(\lambda) = O(\lambda)$. We decompose $\underline{u} = (u_0, \underline{u}_1)$, $\underline{u}_1 \in \mathbb{R}^n$ and introduce the $(n+1) \times (n+1)$ matrix A defined by $A = (A_0, A_1)$, where $A_0 = 0$ and $A_1 = \operatorname{diag}(\alpha_j)$, for $j = 1, ..., n$. Without loss of generality, we can assume that $\underline{f} = (f_0, \underline{f}_1)$ takes the form below

$$f_0(\lambda, \underline{u}) = -u_0^2 + O(|u_0|^3 + \lambda|u_0|^2 + |\underline{u}_1|^2 + |u_0|\,|\underline{u}_1|)$$
$$\underline{f}_1(\lambda, \underline{u}) = O(|\underline{u}|^2) \quad . \tag{1.3a}$$

We also use the following notations:

$$g_0(\lambda, \underline{u}) = -a_0(\lambda)u_0 + (f_0(\lambda, \underline{u}) + u_0^2)$$
$$\underline{g}_1(\lambda, \underline{u}) = -\operatorname{diag}(a_j(\lambda))\underline{u}_1 + \underline{f}_1(\lambda, \underline{u}) \quad . \tag{1.3b}$$

Then, the system (1.1) can be written

$$\partial_\tau u_0 - \partial_{\xi\xi}^2 u_0 + \nu\partial_\xi u_0 = \lambda u_0 - u_0^2 + g_0(\lambda, \underline{u})$$
$$\partial_\tau \underline{u}_1 - \partial_{\xi\xi}^2 \underline{u}_1 + \nu\partial_\xi \underline{u}_1 + A_1\underline{u}_1 = \underline{g}_1(\lambda, \underline{u}) \quad . \tag{1.4}$$

Before we proceed further a few remarks about the generality of (1.4) are appropriate. We could consider, instead of (1.4), a system where $\mathbb{L}(\lambda, \partial_\eta)\underline{u} = (d_{ij}\partial_{\eta_i\eta_j}^2 u_{ij}(\eta, \tau)) + A(\lambda)\underline{u}$, $D = (d_{ij})$ being a symmetric, positive definite $(n+1) \times (n+1)$ matrix. Moreover $A(\lambda)$ could be any symmetric matrix with $\det A(\lambda)$ having a simple zero at $\lambda = 0$, and thus $A(\lambda)$ could be assumed to be of the form $A(\lambda) = (\lambda + O(\lambda^2))E_0 \otimes A_{11}(\lambda)$, where $(E_0)_{ji} = (E_0)_{ij}$, $(E_0)_{ij} = \delta_{0j}$ for $0 \le i \le j = 0, ..., n$ and $A_{11}(\lambda)$ is a symmetric, positive definite $n \times n$-matrix. Under the slight additional assumption that $\underline{e}_0$ is an eigenvector of D as well(that is, $d_{i0} = d_{0j} = 0$ for $i \ne 0, j \ne 0$), one could perform a similar analysis as presented here. Furthermore $\underline{f}$ could be assumed to depend also on $\partial_x\underline{u}$: $\underline{f}(\lambda, \underline{u}, \partial_x\underline{u})$. This generalisation causes a few technical difficulties and some changes in the powers of λ in the subsequent estimates. But the principal result would be still the same.

We return to (1.4) and assume henceforth, that λ is positive. Near $\lambda = 0$, (1.4) has a natural scaling as follows:

$$\underline{u}(\xi, \tau) = \lambda\underline{U}(x, t) \quad , \qquad x = \xi\lambda^{1/2} \quad , \quad t = \tau\lambda \quad ,$$
$$\nu = \gamma\lambda^{1/2} \quad , \quad \gamma \in \mathbb{R}$$

149

which yields

$$\partial_t U_0 - \partial^2_{xx} U_0 + \gamma \partial_x U_0 = U_0 - U_0^2 + G_0(\lambda, \underline{U})$$

$$\partial_t \underline{U}_1 - \partial^2_{xx} \underline{U}_1 + \gamma \partial_x \underline{U}_1 + \frac{1}{\lambda} A_1 \underline{U}_1 = \underline{G}_1(\lambda, \underline{U}) \tag{1.5}$$

where

$$G_0(\lambda, \underline{U}) = \lambda^{-2} g_0(\lambda, \lambda \underline{U}) = -\frac{a_0(\lambda)}{\lambda} U_0 + O(\lambda U_0^2 + |\underline{U}_1|^2 + U_0 |\underline{U}_1|)$$

$$\underline{G}_1(\lambda, \underline{U}) = \lambda^{-2} g_1(\lambda, \lambda \underline{U}) = -\text{diag}\,(\frac{a_j(\lambda)}{\lambda}) \underline{U}_1 + O(|\underline{U}|^2) \quad . \tag{1.6}$$

In the sequel, we shall write $a_0(\lambda) = \lambda^2(a_0 + O(\lambda))$, where a_0 is a constant.

Finally a short remark concerning the scaling in (1.5) should be made. While the scaling of $\underline{u}$, ξ and τ is natural, the scaling of ν as $\nu = \gamma \lambda^{1/2}$ needs an explanation. If we had chosen $\nu = \gamma \lambda^\alpha$, then $\alpha > 1/2$, and λ small would lead to unstable steady solutions connecting a saddle with a focus. If $\alpha < 1/2$, the steady solutions are connections of a saddle with a node and stable, as will be shown. The limit $\lambda \to +0$ leads to a singular bifurcation problem describing the bifurcation of fast waves from the quiescent state(for such problems in an infinite cylindrical domain, see [Sc94]).

It is the system (1.5) which we study for λ near 0. Setting G_0 to 0 yields for U_0 the classical KPP equation. It is well known that this equation has monotone fronts only if $|\gamma| \geq 2$. It is not hard to see that for $|\gamma| < 2$, all steady solutions connecting 1 and 0 are unstable. For $\gamma = 0$ there exists a steady solution connecting 0 with itself, which has a stable manifold of codimension 2. This fact persists for the full system.

The KPP–equation has an elaborate history, which has started with the classical papers [KPP37], [Fi37]. For general results about systems, including the KPP–equation, we also refer the reader to [AW75], for instance. The dynamics of solutions of the KPP–equation in the invariant region $[0, 1]$ is well known cf. [Br83]. In two papers [Sa76], [Sa77], Sattinger has investigated perturbed front–solutions in the 'noncritical' case ($|\gamma| > 2$ here). He has used appropriate weights in the space–variable in order to shift the spectrum of the linearized operator, so that he could prove orbital stability in weighted L^∞–spaces. His basic tool for obtaining the necessary decay in time was an asymptotic representation of the resolvent of the linearization about the front near infinity.

In fact, we could have used his method to show local stability of fronts for (1.5) when the initial conditions are smooth. However, since we are interested in stability without shifts, that we shall treat semi–global results for the perturbed system (1.5) and finally intend to study the critical case $|\gamma| = 2$ in a continuation of this paper [RK95], we decided to go our own way. In the case $|\gamma| > 2$, we use the same weights as Sattinger, but work in Hilbert spaces.

As for the critical situation $|\gamma| = 2$, when the continuous part of the spectrum extends to the point 0 in $\mathbb{C}$, stability for fronts in systems of parabolic equations has been studied by Gardner, Jones and Kapitula in [GJK93] for different situations, and also by Kapitula in [Ka95] using a detailed discussion of the 'Evans' function. In [Ka95], the equality of the dimension of the unstable manifold of one equilibrium with the dimension of the stable manifold of the other equilibrium has to be assumed. It is this requirement which is violated in our case. We shall discuss these facts in more detail in the forthcoming paper [RK95].

A special feature of our analysis is the concept of 'local' perturbations requiring spatial decay properties which exclude $\partial_x \underline{u}^*$ as a member of the corresponding spaces, where $\underline{u}^*$ denotes the front solution. Thus we can obtain asymptotic stability for the individual front. For $|\gamma| > 2$, in appropriate weighted spaces, the time decay is exponential. In the case $|\gamma| = 2$, optimal decay results in time for the scalar KPP equation in this spirit have been obtained by Gallay [Ga94] using, among other methods, renormalization–group arguments. Weaker versions have been proved by Eckmann and Wayne [EW94] by applying energy functionals.

There exists also an extensive literature on long–time behaviour of systems on infinite cylindrical domains in $I\!\!R^n$ arising in flame propagation models(see [BLR92], [Ro92] and the references therein). Their tools in the study of the linearized problem and the linear stability are exponential spatial decay estimates for solutions of second order elliptic equations in a cylinder and various versions of the maximum principle for parabolic equations. In [Ro92], local stability and convergence to a translate of the travelling wave in an adequete weighted Banach space have been proved by adapting the methods of [Sa76] and using the just mentioned careful study of the linearized operator. The existence of a continuum of travelling waves of a KPP type equation on infinite cylindrical domains and their stability in weighted spaces are studied by similar methods in [MR95]. Under some hypotheses which allow the use of the maximum principle, they also give global stability results. Since we do not make such hypotheses here, we cannot apply maximum principles in order to obtain global stability; we shall use perturbation methods instead.

Finally we summarize the structure of this paper. In Section 2 we prove the existence of a front–solution of the system (1.5) when $\gamma \leq -2$ holds. We set $\lambda = \mu^2$ and use a center–manifold reduction to a two–dimensional flow on a manifold modeled over the subspace $\underline{U}_1 = \underline{0}$. There, for $\mu = 0$, the front connects a saddle $e_0^- = (1, 0)$ at $x = -\infty$ with a node $e_0^+ = (0, 0)$ at $x = +\infty$. The transversal intersection of the one–dimensional unstable manifold of e_0^- with the two–dimensional stable manifold of e_0^+ yields the desired result, for $\mu > 0$. We denote this front–solution by $\underline{U}_\mu^*(x)$.

In Section 3 we discuss briefly the effect that exponential weights have on the spectrum of the linearization about $\underline{U}_\mu^*$ and conclude that an ansatz of the form $\underline{U} = \underline{U}_\mu^* + a\underline{W}$ is appropriate to measure the perturbation of $\underline{U}_\mu^*$ where $a(x) = \exp(\frac{\gamma}{2}x)$. The nonlinearity then enforces one to work in the intersection $\mathcal{X}^j$, $j = 0, 1$ of the Sobolev spaces $\mathcal{H}^j(\mathbb{R})$ and $\mathcal{H}^j(\mathbb{R}, a)$ where the latter carries the weight a^2. The decay of the semigroups generated by the linearization L_μ, defined in (3.7), is given in Lemma 3.1.

To prove the local stability of $\underline{U}_\mu^*$ or $\underline{W} = 0$, we show that the system (3.5), which $\underline{W}$ must satisfy, can be reduced in balls of order $O(1/\mu^2)$ to a single nonlocal equation in W_0. The explicit statement of this result is given in Theorem 3.6 for initial conditions in $\mathcal{H}^1(\mathbb{R}) \cap \mathcal{H}^1(\mathbb{R}, a)$. In Theorem 3.9, we prove a similar reduction result, when the size of the first derivatives of the initial data are much larger than these data itself. Finally in both cases, if $\gamma < -2$, we prove local stability results, which are given in Theorems 3.12 and 3.13. It shows that indeed, the long–time behavior of the full system is determined by that of W_0^*, the solution of the "central" equation, which is given in (4.1).

With Section 4 we begin the discussion of some global aspects. As was observed in [Ki92] the front solution of the scalar KPP equation enjoys a semiglobal stability property. This means in terms of W_0, that $W_0^*(t)$, as a solution of (4.1), is asymptotically stable and tends exponentially fast to zero in the space $H^1(\mathbb{R})$ in the case $\gamma < -2$, if only $W_{0|t=0}^* \geq U^0$, $aU^0 = -U_{00}^*$, i.e. if $U_{0|t=0} \geq 0$ in the original scalar KPP equation. If $W_{0|t=0}^* \geq (1 - d)U^0$, where $d > 0$, then $W_0^*(t)$ tends also exponentially fast to zero in the space $H^1(\mathbb{R}, a)$. These facts are proved in Theorem 4.4 and the subsequent propositions.

In the final Section 5 we discuss some of the consequences for System (3.5) of the global stability proved in Section 4 for the central equation. We treat the case $\gamma < -2$ only and show in a first step, that in a short time–interval of order $O(\mu^2|\log \mu|)$, the $\underline{W}_1$ component looses more than a factor of $O(\mu^{-1})$ in amplitude in that interval; see Theorem 5.2. The central component is measured in amplitude by the function $g_0(\mu)$ which becomes unbounded as μ approaches 0. Different forms of $g_0(\mu)$ are discussed depending on the strictness the initial value satisfies the inequality $W_0(0) \geq (1 - \tilde{d}_0)U^0$. If $\tilde{d}_0 > 0$, then $g_0(\mu) = \log(\mu^{-p})$, for some positive $p < 1$, cf Theorem 5.3. If $\tilde{d}_0 > 1/2$ then g_0 can be chosen to be $g_0(\mu) = \mu^{-p_0}$ for some $p_0 \leq 1/2$; see Theorem 5.5.

2 Existence of fronts

Fronts are steady solutions connecting two different equilibria. In this section existence of fronts is shown for (1.4) in the parameter region $0 < \lambda$, $\nu = \gamma\lambda^{1/2}$, $|\gamma| \geq 2$, and λ is assumed to be small. Without loss of generality we may restrict the parameters to $\nu < 0$, $\gamma < -2$ and take $\mu^2 = \lambda$. We show that fronts exist by using a reduction to a centermanifold first. Then, the existence of an orbit connecting $e_\mu^- = (1 + O(\mu^2), 0)$ at $-\infty$ to $e_\mu^+ = (0, 0)$ at $+\infty$ follows from the fact that the unstable manifold of e_μ^- and the stable manifold of e_μ^+ intersect transversally.

We write (1.4) as a first order system for $(\mu, \tilde{u}_0, \tilde{u}_1) \in \mathbb{R} \times \mathbb{R}^{2(n+1)}$, with $\tilde{u}_0 = (u_0, v_0)$,
$v_0 = \partial_\xi u_0$; $\tilde{u}_1 = (u_1, v_1)$, $v_1 = \partial_\xi u_1$, $u = (\tilde{u}_0, \tilde{u}_1)$.

$$\partial_\xi \mu = 0$$
$$\partial_\xi \tilde{u}_0 = \tilde{A}_0(\mu)\tilde{u}_0 - \tilde{g}_0(\mu^2, u) \qquad (2.1)$$
$$\partial_\xi \tilde{u}_1 = \tilde{A}_1(\mu)\tilde{u}_1 - \tilde{g}_1(\mu^2, u)$$

where $\tilde{g}_0 = (0, g_0 - u_0^2)$, $\tilde{g}_1 = (0, g_1)$

$$\tilde{A}_0(\mu) = \begin{pmatrix} 0 & 1 \\ -\mu^2 & \mu\gamma \end{pmatrix} \quad , \quad \tilde{A}_1(\mu) = \begin{pmatrix} 0 & id \\ A_1 & diag(\mu\gamma) \end{pmatrix} \quad . \qquad (2.1a)$$

System (2.1) admits a local center–manifold W_{loc}^c of $\mu = 0$, $\tilde{u} = 0$; it has dimension 3, i.e. there is a C^p–mapping $(\underline{p}_1, \underline{q}_1) : \mathbb{R}^3 \cap V \longrightarrow \mathbb{R}^{2n} \cap V$ such that

$$W_{loc}^c = \left\{ (\mu, \tilde{u}) \mid \tilde{u}_1 = \left(\underline{p}_1(\mu, \tilde{u}_0), \underline{q}_1(\mu, \tilde{u}_0) \right) , (\mu, \tilde{u}_0) \in \mathbb{R}^3 \cap V \right\} , \qquad (2.2)$$

where V is a suitable neighborhood of 0 in $\mathbb{R}^{1+2(n+1)}$ (see [Ca81] or [CH82]). The following facts hold: W_{loc}^c is locally invariant under the flow of (2.1) and

$$\underline{p}_1, \underline{q}_1, D\underline{p}_1, D\underline{q}_1 \text{ vanish for } (\mu, \tilde{\underline{u}}_0) = \underline{0}$$

$$\underline{p}_1(\mu, \underline{0}) = \underline{q}_1(\mu, \underline{0}) = \underline{0}, \ \mu \in (\mathbb{R} \times \underline{0}) \cap V \tag{2.3}$$

$$D_{\mu u_0}^2 \underline{p}_1(0, \underline{0}) = D_{\mu v_0}^2 \underline{p}_1(0, \underline{0}) = \underline{0}$$

holds, where the last equality follows easily from (1.4).

System (2.1) is equivalent to the reduced system

$$\begin{aligned}
\partial_\xi \tilde{\underline{u}}_0 &= \tilde{A}_0(\mu)\tilde{\underline{u}}_0 - \tilde{g}_0\left(\mu^2, \tilde{\underline{u}}_0 + \underline{p}_1(\mu, \tilde{\underline{u}}_0)\right) \\
\tilde{\underline{u}}_1 &= (\underline{p}_1, \underline{q}_1)(\mu, \tilde{\underline{u}}_0)
\end{aligned} \tag{2.4}$$

for solutions $\tilde{\underline{u}}(\xi)$ belonging to some open set $D \subset V$ for all $\xi \in \mathbb{R}$. We describe D by the inequalities $|\mu| < \mu_1$, $|\tilde{\underline{u}}_0| < r$, $|\tilde{\underline{u}}_1| < r$, where $\mu_1 > 0$ and $r > 0$. Now we scale as follows

$$u_0(\xi) = \mu^2 U_0(x) \quad, \quad v_0(\xi) = \mu^2 |\mu| V_0(x) \quad, \quad x = |\mu|\xi \quad, \tag{2.5}$$

which leads to

$$\begin{aligned}
\partial_x U_0 &= V_0 \\
\partial_x V_0 &= \gamma sg(\mu)V_0 - U_0 + U_0^2 + \mu^2 k_0(\mu, U_0, V_0)
\end{aligned} \tag{2.6μ}$$

with

$$\mu^2 k_0(\mu, U_0, V_0) = -\mu^{-4} g_0\left(\mu^2, \mu^2 U_0 + \underline{p}_1(\mu, \mu^2 U_0, \mu^2 |\mu| V_0)\right) \quad . \tag{2.7}$$

Whenever $\left(\mu_1^4 + \mu_1^6\right) R^2 < r^2$, $|\mu| < \mu_1$ and $|U_0|^2 + |V_0|^2 < R^2$ holds, (2.6μ) and (2.1) are equivalent.

Let us estimate k_0 for small $|\mu|$. Since the change of $sg(\mu)$ just inverts the orientation of x, we may assume μ to be positive, what we will do henceforth. Applying the integral Taylor formula about $\underline{U}_0 := (U_0, V_0) = \underline{0}$ to $\underline{p}_1(\mu, \mu^2 U_0, \mu^3 V_0)$, we obtain $\underline{p}_1(\mu, \mu^2 U_0, \mu^3 V_0) = \mu^4 \underline{p}_1^*(\mu, U_0, V_0)$, where $\underline{p}_1^*$ is a C^{p-2}-function with $\underline{p}_1^*(\mu, \underline{0}) = \underline{0}$. Finally, applying that formula at $\mu = 0$, $\underline{U}_0 = \underline{0}$ to g_0, we conclude from (2.7) that $k_0(\mu, \underline{U}_0)$ is a C^{p-3}-function, defined for $0 \le \mu < \mu_1$, $|\underline{U}_0| < R$, satisfying $k_0(\mu, \underline{0}) = 0$.

The existence of a front solution of (2.6μ), will follow from the existence of a solution of

$$\begin{aligned}
\partial_x U_0 &= V_0 \\
\partial_x V_0 &= \gamma V_0 - U_0 + U_0^2
\end{aligned} \tag{2.8}$$

via perturbation arguments. The equilibria are $e_0^- = (1, 0)$, $e_0^+ = (0, 0)$. e_0^- is a saddle point with eigenvalues $\sigma_{\pm}^- = \frac{1}{2}(\gamma \pm \alpha)$, where $\alpha = (\gamma^2 + 4)^{1/2}$; e_0^+ is a stable node with eigenvalues $\sigma_{\pm}^+ = \frac{1}{2}(\gamma \pm \beta)$, where $\beta = (\gamma^2 - 4)^{1/2}$. Observe that $\gamma \leq -2$ has to hold to obtain a front. A simple phase plane analysis shows the existence of a solution $\underline{U}_{00}^* = (U_{00}^*, V_{00}^*)$ of (2.8)–unique up to translations in x–connecting e_0^- and e_0^+ (see [Sa76]). It converges to its respective equilibria like $\exp(\sigma_+^- x)$ as $x \longrightarrow -\infty$, and like $\exp(\sigma_+^+ x)$ for $\gamma < -2$ (resp. xe^{-x} for $\gamma = -2$) as $x \longrightarrow +\infty$. Moreover, $U_{00}^*(x) \subset (0, 1)$ holds for $x \in I\!\!R$.

The 1–d unstable manifold $W^u(e_0^-)$ and the 2–d stable manifold $W^s(e_0^+)$ intersect transversally in $I\!\!R^2$. This being a structurally stable property, we expect it to imply the existence of an orbit $\underline{U}_{0\mu}^*$ of (2.6μ), close to $\underline{U}_{00}^*$, connecting the equilibria $e_\mu^- = (1 + O(\mu^2), 0)$ at $-\infty$ and $e_\mu^+ = \underline{0}$ at $+\infty$. This will result from the subsequent Lemmata.

The unique existence of the equilibria $e_\mu^- = (1 + O(\mu^2), 0)$ as a saddle and $e_\mu^+ = (0, 0)$ as a stable node follows immediately from the implicit function theorem applied to the function $-l + l^2 + \mu^2 k_0(\mu, l, 0)$ at $(\mu, l) = (0, 1)$ and $(0, 0)$ successively, and from the fact that $k_0(\mu, \underline{0}) = 0$. The corresponding eigenvalues are denoted by $\sigma_{\pm}^-(\mu)$, $\sigma_{\pm}^+(\mu)$. For small positive μ, $\sigma_{\pm}^-(\mu)$ are real numbers of opposite signs, while $\mathrm{Re}\,\sigma_{\pm}^+(\mu)$ is negative. All eigenvalues are simple, if $\gamma < -2$. Moreover we have

$$\sigma_{\pm}^-(\mu) - \sigma_{\pm}^- = O(\mu^2)$$

$$\sigma_{\pm}^+(\mu) - \sigma_{\pm}^+ = \begin{cases} O(\mu^2) & \text{if } \gamma < -2 \\ O(\mu) & \text{if } \gamma = -2 \end{cases} \qquad (2.9)$$

Lemma 2.1 *Given the local groups $T_\mu(x)$, resp. $T_0(x)$, defined by (2.6μ) and (2.8), there exists a positive, nondecreasing function $K(\cdot)$ on $[0, \infty)$ such that, for every positive x_0, R_0 and for any $\underline{U}_0^\mu$, $\underline{U}_0^0 \in I\!\!R^2$,*

$$\max_{x \in [0, x_0]} \left\{ \max\left(|T_\mu(x)\underline{U}_0^\mu|, |T_0(x)\underline{U}_0^0| \right) \right\} < R_0 \quad ,$$

implies

$$|T_\mu(x)\underline{U}_0^\mu - T_0(x)\underline{U}_0^0| \leq K(R_0)\left(\mu^2 + |\underline{U}_0^\mu - \underline{U}_0^0|\right) \cdot \exp\left(K(R_0)x\right) \qquad (2.10)$$

for $x \in [0, x_0]$.

The elementary proof is left to the reader.

155

Since $(0,0) = \underline{0}$ is a stable equilibrium of T_μ, $W^s_{loc,\mu}(\underline{0})$ is a neighborhood of $\underline{0}$. The vectorfields, defined by (2.6μ) and (2.8), are C^1-close, so we could appeal to known results to conclude that this neighborhood can be chosen independently of μ and that $W^u_\mu(e^-_\mu) \cap W^s_\mu(e^+_\mu) \neq \emptyset$ for small μ. But, as we need more precise estimates, we give a direct elementary proof here.

Lemma 2.2

(i) *Given δ, with $0 < \delta < -\frac{1}{2}\,Re\,\sigma^+_+$. Then there are positive constants c_1, c_2, μ_2 and ρ_2, such that $T_\mu(x)\underline{U}^0_0$ exists for all $x \geq 0$, whenever $0 \leq \mu < \mu_2$ and $|\underline{U}^0_0| < \rho_2$. Moreover, for all $x \geq 0$*

$$|T_\mu(x)\underline{U}^0_0| \leq c_1|\underline{U}^0_0|\exp\left[\left(Re\,\sigma^+_+ + \frac{\delta}{2}\right)x\right] \tag{2.11}$$

holds.

(ii) *Furthermore, for $\underline{U}^1_0$, $\underline{U}^2_0$, with $|\underline{U}^j_0| < \rho_2$, we have*

$$|T_\mu(x)\underline{U}^1_0 - T_0(x)\underline{U}^2_0| \leq c_2\left(\mu^2 + |\underline{U}^1_0 - \underline{U}^2_0|\right)\exp\left(Re\,\sigma^+_+ + \delta\right)x \tag{2.12}$$

for all $x \geq 0$. Here, c_1, c_2, μ_2, ρ_2 depend on δ but not on μ.

Proof. Denote by e^{C_0} the linear group generated by the linearized system of (2.8) about $\underline{U}_0 = 0$. There exists a constant $c_1 \geq 1$ such that

$$\left|e^{C_0 x}\underline{U}^0_0\right| \leq c_1\exp\left\{\left(Re\,\sigma^+_+ + \frac{\delta}{4}\right)x\right\}|\underline{U}^0_0| \tag{2.13}$$

for $x \geq 0$. From (2.6μ) we obtain

$$\underline{U}^\mu_0(x) =: T_\mu(x)\underline{U}^0_0 = e^{C_0 x}\underline{U}^0_0 + \int_0^x e^{C_0(x-s)}\underline{F}^\mu\left(\underline{U}^\mu_0(s)\right)ds \tag{2.14}$$

where

$$\underline{F}^\mu(\underline{U}_0) = \left(0,\, U^2_0 + \mu^2 k_0(\mu, \underline{U}_0)\right)^t\;.$$

Clearly, there are suitable positive constants ρ_1, μ_2, $\widetilde{K}$, such that

$$|\underline{F}^\mu(\underline{U}_0)| \leq \left(\rho_1 + \mu^2\widetilde{K}\right)|\underline{U}_0| \tag{2.15}$$

156

holds, when $|\underline{U}_0| < \rho_1$, $\mu < \mu_2$ (observe that $k_0(\mu, \underline{0}) = 0$). In addition, we require

$$c_1 \left(2\rho_1 + \mu_2^2 \widetilde{K} \right) < \frac{\delta}{4} \quad . \tag{2.16}$$

Now choose $0 < 2c\rho_2 < \rho_1$ and set for $|\underline{U}_0^0| \leq \rho_2$

$$x_0 := \sup \left\{ x \geq 0 \,\big|\, |T_\mu(s)\underline{U}_0^0| < \rho_1 \,; \, 0 \leq s \leq x \right\} \leq \infty \quad .$$

We infer from (2.13)–(2.15), for $0 \leq x < x_0$,

$$|\underline{W}(x)| \leq c_1 \left(|\underline{U}_0^0| + (\rho_1 + \mu^2 \widetilde{K}) \int\limits_0^x |\underline{W}(s)| ds \right)$$

where

$$\underline{U}_0^\mu(x) = \underline{W}(x) \exp \left(\operatorname{Re} \sigma_+^+ + \frac{\delta}{4} \right) x$$

Applying the Gronwall Lemma and using (2.16) we obtain

$$|\underline{U}_0^\mu(x)| \leq c_1 |\underline{U}_0^0| \exp \left[\left(\operatorname{Re} \sigma_+^+ + \frac{\delta}{2} \right) x \right] \leq \frac{\rho_1}{2} \quad , \tag{2.17}$$

for $0 \leq x < x_0$. This implies $x_0 = +\infty$ and thus part (i) of the assertion.
Statement (ii) is proved in a similar way. We write

$$\underline{U}_0^\mu(x) - \underline{U}_0^0(x) = e^{C_0 x} \left(\underline{U}_0^1 - \underline{U}_0^2 \right) + \int\limits_0^x e^{C_0(x-s)} G(s) ds \quad , \tag{2.18}$$

where

$$G(s) := \left(0, \, (U_0^\mu - U_0^0) (U_0^\mu + U_0^0) + \mu^2 k_0 \left(\mu, \underline{U}_0^\mu \right) \right)^t \quad .$$

Due to (2.15) and (2.17), we have, for $s \geq 0$,

$$|G(s)| \leq 2\rho_1 \left| \underline{U}_0^\mu - \underline{U}_0^0 \right| + \mu^2 \widetilde{K} c_1 \rho_2 \exp \left[\left(\operatorname{Re} \sigma_+^+ + \frac{\delta}{2} \right) s \right] \tag{2.19}$$

whenever $\mu < \mu_2$, $|\underline{U}_0^j| < \rho_2$, $j = 1, 2$. Arguing as above and using (2.13), (2.16), (2.18), (2.19), we obtain (2.12). ∎

In the following Lemma we are going to compare the local unstable manifolds $W_{loc,\mu}^u \left(e_\mu^- \right)$ and $W_{loc,0}^u \left(e_0^- \right)$, where $|e_\mu^- - e_0^-| \leq c\mu^2$, $e_0^- = (1, 0)$. We introduce the following semidistance on $\mathbb{R}^2$. If B_1, B_2 denote two subsets of $\mathbb{R}^2$, we define

$$d(B_1, B_2) := \sup_{B_1} \inf_{B_2} |b_1 - b_2| \quad .$$

157

Lemma 2.3

(i) *There are three positive constants μ_3, ρ_3, c_3 and a neighborhood V_μ of e_μ^- in $\mathbb{R}^2$ for $\mu \in [0, \mu_3]$, such that $W_\mu^u(e_\mu^-; V_\mu)$ is a C^{p-3}-submanifold of $\mathbb{R}^2$ of dimension 1. Moreover, for each μ, V_μ contains the ball of radius ρ_3 about e_0^-, and the following inequality is valid*

$$d\left(W_\mu^u(e_\mu^-; V_\mu), W_0^u(e_0^-; V_0)\right) + d\left(W_0^u(e_0^-; V_0), W_\mu^u(e_\mu^-; V_\mu)\right) \le c_3 \mu^2 \quad (2.20)$$

(ii) *For any $\delta > 0$, with $0 < \delta < \frac{1}{2}\sigma_+^-$, there exist positive constants c_4, μ_4, ρ_4, with $0 < \rho_4 \le \rho_3$, such that, if $\underline{U}_0^1 \in W_\mu^u(e_\mu^-; V_\mu) \cap B(e_\mu^-, \rho_4)$ and $\underline{U}_0^2 \in W_0^u(e_0^-; V_0) \cap B(e_0^-, \rho_4)$, we have for all $x \ge 0$ and $\mu \in [0, \mu_4]$*

$$\left| T_\mu(-x)\underline{U}_0^1 - T_0(-x)\underline{U}_0^2 \right| \le c_4 \left[\mu^2 + (\mu^2 + |\underline{U}_0^1 - \underline{U}_0^2|) \cdot e^{-(\sigma_+^- - \delta)x} \right] \quad (2.21)$$

Proof.

(i) Since the vector fields in (2.6μ) and (2.8) are C^{p-3}-close of order μ^2, (2.20) is a direct consequence of general results in [PaM82].

(ii) Set $e_\mu^- = (e_{\mu 0}^-, 0)$ and introduce the transformation

$$U_0 = e_{\mu 0}^- + u_\mu + v_\mu$$

$$V_0 = \sigma_-^-(\mu)u_\mu + \sigma_+^-(\mu)v_\mu \quad ,$$

then $W_\mu^u(e_\mu^-; V_\mu)$ is given by a graph $(h_\mu(v_\mu), v_\mu)$, where h_μ is a C^{p-3}-function in a neighborhood of 0 in $\mathbb{R}$. Moreover for some $\tilde{\rho}_3 > 0$ and all $\mu \in [0, \mu_3]$

$$\|h_0 - h_\mu\|_{C^1(B(0,\tilde{\rho}_3), \mathbb{R})} \le C\mu^2 \quad . \quad (2.22)$$

Thus, the comparison of the flows on the local unstable manifolds reduces to comparing the flows of the system

$$\partial_x v_\mu = \sigma_+^-(\mu)v_\mu + \frac{1}{\sigma_+^-(\mu) - \sigma_-^-(\mu)}((v_\mu + h(v_\mu))^2 +$$

$$+ \mu^2 \overline{G}(\mu, h_\mu(v_\mu), v_\mu)) \quad (2.23)$$

$$\partial_x v_0 = \sigma_+^- v_0 + \frac{1}{\sigma_+^- - \sigma_-^-}(v_0 + h_0(v_0))^2 \quad ,$$

where $\overline{G}$ is a C^{p-5}-function on a neighborhood of 0 in $\mathbb{R}^3$ independent of μ, and $\overline{G}(\mu, 0, 0) = 0$ holds. Arguing as in the proof of Lemma 2.2, we conclude the validity of similar estimates as in that Lemma for $|v_\mu(-x)|$ and $|v_\mu(-x) - v_0(-x)|$. Using then (2.22), we obtain an estimate similar to (2.12) for $|h_\mu(v_\mu(-x)) - h_0(v_0(-x))|$. Finally, returning to (U_0, V_0), we obtain (2.21). ∎

Now we show the existence of a front solution of (2.6μ) connecting e_μ^- with e_μ^+. We denote this solution by $\underline{U}_{\mu 0}^*$. Choose some $\underline{a}_0 = (a_0, b_0) \in W_0^u(e_0^-; V_0) \cap B(e_0^-; \rho_4/2)$ and x_0, so that $\underline{a}_0 = \underline{U}_{00}^*(x_0)$. Let $x_2 \geq 0$ be any real number satisfying

$$T_0(x_2)\underline{a}_0 \in B(\underline{0}; \rho_2/2) \quad , \tag{2.24}$$

where ρ_2 is taken from Lemma 2.2. Due to Lemma 2.3, there exists a positive constant μ_5, and for each $\mu \in (0, \mu_5]$ at least one element $\underline{a}_\mu \in W_\mu^u(e_\mu^-; V_\mu)$ with $|\underline{a}_\mu - \underline{a}_0| < c_3\mu^2 < \rho_4/2$. Thus $\underline{a}_\mu \in W_\mu^u(e_\mu^-, V_\mu) \cap B(e_0^-; \rho_4)$ holds. Lemma 2.1 serves to compare $T_\mu(x)\underline{a}_\mu$ and $T_0(x)\underline{a}_0$ for $0 \leq x \leq x_2$. Choose

$$R_0 \geq 2\max\left(\rho_2, \sup_{x \in \mathbb{R}} |\underline{U}_{00}^*(x)|\right)$$

and $0 < \mu_6 \leq \mu_5$ such that

$$K(R_0)\mu_6^2(1 + c_3)\exp(K(R_0)x_2) < \rho_2/2$$

$$\sup_{x \in \mathbb{R}} |\underline{U}_{00}^*(x)| + K(R_0)\mu_6^2(1 + c_3)\exp(K(R_0)x_2) < R_0 \tag{2.25}$$

holds. Then we obtain for x, μ with $0 \leq x \leq x_2, 0 < \mu < \mu_6$

$$\left|T_\mu(x)\underline{a}_\mu - T_0(x)\underline{a}_0\right| \leq K(R_0)\mu^2(1 + c_3)\exp(K(R_0)x_2) \quad . \tag{2.26}$$

The estimates (2.25), (2.26) and property (2.24) imply that $T_\mu(x_2)\underline{a}_\mu \in B(\underline{0}; \rho_2)$. Hence, by Lemma 2.2, $T_\mu(x)\underline{a}_\mu \to \underline{0}$ as $x \to \infty$. We conclude via Lemmata 2.2, 2.3 and by (2.26) for $x \in \mathbb{R}$

$$\left|T_\mu(x)\underline{a}_\mu - T_0(x)\underline{a}_0\right| \leq C\mu^2 \quad . \tag{2.27}$$

Set $\underline{U}_{\mu 0}^*(x) = T_\mu(x - x_0)\underline{a}_\mu$. From (2.27) and from (2.6μ) and (2.8) we deduce for $\mu \in (0, \mu_6)$

$$\|\underline{U}_{\mu 0}^* - \underline{U}_{00}^*\|_{C_b^2(\mathbb{R}, \mathbb{R})} < C\mu^2 \quad . \tag{2.28}$$

159

We remark that $\underline{U}^*_{\mu 0} \simeq e^-_\mu - \underline{c}_{-\infty} \exp(\sigma^-_+(\mu)x)$ as $x \to -\infty$; $\underline{U}^*_{\mu 0}(x) \simeq \underline{c}_\infty \exp(\sigma^+_+(\mu)x)$ as $x \to +\infty$ is valid for $\gamma < -2$. In the case $\gamma = -2$ the asymptotic behavior depends on properties of the system, and will be discussed in [RK95]. Now define

$$\underline{U}^*_\mu(x) = \left(U^*_{\mu 0}(x), \frac{1}{\mu^2} \underline{p}_1 \left(\mu, \mu^2 U^*_{\mu 0}(x), \mu^3 V^*_{\mu 0}(x) \right) \right) \quad . \tag{2.29}$$

$\underline{U}^*_\mu$ is a stationary solution of (1.5), where $\lambda = \mu^2$, connecting $\left(e^-_{\mu 0}, \frac{1}{\mu^2} \underline{p}_1(\mu, \mu^2 e^-_{\mu 0}, 0) \right)$ and $\underline{0}$. Due to the property

$$\underline{p}_1 \left(\mu, \mu^2 U_0, \mu^3 V_0 \right) = \mu^4 \underline{p}^*_1 \left(\mu, U_0, V_0 \right) \quad ,$$

to (2.28) and the boundedness of $U^*_{00}(x)$ in $C^2(I\!\!R)$, we have shown, for $0 < \mu < \mu_6$

$$\|\underline{U}^*_{\mu 1}\|_{C^2(I\!\!R, I\!\!R^n)} \leq C\mu^2 \quad . \tag{2.30}$$

Theorem 2.4 *For* $\gamma \leq -2$, *there exist a positive number* μ_0, *and for each* $\mu \in (0, \mu_0]$, *a stationary solution* $\underline{U}^*_\mu(x)$ *of the system* (1.5). *It is given by* (2.29). *Moreover, there are positive constants* C^* *and* C^*_0 *such that*

$$\|\underline{U}^*_\mu\|_{C^2_b(I\!\!R; I\!\!R^{n+1})} \leq C^* \tag{2.31}$$

$$\|\underline{U}^*_\mu - (U^*_{00}, 0)\|_{C^2_b(I\!\!R; I\!\!R^{n+1})} \leq C^*_0 \mu^2 \tag{2.32}$$

holds.

Remark. From Lemma 2.2 and from (2.26), we deduce that, for $x \geq x_2$,

$$\begin{aligned} |T_\mu(x)\underline{a}_\mu - T_0(x)\underline{a}_0| &\leq c_2 \left[\mu^2 + K(R_0)\mu^2 (1 + c_3) \exp(K(R_0)x_2) \right] \cdot \\ &\quad \exp\left((\operatorname{Re}\sigma^+_+ + \delta)(x - x_2) \right) \quad . \end{aligned} \tag{2.33}$$

It follows from (2.33) and the definition of $\underline{U}^*_{\mu 0}(x)$ that, for every positive constant d, there exists $x_d > 0$, such that for $\mu \in (0, \mu_0]$ and for $x \geq x_d$,

$$|\underline{U}^*_{\mu 0}(x) - \underline{U}^*_{00}(x)| \leq d\mu^2 \quad . \tag{2.34}$$

3 Local stability of fronts

In this section we study the local stability of the fronts $\underline{U}^*_\mu$, which have been constructed previously. Based on a reduction procedure we show that the system inherits

its long–time behavior from the scalar equation. In this way we can prove local exponential asymptotic stability when $\gamma < -2$ holds (Theorems 3.12 and 3.13).

Even for the limiting case $\gamma = -2$ asymptotic stability prevails, but the proof of that is postponed to [RK95], the second part of this paper.

The stability results are shown for two different classes of initial conditions which differ by the degree of their regularity. In order to obtain asymptotic stability for every individual front we have to exclude $\partial_x \underline{U}_\mu^*$ from the space of admissible functions. After having defined the appropriate semigroups which are generated by the linearization about $\underline{U}_\mu^*$ and studied their decay for large times we prove the reduction result and the local asymptotic stability.

We start with a simple equation to explain effects that weights can have on the long–time behavior of solutions. We take the formal limits of (1.5) and (1.6) as $\mu \to 0+$ and set $\underline{U} = \underline{U}_0^* + \underline{V}$; then V_0 satisfies

$$\partial_t V_0 + \widehat{L}_0 V_0 + V_0^2 = 0 \tag{3.1}$$

where

$$\widehat{L}_0 = -\partial_{xx}^2 + \gamma \partial_x + (2U_{00}^* - 1) \quad .$$

Consider $\widehat{L}_0$ as a linear operator in $L^2(\mathbb{R})$. Its essential spectrum $\sigma_e(\widehat{L}_0)$, in the sense of Krein, is contained in the parabolic region

$$P = \left\{ \zeta = \zeta_r + i\zeta_i \in \mathbb{C} \,|\, \zeta_r - \frac{1}{\gamma^2}\zeta_i^2 + 1 \geq 0 \right\} \tag{3.2}$$

and its boundary ∂P satisfies $\partial P \subset \sigma_e(\widehat{L}_0) \subset P$ ([He80]). Therefore $\widehat{L}_0$ has spectral points in $\mathbb{C}^-$, the negative halfplane of $\mathbb{C}$, and $\exp(-\widehat{L}_0 t)$ is linearly unstable. The introduction of an exponential weight as in Sattinger [Sa76] may shift the spectrum to the right. Its effect is seen, when we choose $\gamma_0 \in \mathbb{R}$ and $V_0 = W_0 \exp(\gamma_0 x)$. Then (3.1) becomes

$$\partial_t W_0 + \widetilde{L}_0 W_0 + e^{\gamma_0 x} W_0^2 = 0 \tag{3.3}$$

where $\widetilde{L}_0 = -\partial_{xx}^2 + (\gamma - 2\gamma_0)\partial_x + \widetilde{b}(x)$ and $\widetilde{b}(x) = 2U_{00}^* - 1 + \gamma_0(\gamma - \gamma_0)$. Similarly, as for (3.1), we conclude that the essential spectrum $\sigma_e(\widetilde{L}_0)$ belongs to $\mathbb{C} - \mathbb{C}^-$, if $\gamma_0 \in [\frac{\gamma}{2} - \theta, \frac{\gamma}{2} + \theta]$, where $\theta = (\frac{\gamma^2}{4} - 1)^{1/2}$ (and that $\sigma_e(\widetilde{L}_0)$ contains 0 if $\gamma_0 = \frac{\gamma}{2} \pm \theta$). Set $\alpha = \frac{(\gamma - 2\gamma_0)}{2}$. The fundamental solution of $\partial_t + \widetilde{L}_0$ at $x = \mp\infty$ is given by $G(x, t) \exp\left(\alpha x - (\frac{\gamma^2}{4} \pm 1)t\right)$, where $G(x, t)$ is the Gaussean $(4\pi t)^{-\frac{1}{2}} \exp(-\frac{x^2}{4t})$. Hence the fundamental solution modifies the diffusive part, represented by G, by a

161

"convective" term, which travels with speed $\frac{(\frac{\gamma^2}{4} \pm 1)}{\alpha}$ at $x = \mp \infty$ relative to the wave, if $\alpha \neq 0$.

Therefore, whenever α is nonzero (i.e. $\gamma_0 \neq \frac{\gamma}{2}$), the solutions of $(\partial_t + \widetilde{L}_0)W_0 = 0$ have a drift, which is asymptotically given by the wave $\exp\left(\alpha x - (\frac{\gamma^2}{4} \pm 1)t\right)$. Thus, an artificial time dependence is introduced. Such an effect is called convective. To avoid convective effects we have therefore to choose $\gamma_0 = \frac{\gamma}{2}$.

Another argument supporting this choice proceeds as follows. Take $\gamma < -2$. Assume we require that W_0 belongs to $H^1(\mathbb{R})$. Due to the presence of the non linear term, we also need to require that $e^{\gamma_0 x} W_0$ belongs to $H^1(\mathbb{R})$. Then this restricts the functional space in which we are studying the stability. The larger $|\gamma_0|$ is, the smaller is our functional space. Considerations of a right balance between the decay in time and the smallness of the functional space lead us to choose γ_0 in $[\frac{\gamma}{2}, \frac{\gamma}{2} + \theta]$ only. Among these values of γ_0, only the limit choices $\gamma_0 = \frac{\gamma}{2}$ or $\gamma_0 = \frac{\gamma}{2} + \theta$ are really interesting. In this paper, we choose $\gamma_0 = \frac{\gamma}{2}$, for which the time decay is optimal and of exponential order if $\gamma < -2$ and is also independent of any shift. If $\gamma_0 = \frac{\gamma}{2} + \theta$, the time decay for the solutions of (3.3) is of polynomial order. Moreover the choice of this weight for studying the perturbations in the system (1.5) can lead either to instability, or to stability with exponential decay, or to stability with polynomial decay. This case will be studied in detail in the part two of this paper([RK95]). One might think that the choice of a C^1 weight–function like $\tilde{a}(x) = 1$ for x near $-\infty$ and $e^{\frac{\gamma}{2}x}$ near $+\infty$ is less restrictive. But it is not true. Actually our choice of the space X^k in (3.8) is equivalent to the choice of the weighted Sobolev space $H^k(\tilde{a})$. Moreover, proving decay estimates successively in $H^k(\mathbb{R})$ and in $H^k(a)$ leads to simpler and nicer proofs.

As we have just explained it, we introduce the weight $a = \exp(\gamma x/2)$ and make the ansatz

$$\underline{U} = \underline{U}_\mu^* + a\underline{W} \quad , \quad a(x) = e^{\frac{\gamma}{2}x} \quad . \tag{3.4}$$

From (1.5) we obtain

$$\partial_t \underline{W} + L_\mu \underline{W} = \underline{Q}(\underline{W}) + \underline{H}(\mu, \underline{W}) \tag{3.5}$$

where

$$\underline{Q}(\underline{W}) = \begin{pmatrix} Q(\underline{W}) \\ \underline{0} \end{pmatrix} \quad , \quad Q(\underline{W}) = -aW_0^2$$

$$\underline{H}(\mu, \underline{W}) = a^{-1}\mu^{-4} \left\{ \underline{g}\left(\mu^2, \mu^2 \left(\underline{U}_\mu^* + a\underline{W}\right)\right) - \underline{g}^* \right\} = \tag{3.6}$$

$$= \mu^{-2} D_{\underline{u}}\underline{g}^* \cdot \underline{W} + \int_0^1 (1-s) D_{\underline{u}\underline{u}}^2 \underline{g}\left(\mu^2, \mu^2 \left(\underline{U}_\mu^* + as\underline{W}\right)\right)(\underline{W}, a\underline{W})ds$$

Here we have used the notation $D_{\underline{u}}\underline{g}^* = D_{\underline{u}}\underline{g}(\mu^2, \mu^2\underline{U}_\mu^*)$ and $\underline{g}^* = \underline{g}(\mu^2, \mu^2\underline{U}_\mu^*)$. We make use of similar notations throughout the paper.

The linear operator L_μ is given by

$$L_\mu \underline{W} = -\partial_{xx}^2 \underline{W} + B(x, \mu)\underline{W} = \begin{pmatrix} L_{\mu 0} W_0 \\ L_{\mu 1} \underline{W}_1 \end{pmatrix} \tag{3.7}$$

where

$$B(x, \mu)\underline{W} = \begin{pmatrix} b(x, \mu)W_0 \\ B_1(x, \mu)\underline{W}_1 \end{pmatrix}$$

$$b(x, \mu) = 2U_{\mu 0}^*(x) + \frac{\gamma^2}{4} - 1 \tag{3.7a}$$

$$B_1(x, \mu) = \mu^{-2}A_1 + \frac{\gamma^2}{4}\mathbb{I}_1 \quad .$$

The following inequalities are valid

$$b(x, \mu) \geq b_0 := \frac{\gamma^2}{4} - 1$$

$$B_1(x, \mu) \geq \left(\frac{\alpha}{\mu^2} + \frac{\gamma^2}{4}\right)\mathbb{I}_1 \quad , \quad \alpha = \min_j(\alpha_j) > 0 \quad . \tag{3.7b}$$

It is clear from (3.6) that we have to choose two weights in order to be able to control the nonlinearity. The weights have to differ by the factor $a = \exp(\gamma x/2)$. To be specific we choose

$$H^k = H^k(\mathbb{R}), \|\cdot\|_k , k \in \mathbb{N}$$

$$H^k(a) = H^k(\mathbb{R}, a), \|\cdot\|_{k, a} = \left(\sum_{j=0}^k \left\| a\frac{\partial^j \cdot}{\partial x^j} \right\|_0^2\right)^{1/2} \tag{3.8}$$

$$X^k = H^k \cap H^k(a), \|\cdot\|_{X^k} \quad .$$

Vector–valued spaces are marked with script letters, e.g. $\mathcal{X}^k$. $H^k(\mathbb{R})$ denotes the usual Sobolev space $W^{k,2}(\mathbb{R})$, and $H^k(\mathbb{R}, a) = \{v \in H^k_{loc}(\mathbb{R}) \mid a\frac{\partial^j v}{\partial x^j} \in L^2(\mathbb{R}), 0 \le j \le k\}$ and the norm is as defined in (3.8).

Next, we discuss the semigroup generated by $-L_\mu$ in $\mathcal{X}^0$ for $t \ge 0$. We do this by intersecting the holomorphic semigroups generated by $-L_\mu$ in $\mathcal{H}^0(a)$ and in $\mathcal{H}^0$. In fact we take, for later use, a slight modification of L_μ, which is defined as follows

$$L_\mu^d \underline{W} = L_\mu \underline{W} - (1-d) \begin{pmatrix} U_{\mu 0}^* W_0 \\ 0 \end{pmatrix} = \begin{pmatrix} L_{\mu 0}^d W_0 \\ L_{\mu 1} \underline{W}_1 \end{pmatrix} \tag{3.9}$$

for $0 < d \le 1$. Remark that $L_\mu^1 = L_\mu$ and that, in L_μ^d, $b(x, \mu)$ is replaced by

$$b^d = (1+d)U_{\mu 0}^* + b_0 \quad .$$

Inequality (3.7b) is still valid for b^d.

Lemma 3.1 *Let* $\gamma \le -2$, μ_0 *as in Section 2, and* $\mu \in (0, \mu_0]$. *Then the operator* $-L_\mu^d$ *generates a holomorphic semigroup in* $\mathcal{X}^0$ *for* $t \ge 0$.

Moreover, for every $\delta \in (0, d)$, *there exist positive constants* μ_1, C_{30}, $\mu_1 \le \mu_0$, *depending solely on* γ, δ, d, *such that the following estimates are valid for* $t > 0$, $j = 0, 1$:

$$\left\| \exp\left(-L_{\mu 0}^d t\right) \right\|^2_{\mathcal{L}(H^j, H^j)} \le C_{30} e^{-2b_0 t} \tag{3.10a}$$

$$\left\| \exp\left(-L_{\mu 1} t\right) \right\|^2_{\mathcal{L}(\mathcal{H}^j, \mathcal{H}^j)} \le e^{-\left(\frac{\gamma^2}{2} + \frac{2\alpha}{\mu^2}\right)t} \tag{3.10b}$$

$$\left\| \exp\left(-L_{\mu 0}^d t\right) \right\|^2_{\mathcal{L}(X^j, X^j)} \le C_{30} \left(e^{-2b_0 t} + e^{-2(d-\delta)t}\right) \tag{3.11a}$$

$$\left\| \exp\left(-L_{\mu 1} t\right) \right\|^2_{\mathcal{L}(\mathcal{X}^j, \mathcal{X}^j)} \le e^{-\frac{2\alpha}{\mu^2} t} \tag{3.11b}$$

$$\left\| \exp\left(-L_{\mu 0}^d t\right) \right\|^2_{\mathcal{L}(X^0, X^1)} \le C_{30} \left(1 + t^{-\frac{1}{2}}\right)^2 \left(e^{-2b_0 t} + e^{-2(d-\delta)t}\right) \tag{3.11c}$$

$$\left\| \exp\left(-L_{\mu 1} t\right) \right\|^2_{\mathcal{L}(\mathcal{X}^0, \mathcal{X}^1)} \le C_{30} \left(1 + t^{-\frac{1}{2}}\right)^2 e^{-\frac{2\alpha}{\mu^2} t} \tag{3.11d}$$

The inequalities (3.10a), (3.10b), (3.11b) *and* (3.11d) *hold also in the case* $d = 0$.

Proof. The operator L_μ^d is selfadjoint in $\mathcal{H}^0$. Therefore it generates a holomorphic semigroup in $\mathcal{H}^0$. The estimate (3.10a) follows from (3.5) by taking the scalar product of

$$\partial_t W_0 - \partial_{xx}^2 W_0 + b^d(x, \mu) W_0 = 0$$

by W_0, $-W_{0xx}$, respectively. We obtain

$$\frac{d}{dt}\|W_0\|_0^2 + 2\|W_{0x}\|_0^2 + 2\int_{\mathbb{R}} b^d |W_0|^2 = 0 \tag{3.12a}$$

$$\frac{d}{dt}\|W_{0x}\|_0^2 + 2\|W_{0xx}\|_0^2 + 2\int_{\mathbb{R}} b^d |W_{0x}|^2 - \int_{\mathbb{R}} b_{xx}^d |W_0|^2 = 0 \quad . \tag{3.12b}$$

Applying the Gronwall Lemma to (3.12a) and using (3.7b) yields the estimate (3.10a) for $j = 0$ with C_{30} replaced by 1. If we integrate the inequality (3.12a) between t and $t + \theta$, $\theta \geq 0$, we also have

$$\int_t^{t+\theta} \|W_{0x}\|_0^2 \leq \|W_0(t)\|_0^2 \quad . \tag{3.12c}$$

Applying the uniform Gronwall Lemma to (3.12b) and taking into account (3.12c) as well as (3.10a) implies for $t > 0$, $\tau = \max(0, t-1)$

$$\begin{aligned}
\|W_{0x}(t)\|_0^2 &\leq \frac{1}{t-\tau}\int_\tau^t \|W_{0x}\|_0^2 ds + c_1 \int_\tau^t \|W_0\|_0^2 ds \\
&\leq c_2 \left(1 + (t-\tau)^{-1}\right) e^{-2b_0 t}\|W_0(0)\|_0^2 \quad .
\end{aligned} \tag{3.12d}$$

We thus have proved the estimate (3.10a) for $j = 1$, in the case $t \geq 1$. To obtain the estimate (3.10a) for $j = 1$ and $0 \leq t \leq 1$, we simply apply the usual Gronwall Lemma to (3.12b).

Estimate (3.10b) follows similarly from

$$\partial_t \underline{W}_1 + L_{\mu 1}\underline{W}_1 = \partial_t \underline{W}_1 - \partial_{xx}^2 \underline{W}_1 + \left(\frac{\gamma^2}{4} + \frac{1}{\mu^2}A_1\right)\underline{W}_1 = 0$$

via scalar multiplication by $\underline{W}_1$ and $-\partial_{xx}^2 \underline{W}_1$ successively. Again the uniform Gronwall Lemma implies that, for $t > 0$, $\tau = \max(0, t-1)$,

$$\begin{aligned}
\|\underline{W}_{1x}(t)\|_0^2 &\leq \frac{1}{t-\tau}\int_\tau^t \|\underline{W}_{1x}\|_0^2 ds + c_3 \int_\tau^t \|\underline{W}_1\|_0^2 ds \\
&\leq c_4 \left(1 + (t-\tau)^{-1}\right) e^{-\left(\frac{\gamma^2}{2} + \frac{2\alpha}{\mu^2}\right)t}\|\underline{W}_1(0)\|_0^2 \quad .
\end{aligned} \tag{3.13}$$

To prove the various estimates (3.11), we have to show first that L_μ^d generates a holomorphic semigroup in $\mathcal{H}^0(a)$. We complexify $\mathcal{H}^0(a)$ and write

$$L_\mu^d \underline{W} = -\frac{1}{a^2}\left(a^2\underline{W}_x\right)_x + \gamma\underline{W}_x + \begin{pmatrix} b^d(x,\mu)W_0 \\ B_1(x,\mu)\underline{W}_1 \end{pmatrix} . \tag{3.14}$$

This defines a closed linear operator in $\mathcal{H}^0(a)$ with domain in $\mathcal{H}^2(a)$. Set $\tilde{L}_\mu = L_\mu^d + (\frac{\gamma^2}{4}+2)I$. Take the (complex) scalar product of $-\tilde{L}_\mu\underline{W} + \zeta\underline{W} = \underline{F} \in \mathcal{H}^0(a)$ with $a^2\underline{W}$, $\zeta = \zeta_r + i\zeta_i \in \mathbb{C}$, to obtain

$$\|\underline{W}_x\|_{0,a}^2 + (1-\zeta_r)\|\underline{W}\|_{0,a}^2 \le \|\underline{F}\|_{0,a}\|\underline{W}\|_{0,a}$$

$$|\zeta_i|\,\|\underline{W}\|_{0,a}^2 \le |\mathrm{Im}\int_{I\!R} a^2\underline{F}\cdot\overline{\underline{W}}| + |\gamma|\,|\int_{I\!R} a^2\underline{W}_x\cdot\overline{\underline{W}}|$$

or $|\zeta_i|\,\|\underline{W}\|_{0,a} \le (|\gamma|+1)\,\|\underline{F}\|_{0,a}$. This implies, for the resolvent $\tilde{R}$ of $\tilde{L}_\mu$, the estimate $\|\tilde{R}(\zeta)\underline{F}\|_{0,a} \le c\|\underline{F}\|_{0,a}/|\zeta-1|$ for $\zeta_r < 1$ and all $\zeta_i \ne 0$. Hence, the condition in [Ka76], p. 487f is fulfilled: $-\tilde{L}_\mu$, and thus also $-L_\mu^d$, generates a holomorphic semigroup for $t \ge 0$ in $\mathcal{H}^0(a)$. Now, take $\underline{W}^0 \in \mathcal{X}^0 = \mathcal{H}^0 \cap \mathcal{H}^0(a)$. Then $\exp(-L_\mu^d t)\underline{W}^0$ belongs to $\mathcal{X}^0$ for all $t > 0$. This follows from the unique solvability of the initial value problem of $\partial_t\underline{W} + L_\mu^d\underline{W} = 0$ in $\mathcal{H}^0$ as well as in $\mathcal{H}^0(a)$. Hence $-L_\mu^d$ generates a holomorphic semigroup in $\mathcal{X}^0$. The reader may verify also, that $\exp(-L_\mu^d t)$ in $\mathcal{X}^0$ is generated by $-L_\mu^d$ in $\mathcal{H}^0(1+a)$.

Now return to the real Banach spaces. To verify (3.11a) we take the scalar product of (3.15) with a^2W_0, where

$$\partial_t W_0 - \frac{1}{a^2}\left(a^2 W_{0x}\right)_x + \gamma W_{0x} + b^d W_0 = 0 \tag{3.15}$$

and obtain

$$\frac{d}{dt}\|W_0\|_{0,a}^2 + 2\|W_{0x}\|_{0,a}^2 + 2\int_{I\!R} a^2 b^d W_0^2 + 2\gamma\int_{I\!R} a^2 W_{0x}W_0 = 0 . \tag{3.16a}$$

Using the estimate

$$|2\gamma\int_{I\!R} a^2 W_{0x}W_0| \le 2\|W_{0x}\|_{0,a}^2 + \frac{\gamma^2}{2}\|W_0\|_{0,a}^2$$

leads to

$$\frac{d}{dt}\|W_0\|_{0,a}^2 + 2\int_{I\!R} a^2\left((1+d)U_{\mu 0}^* - 1\right)W_0^2 \le 0 . \tag{3.16b}$$

We know from Section 2, that $U_{00}^*(-\infty) = 1$ and that $\|U_{\mu 0}^* - U_{00}^*\|_{C_b^2(\mathbb{R})} \le c\mu^2$ holds. For $\delta \in (0, d)$, choose $\beta = \beta(\delta, d)$ such that

$$(1+d)U_{00}^*(\beta) - 1 \ge d - \frac{\delta}{2} \quad .$$

Then, take a positive constant $\mu_1 \le \mu_0$, $\mu_1(\delta, d)$, to have

$$(1+d)U_{\mu 0}^*(x) - 1 \ge d - \frac{3}{4}\delta \tag{3.17}$$

for $x \le \beta$. Then we conclude

$$\frac{d}{dt}\|W_0\|_{0,a}^2 + 2\left(d - \frac{3}{4}\delta\right)\|W_0\|_{0,a}^2 \le 2\left(1 + \left(d - \frac{3}{4}\delta\right)\right)e^{\beta\gamma}\int_{\mathbb{R}} W_0^2 \le$$

$$\le 4e^{\beta\gamma}e^{-2b_0 t}\|W_0(0)\|_0^2$$

by using (3.16b), (3.17), and (3.10a), for $j = 0$. This yields (3.11a) via the Gronwall Lemma.

Due to the estimate

$$\left|2\gamma\int_{\mathbb{R}} a^2 W_{0x}W_0\right| \le \frac{\gamma^2 + d}{2}\|W_0\|_{0,a}^2 + \frac{2\gamma^2}{\gamma^2 + d}\|W_{0x}\|_{0,a}^2 \quad ,$$

we also derive from (3.10a), (3.16a) and (3.17) that

$$\frac{d}{dt}\|W_0\|_{0,a}^2 + \frac{2d}{\gamma^2 + d}\|W_{0,x}\|_{0,a}^2 + \frac{3}{2}(d - \delta)\|W_0\|_{0,a}^2$$

$$\le 4e^{\beta\gamma}e^{-2b_0 t}\|W_0(0)\|_0^2 \quad .$$

Integrating the previous inequality and using the estimate (3.11a) for $j = 0$ yields

$$\int_t^{t+\theta} \|W_{0x}\|_{0,a}^2 ds \le \frac{\gamma^2 + d}{2d}\left(\|W_0(t)\|_{0,a}^2 + 4e^{\beta\gamma}\theta e^{-2b_0 t}\|W_0(0)\|_0^2\right) \tag{3.16c}$$

$$\le c_5(\theta + 1)\left(e^{-2b_0 t} + e^{-2(d-\delta)t}\right)\|W_0(0)\|_{X^0}^2$$

for all $t \ge 0$, $\theta \ge 0$.

To prove (3.11a) for $j = 1$, take the scalar product of (3.15) with $-(a^2 W_{0x})_x$. Observe that $2a_x/a = \gamma$ holds and obtain

$$\frac{d}{dt}\|W_{0x}\|_{0,a}^2 + 2\|W_{0xx}\|_{0,a}^2 - \gamma^2\|W_{0x}\|_{0,a}^2 + 2\int_{\mathbb{R}} a^2 b^d W_{0x}^2 = \int_{\mathbb{R}} \left(a^2 b_x^d\right)_x W_0^2 \quad .$$

From $\|W_{0xx} + \frac{\gamma}{2} W_{0x}\|_{0,a}^2 \geq 0$ we derive

$$\|W_{0xx}\|_{0,a}^2 \geq \frac{\gamma^2}{4} \|W_{0x}\|_{0,a}^2 \quad ,$$

whence it follows, via (3.7a),

$$\frac{d}{dt}\|W_{0x}\|_{0,a}^2 + 2 \int_{\mathbb{R}} \left((1+d)U_{\mu0}^* - 1\right) a^2 |W_{0x}|^2 \leq \int_{\mathbb{R}} a^2 \left(b_{xx}^d + \gamma b_x^d\right) W_0^2 \quad .$$

Proceeding as in the case $j = 0$, we conclude

$$\frac{d}{dt}\|W_{0x}\|_{0,a}^2 + 2\left(d - \frac{3}{4}\delta\right) \|W_{0x}\|_{0,a}^2 \leq c_6 e^{\beta\gamma} \int_{\mathbb{R}} W_{0x}^2 + \int_{\mathbb{R}} a^2 \left(b_{xx}^d + \gamma b_x^d\right) W_0^2$$

or

$$\frac{d}{dt}\|W_{0x}\|_{0,a}^2 + 2\left(d - \frac{3}{4}\delta\right) \|W_{0x}\|_{0,a}^2 \leq c_7 \left(e^{-2b_0 t} + e^{-2(d-\delta)t}\right) \|W_0(0)\|_{X^1}^2$$

which implies (3.11a) for $j = 1$ via the Gronwall Lemma. If we apply the uniform Gronwall Lemma to the first inequality and take into account the estimates (3.12c), (3.10a) and (3.16c), we derive, for $t > 0$ and $\tau = \max(0, t-1)$,

$$\begin{aligned}
\|W_{0x}\|_{0,a}^2 &\leq \frac{1}{t-\tau}\int_\tau^t \|W_{0x}\|_{0,a}^2 ds + \|W_0(0)\|_{X^0}^2 \left(\int_\tau^t c_8 \cdot \left(e^{-2b_0 s} + e^{-2(d-\delta)s}\right) ds\right) \\
&\leq c_9 \left(1 + (t-\tau)^{-1}\right) \left(e^{-2b_0 t} + e^{-2(d-\delta)t}\right) \|W_0(0)\|_{X^0}^2 \quad ,
\end{aligned}$$

which, together with (3.12d) as well as (3.10a) and (3.11a) for $j = 0$, implies the estimate (3.11c).

Estimate (3.11b) is easier. Use for $\partial_t \underline{W}_1 + L_{\mu 1}\underline{W}_1 = 0$ the form

$$\partial_t \underline{W}_1 - \frac{1}{a^2}\left(a^2 \underline{W}_{1x}\right)_x + \gamma \underline{W}_{1x} + B_1 \underline{W}_1 = 0 \quad . \tag{3.18}$$

Take the scalar product with $a^2 \underline{W}_1$ to obtain

$$\frac{d}{dt}\|\underline{W}_1\|_{0,a}^2 + 2\|\underline{W}_{1x}\|_{0,a}^2 + \gamma \int_{\mathbb{R}} a^2 (|\underline{W}_1|^2)_x + 2\int_{\mathbb{R}} a^2 (B_1\underline{W}_1, \underline{W}_1) = 0 \quad .$$

Estimate the third term in the above equation as follows

$$\gamma \int_{\mathbb{R}} a^2 (|\underline{W}_1|^2)_x \geq -\frac{\gamma^2}{2}\|\underline{W}_1\|_{0,a}^2 - 2\|\underline{W}_{1x}\|_{0,a}^2$$

and use $B_1 \geq \left(\dfrac{\gamma^2}{4} + \dfrac{\alpha}{\mu^2} \right) I_1$ to conclude

$$\frac{d}{dt} \|\underline{W}_1\|_{0,a}^2 + \frac{2\alpha}{\mu^2} \|\underline{W}_1\|_{0,a}^2 \leq 0 \qquad (3.19)$$

whence (3.11b) follows for $j = 0$. The case $j = 1$ is obtained quite similarly. We multiply (3.18) by $-(a^2 \underline{W}_{1x})_x$ and use the same estimate as above to verify (3.19) for $\|\underline{W}_{1x}\|_{0,a}^2$. As in the case of the operator $L_{\mu 0}^d$, we use the uniform Gronwall Lemma to prove (3.11d). ■

Now we turn to the nonlinear system (3.5) and discuss first the properties of $\underline{g}$ in (1.4) as a mapping in $\mathcal{X}^1$. Let $\underline{u}$ be an element of $\mathcal{X}^1$. It follows immediately from the representations of $\underline{u}$ and $a\underline{u}$ by their indefinite integrals that $\lim\limits_{x \to \pm\infty} (a^j \underline{u})(x) = 0$, $j = 0, 1$, and that they are bounded and uniformly continuous functions on $I\!\!R$. Moreover, we have the estimates

$$\begin{aligned}
\|\underline{u}\|_\infty &\leq \sqrt{2} \|\underline{u}\|_0^{1/2} \|\underline{u}_x\|_0^{1/2} \leq \|\underline{u}\|_1 \\
\|a\underline{u}\|_\infty &\leq \sqrt{2} \|\underline{u}\|_{0,a}^{1/2} \|\underline{u}_x\|_{0,a}^{1/2} \leq \|\underline{u}\|_{1,a}
\end{aligned} \qquad (3.20)$$

where $\|\cdot\|_\infty$ denotes the L^∞-norm.

The estimates are simple consequences of

$$a^{2j} |\underline{u}|^2(x) = \int\limits_{-\infty}^{x} \partial_s \left(a^{2j} \underline{u} \cdot \underline{u} \right)(s) ds \quad , \quad j = 0, 1 \quad ,$$

the Cauchy–Schwarz inequality and the fact that $2aa_x = a^2\gamma$, $\gamma \leq -2$ holds.

Therefore we have a continuous embedding $J : \mathcal{X}^1 \to C_{b,u}^0(I\!\!R, I\!\!R^{n+1})$, where we denote by $C_{b,u}^0(I\!\!R, I\!\!R^{n+1})$ the space of bounded and uniformly continuous mappings from $I\!\!R$ into $I\!\!R^{n+1}$. In fact, $J\mathcal{X}^1$ lies in the closed subspace defined by $\underline{u}(\pm\infty) = 0$, $a\underline{u}(\pm\infty) = 0$.

Lemma 3.2 *Let* $\underline{g} \in C^p(I\!\!R \times I\!\!R^{n+1}, I\!\!R^{n+1})$ *for some integer* $p \geq 2$, *and assume* $\underline{g}(\lambda, \underline{0}) = \underline{0}$. *Then* $\underline{g}(\lambda, \cdot) J : R \times \mathcal{X}^1 \to \mathcal{X}^1$ *is a* C^{p-2} *mapping.*

Proof. We assume that $\underline{u}(x)$ belongs to a bounded set $\mathcal{B}_1$ of $\mathcal{X}^1$. We write

$$\underline{g}(\lambda, \underline{u}(x)) = \int\limits_0^1 \nabla_{\underline{u}} \underline{g}(\lambda, s\underline{u}(x)) \cdot \underline{u}(x) ds$$

where we use $\nabla_{\underline{u}}$ to denote the derivative in $\mathbb{R}^{n+1}$ and by $D_{\underline{u}}$ the F–derivative in $\mathcal{X}^1$. We have, since (3.20) holds, $|\nabla_{\underline{u}}g(\lambda, s\underline{u}(x))| \leq c$ for $x \in \mathbb{R}$, $s \in [0, 1]$, and $|\lambda| \leq \lambda_0$. If $\underline{u} \in \mathcal{B}_1$ we obtain

$$\|a^j g(\lambda, \underline{u})(x)\|_0 \leq C\|a^j \underline{u}(x)\|_0 \quad , \tag{3.21a}$$

$$\begin{aligned}
\|a^j \partial_x g(\lambda, \underline{u})\|_0 &= \| \int_0^1 [\nabla^2_{\underline{u}\underline{u}}g(\lambda, s\underline{u}(x))\left(\underline{u}(x), a^j(x)\partial_x\underline{u}(x)\right) \\
&\quad + \nabla_{\underline{u}}g(\lambda, s\underline{u}(x)) \cdot \partial_x\underline{u}(x)a^j(x)]ds\|_0 \\
&\leq C_1[\|\underline{u}\|^2_{\mathcal{X}^1} + \|\underline{u}\|_{\mathcal{X}^1}] \quad , \quad j = 0, 1
\end{aligned} \tag{3.21b}$$

where the constants C and C_1 depend only on $\mathcal{B}_1$. Higher derivatives can be calculated and estimated similarly up to order $p-2$ (in fact up to order p in λ; but we neglect this unimportant detail). Hence the assertion follows.

We shall use the fact that $\mathcal{X}^1$ is a Banach–algebra, which follows from (3.20), and calculate F–derivatives in $\mathcal{X}^1$ as we do with classical derivatives. We suppress the explicit notations of the embedding J from now on.

We estimate the nonlinearities Q and $\underline{H}$ in (3.5) and (3.6) by using Lemma 3.2. It will be necessary to control the dependence on $\underline{W}$ and μ in detail. Therefore we apply a suitable Taylor expansion for $\underline{H}$, which is an extension of the one given in (3.6).

$$\begin{aligned}
\underline{H}(\mu, \underline{W}) &= \mu^{-2}D_{\underline{u}}g^*\underline{W} + \int_0^1 (1-s)D^2_{\underline{u}\underline{u}}g\left(\mu^2, \mu^2\left(\underline{U}^*_\mu + sa\underline{W}\right)\right) ds\,(\underline{W}, a\underline{W}) \\
&= \mu^{-2}D_{\underline{u}}g^*\underline{W} + \frac{1}{2}D^2_{\underline{u}\underline{u}}g(0, \underline{0})(\underline{W}, a\underline{W}) + \mu^2 \int_0^1 \int_0^1 (1-s)\cdot \\
&\quad \cdot\Big\{ D^3_{\underline{u}\underline{u}\lambda}g\left(\sigma\mu^2, \sigma\mu^2\left(\underline{U}^*_\mu + sa\underline{W}\right)\right)(\underline{W}, a\underline{W}) + \\
&\quad + D^3_{\underline{u}\underline{u}\underline{u}}g\left(\sigma\mu^2, \sigma\mu^2\left(\underline{U}^*_\mu + sa\underline{W}\right)\right)(\underline{W}, a\underline{W}, \underline{U}^*_\mu + sa\underline{W}) \Big\}ds\,d\sigma \quad .
\end{aligned} \tag{3.22}$$

In the following we denote by $\| \cdot \|_{k, a^j}$ the norms $\| \cdot \|_k$ and $\| \cdot \|_{k, a}$ for $j = 0, 1$ respectively.

Lemma 3.3 Let $g \in C^p(\mathbb{R} \times \mathbb{R}^{n+1}, \mathbb{R}^{n+1})$, $p \geq 4$, be as in (1.3b), and let μ_0 and R be positive constants. Assume that $\underline{W} \in \mathcal{X}^1$ satisfies

$$\|a^l\underline{W}\|_\infty \leq \frac{2R}{\mu^2} \quad , \quad \text{for some} \quad \mu \in (0, \mu_0] \quad \text{and} \quad l = 0, 1 \quad .$$

Then the following estimates are valid for j, $k = 0, 1$.

$$C_{32}^{-1}\|\underline{Q}(\underline{W})\|_{k, a^j} \leq \|aW_0\|_\infty \|W_0\|_{k, a^j} \tag{3.23a}$$

$$C_{32}^{-1}\|\underline{H}_0(\mu, \underline{W})\|_{0, a^j} \leq \|W_0\|_{0, a^j} \left(\mu^2(1 + \|aW_0\|_\infty) + \|a\underline{W}_1\|_\infty\right) + \tag{3.23b}$$

$$+\|\underline{W}_1\|_{0, a^j} \left(1 + \|aW_0\|_\infty + \sigma\|a\underline{W}_1\|_\infty\right) + \mu^2 \|\underline{W}\|_{0, a^j} \|a\underline{W}\|_\infty^2$$

$$C_{32}^{-1}\|\partial_x \underline{H}_0(\mu, \underline{W})\|_{0, a^j} \leq \|W_0\|_{1, a^j} \left(\mu^2(1 + \|a\underline{W}_0\|_\infty) + \|a\underline{W}_1\|_\infty\right) + \tag{3.23c}$$

$$+\|\underline{W}_1\|_{1, a^j} \left(1 + \|aW_0\|_\infty + \sigma\|a\underline{W}_1\|_\infty\right) + \mu^2 \|\underline{W}\|_{1, a^j} \|a\underline{W}\|_\infty^2$$

$$C_{33}^{-1}\|\underline{H}_1(\mu, \underline{W})\|_{k, a^j} \leq \|\underline{W}\|_{k, a^j} \left(1 + \|a\underline{W}\|_\infty + \mu^2 \|a\underline{W}\|_\infty^2\right) \; . \tag{3.23d}$$

The positive constants C_{32}, C_{33} depend on R and μ_0 only. They are chosen to be continuous and increasing in each of these variables separately. The constant σ is defined as follows

$$\sigma = \begin{cases} 1 & \text{if } D^2_{\underline{u}_1 \underline{u}_1} g_0(0, \underline{0}) \neq 0 \\ \mu^2 & \text{if } D^2_{\underline{u}_1 \underline{u}_1} g_0(0, \underline{0}) = 0 \end{cases} \tag{3.23e}$$

and has been introduced to study the influence of 'critical' terms of the nonlinearity on the global stability of the fronts.

The proof of (3.23a) is a direct consequence of the definition of $\underline{Q}(\underline{W}) = (-aW_0^2, \underline{0})^t$. The rest of the assertion follows from (3.22), the fact that $\underline{U}^*_{\mu 1}/\mu^2$ is bounded, Lemma 3.2 and the special form of g_0 in (1.3b). The lengthy, but straightforward calculations are left to the reader.

We shall need the previous estimates to prove subsequently a reduction procedure via a fixed point argument. To this purpose explicit bounds for the Lipschitz–continuity of $\underline{H}$ and $\underline{Q}$ have to be calculated. They follow from suitable Taylor–expansions which are direct consequences of (3.22).

Lemma 3.4 *Assume that the hypotheses of the previous Lemma hold. Moreover, let $\underline{\widetilde{W}} \in \mathcal{X}^1$ satisfy $\|a^l\underline{\widetilde{W}}\|_\infty \leq 2R/\mu^2$, for $l = 0, 1$. Then there are positive constants $\widetilde{C}_{32}$, $\widetilde{C}_{33}$, having the same properties as those in Lemma 3.3, such that the following inequalities hold for j, $k = 0, 1$:*

$$\widetilde{C}_{32}^{-1}\|\underline{Q}(\underline{W}) - \underline{Q}(\underline{\widetilde{W}})\|_{k, a^j} \leq \|W_0 - \widetilde{W}_0\|_{k, a^j} \left(\|aW_0\|_\infty + \|a\widetilde{W}_0\|_\infty\right) + \tag{3.24a}$$

$$+k\|a^j(W_0 - \widetilde{W}_0)\|_\infty \left(\|aW_{0x}\|_0 + \|a\widetilde{W}_{0x}\|_0\right)$$

$$\widetilde{C}_{32}^{-1}\|\underline{H}_0(\mu, \underline{W}) - \underline{H}_0(\mu, \underline{\widetilde{W}})\|_{0, a^j} \leq \|W_0 - \widetilde{W}_0\|_{0, a^j} \; . \tag{3.24b}$$

$$\cdot \left\{ \mu^2 \left(1 + \|aW_0\|_\infty + \|a\widetilde{W}_0\|_\infty \right) + \|a\underline{W}_1\|_\infty + \|a\widetilde{\underline{W}}_1\|_\infty \right\} +$$

$$+ \|\underline{W}_1 - \widetilde{\underline{W}}_1\|_{0,\,a^j} \left\{ 1 + \|aW_0\|_\infty + \|a\widetilde{W}_0\|_\infty + \sigma \left(\|a\underline{W}_1\|_\infty + \|a\widetilde{\underline{W}}_1\|_\infty \right) \right\}$$

$$+ \mu^2 \|\underline{W} - \widetilde{\underline{W}}\|_{0,\,a^j} \left(\|a\underline{W}\|_\infty^2 + \|a\widetilde{\underline{W}}\|_\infty^2 \right)$$

$$\widetilde{C}_{32}^{-1} \|\partial_x \left(H_0(\mu,\,\underline{W}) - H_0(\mu,\,\widetilde{\underline{W}}) \right) \|_{0,\,a^j} \leq \|W_0 - \widetilde{W}_0\|_{1,\,a^j} \cdot \qquad (3.24\text{c})$$

$$\cdot \left\{ \mu^2 \left(1 + \|aW_0\|_\infty + \|a\widetilde{W}_0\|_\infty \right) + \|a\underline{W}_1\|_\infty + \|a\widetilde{\underline{W}}_1\|_\infty \right\} +$$

$$+ \|a^j \left(W_0 - \widetilde{W}_0 \right)\|_\infty \left\{ \mu^2 \left(\|aW_{0x}\|_0 + \|a\widetilde{W}_{0x}\|_0 \right) + \|a\underline{W}_{1x}\|_0 + \|a\widetilde{\underline{W}}_{1x}\|_0 \right\} +$$

$$+ \|\underline{W}_1 - \widetilde{\underline{W}}_1\|_{1,\,a^j} \left\{ 1 + \|aW_0\|_\infty + \|a\widetilde{W}_0\|_\infty + \sigma \left(\|a\underline{W}_1\|_\infty + \|a\widetilde{\underline{W}}_1\|_\infty \right) \right\} +$$

$$+ \|a^j (\underline{W}_1 - \widetilde{\underline{W}}_1)\|_\infty \left\{ \|aW_{0x}\|_0 + \|a\widetilde{W}_{0x}\|_0 + \sigma \left(\|a\underline{W}_{1x}\|_0 + \|a\widetilde{\underline{W}}_{1x}\|_0 \right) \right\} +$$

$$+ \mu^2 \|\underline{W} - \widetilde{\underline{W}}\|_{1,\,a^j} \left(\|a\underline{W}\|_\infty^2 + \|a\widetilde{\underline{W}}\|_\infty^2 \right) +$$

$$+ \mu^2 \|a^j (\underline{W} - \widetilde{\underline{W}})\|_\infty \left(\|a\underline{W}\|_\infty \|a\underline{W}_x\|_0 + \|a\widetilde{\underline{W}}\|_\infty \cdot \|a\widetilde{\underline{W}}_x\|_0 \right)$$

$$\widetilde{C}_{33}^{-1} \|\underline{H}_1(\mu,\,\underline{W}) - \underline{H}_1(\mu,\,\widetilde{\underline{W}})\|_{k,\,a^j} \leq \|\underline{W} - \widetilde{\underline{W}}\|_{k,\,a^j} \cdot \qquad (3.24\text{d})$$

$$\cdot \left\{ 1 + \|a\underline{W}\|_\infty + \|a\widetilde{\underline{W}}\|_\infty + \mu^2 \left(\|a\underline{W}\|_\infty^2 + \|a\widetilde{\underline{W}}\|_\infty^2 \right) \right\} +$$

$$+ k \|a^j (\underline{W} - \widetilde{\underline{W}})\|_\infty \left\{ \|\underline{W}\|_{k,\,a} + \|\widetilde{\underline{W}}\|_{k,\,a} + \mu^2 \left(\|a\underline{W}\|_\infty \|\underline{W}\|_{k,\,a} + \|a\widetilde{\underline{W}}\|_\infty \|\widetilde{\underline{W}}\|_{k,\,a} \right) \right\}$$

The next step in our analysis is a reduction of Equation (3.5) to a (nonlocal) evolution equation in X^1. This is based on a simple, well known Lemma.

Lemma 3.5 *Let X and Y denote two Banach spaces, and $D \subset X$, $P \subset Y$ be nonempty closed sets. Given a mapping $F : D \times P \to D$, such that $F(x, \cdot) : P \to D$ and $F(\cdot, p) : D \to D$ are Lipschitz-continuous uniformly in D and P respectively. Moreover, the Lipschitz-constant α of $F(\cdot, p)$ satisfies $\alpha < 1$.*

Then there exists a unique Lipschitz-continuous family of fixed-points $x^*(p)$, $p \in P$, of the mapping $F(\cdot, p)$. If the interior of P and D are not empty, and if $F \in C^k(D \times P, D)$, $k \geq 1$, holds, then $x^*(p)$ is C^k in some neighborhood of p^*, if $p^* \in \overset{\circ}{P}$, $x^*(p^*) \in \overset{\circ}{D}$.

Proof. The existence of a fixed point for every $p \in P$ follows from Banach's fixed point theorem. The family $x^*(p)$ derives its existence from Zorn's Lemma.

Furthermore

$$\|x^*(p) - x^*(q)\|_X = \|F(x^*(p),\, p) - F(x^*(q),\, p)+$$
$$+F(x^*(q),\, p) - F(x^*(q),\, q)\|_X$$

hence

$$\|x^*(p) - x^*(q)\|_X(1 - \alpha) \le c\|p - q\|_Y \quad .$$

The regularity assertion is an immediate consequence of the implicit function theorem.
∎

We set: $\mathcal{X}^j = X_0^j \oplus \mathcal{X}_1^j \equiv X^j \oplus \mathcal{X}_1^j$.

Now define for $\mu \in (0,\, \mu_0]$, $t \ge 0$,

$$K_j(\mu,\, t) = \exp(-L_{\mu j}t) \quad , \quad j = 0,\, 1 \tag{3.25}$$

where $K_0 \in \mathcal{L}(X_0^1,\, X_0^1)$, resp. $K_1 \in \mathcal{L}(\mathcal{X}_1^1,\, \mathcal{X}_1^1)$. The $\underline{W}_1$–part of (3.5) can be written

$$\underline{W}_1(t) = K_1(\mu,\, t)\underline{W}_1^0 + \int_0^t K_1(\mu,\, t - s)\underline{H}_1(\mu,\, W_0(s),\, \underline{W}_1(s))ds \tag{3.26}$$

where $\underline{W}_1(0) = \underline{W}_1^0 \in \mathcal{X}_1^1$. From (3.10b) and (3.11b) we deduce

$$\|K_1(\mu,\, t)\underline{W}_1^0\|_{\mathcal{X}_1^j} \le e^{-\frac{\alpha}{\mu^2}t}\|\underline{W}_1^0\|_{\mathcal{X}_1^j} \le \mu^2(\alpha e t)^{-1}\|\underline{W}_1^0\|_{\mathcal{X}_1^j} \quad .$$

To avoid the singularity at $t = 0$ we write $\underline{W}_1(t) = K_1(\mu,\, t)\underline{W}_1^0 + \underline{M}_1(t)$ and obtain

$$\underline{M}_1(t) = \int_0^t K_1(\mu,\, t - s)\underline{H}_1\left(\mu,\, W_0(s),\, K_1(\mu,\, s)\underline{W}_1^0 + \underline{M}_1(s)\right) ds \quad . \tag{3.27}$$

To simplify notation, set $\mathcal{F}(\mu,\, W_0,\, \underline{W}_1^0,\, \underline{M}_1)$ for the right–hand side of (3.27). Apply Lemma 3.5 to the fixed–point problem

$$\underline{M}_1 = \mathcal{F}\left(\mu,\, W_0,\, \underline{W}_1^0,\, \underline{M}_1\right) \tag{3.28}$$

where $\mu \in (0,\, \mu_0]$ is fixed; W_0 and $\underline{W}_1^0$ are considered as parameters.

To control the time behavior of $\underline{W}$, we introduce a continuous function $\rho(t)$, $t \ge 0$, $\rho(t) \ge 1$. Assume furthermore that for $m,\, j = 0,\, 1$; $0 \le m \le j \le 1$, $t \ge 0$ and

173

$\mu \in (0, \mu_0]$

$$\mu^{-m+j-2} \| \int_0^t K_1(\mu, t-s)\rho(t)\rho^{-1}(s)ds \|_{\mathcal{L}(\mathcal{X}^m, \mathcal{X}^j)} \leq k^* \tag{3.29}$$

$$\rho(t)\exp\left(-\frac{\alpha}{16\mu_0^2}t\right) \leq k_1^*$$

holds for some $k^*(\mu_0)$, $k_1^*(\mu_0)$. The reader may verify the validity of (3.29) for $\rho(t) = \exp(\epsilon t)$, if $0 < \epsilon < \alpha/16\mu_0^2$, and for $\rho(t) = 1 + t^\beta$, if $\beta > 0$. The first case will be applied later.

The Banach spaces to be used are defined as follows

$$\mathcal{Y}_\rho^0 = \left\{ \underline{W} \in C_b^0\left([0, \infty), \mathcal{X}^0\right); \ |\underline{W}|_{\rho, 0} = \sup_{t \geq 0} \rho(t)\|\underline{W}(t)\|_{\mathcal{X}^0} < \infty \right\}$$

$$\mathcal{Y}_\rho^1 = \left\{ \underline{W} \in C_b^0\left([0, \infty), \mathcal{X}^1\right); \ |\underline{W}|_{\rho, 1} = \sup_{t \geq 0} \rho(t)\|\underline{W}(t)\|_{\mathcal{X}^1} < \infty \right\} \tag{3.30}$$

$$\mathcal{Y}_\rho^j = Y_{0\rho}^j \oplus \mathcal{Y}_{1\rho}^j \ , \quad j = 0, 1$$

If $\underline{W}_1^0 \in \mathcal{X}_1^j$ then $K_1\underline{W}_1^0 \in \mathcal{Y}_{1\rho}^j$. Therefore, $\underline{W} \in \mathcal{Y}_\rho^j$ implies $W_0 \in Y_{0\rho}^j$, $\underline{M}_1 \in \mathcal{Y}_{1\rho}^j$, where

$$\underline{W} = W_0 + K_1\underline{W}_1^0 + \underline{M}_1 \ . \tag{3.31}$$

We shall prove two different reduction results, which will enable us to show the local stability of the fronts by just working in the reduced space X_0^1. This is the counterpart of the center–manifold reduction for the steady solutions in the nonsteady case. Since we have now two unbounded directions, the reduced equation will contain nonlocal terms [Mi92].

Besides the estimates in Lemma 3.1, a central role in the proof of the reduction is played by the estimates of $\underline{H}_1$ in Lemmas 3.3 and 3.4. In view of (3.20) the estimates on $\|W\|_\infty$ and $\|aW\|_\infty$ give us some freedom in the weight we put on $\|\underline{W}\|_{\mathcal{X}^0}$ resp. $\|\underline{W}\|_{\mathcal{X}^1}$. We exploit this by giving two different reduction results. The first one is valid, if $\underline{W}$ stays in a ball of radius $\delta\mu^{-2}$ in $\mathcal{X}^1$: $\underline{W} \in B(0, \delta\mu^{-2}) \subset \mathcal{X}^1$, while the second one holds if $\underline{W} \in (B(0, \delta\mu^{l-1}) \subset \mathcal{X}^0) \cap (B(0, \delta\mu^{-l-1}) \subset \mathcal{X}^1)$, for some $l \geq 0$. This then implies that $\|a^j\underline{W}\|_\infty = O(\delta\mu^{-1})$, for $j = 0, l \geq 0, 1$. It is useful when the considered solutions have some roughness for small times and are smooth for large times.

A specific example may clarify the situation. We consider the following initial conditions $\underline{W}^0 = (W_0^0, \ldots, W_n^0)$ where

$$W_i^0(x) = \delta_i \mu^{-p_i} e^{-\frac{\gamma}{2}x - |x|\mu^{-2l}}$$

for δ_i, $l \geq 0$, $\mu > 0$, $p_i \in \mathbb{R}$. If μ is sufficiently small, W_i^0 belongs to X^1. Moreover we have

$$
\begin{aligned}
\|W_i^0\|_0 &= \delta_i \mu^{-p_i+l} \left(1 - \frac{\gamma^2}{4}\mu^{4l}\right)^{1/2} \\
\|W_{ix}^0\|_0 &= \delta_i \mu^{-p_i-l} \\
\|W_i^0\|_{0,a} &= \delta_i \mu^{-p_i+l} \\
\|W_{ix}^0\|_{0,a} &= \delta_i \mu^{-p_i-l} \left(1 + \frac{\gamma^2}{4}\mu^{4l}\right)^{1/2}
\end{aligned}
$$

and

$$\|W_i^0\|_{L^\infty(\mathbb{R}) \cap L^\infty(a)} = \delta_i \mu^{-p_i} \quad .$$

From now on we fix a large positive number R for the whole section. Define

$$c^* = \max(C_{33}, \widetilde{C}_{33}) \tag{3.32}$$

and introduce μ_1^*, $\delta^* > 0$ such that

$$3\delta^* < R \quad , \quad \mu_1^* \leq \inf\left(\frac{1}{2}, \frac{1}{4k^*c^*}, \left[\frac{1}{2k^*c^*(2 + k_1^*)}\right]\right) \tag{3.33a}$$

$$9\delta^{*^2} + 3\sqrt{2}\delta^* \leq \frac{1}{4(2 + k_1^*)k^*c^*} \tag{3.33b}$$

holds. Set $\mu_0^* = \min(\mu_0, \mu_1^*)$.

Theorem 3.6 *For every $\mu \in (0, \mu_0^*]$, $W_0 \in Y_{0\rho}^1$, $\underline{W}_1^0 \in \mathcal{X}_1^1$, satisfying $|W_0|_{\rho,1} \leq \delta^*/\mu^2$, $\|\underline{W}_1^0\|_{\mathcal{X}^1} \leq \delta^*/\mu^2$, there exists in $cl\, B(0, \delta^*/\mu^2) \subset \mathcal{Y}_{1\rho}^1$ a unique fixed point $\underline{M}_1^*(\mu, W_0, \underline{W}_1^0)$ of the mapping $\mathcal{F}(\mu, W_0, \underline{W}_1^0, \cdot)$ defined in (3.27). This fixed point is a Lipschitz–continuous function of $(W_0, \underline{W}_1^0)$ with a constant C_{34}^* independent of μ.*

Proof. We apply Lemma 3.5 to $\mathcal{F}$ in the closed set $D = cl\, B(0, \delta^*/\mu^2) \subset \mathcal{Y}_{1\rho}^1$. As P, we choose $P = cl\, B(0, \delta^*/\mu^2) \times cl\, B(0, \delta^*/\mu^2) \subset Y_{0\rho}^1 \times \mathcal{X}_1^1$. Under the assumptions of the theorem, $\mathcal{F}$ maps the ball $B(0, \delta^*/\mu^2) \subset \mathcal{Y}_{1\rho}^1$ into itself. Indeed, according to Lemma 3.1, to (3.23d), (3.29),(3.33a) and (3.33b), we have

$$\mu^2 \,|\,\mathcal{F}(\mu, W_0, \underline{W}_1^0, \underline{M}_1)\,|_{\rho,1} \;\leq\; \mu^2 \sup_{t\geq 0}\left\{\int\limits_0^t \|K_1(\mu, t-s)\rho(t)\rho^{-1}(s)\|_{\mathcal{L}(\mathcal{X}^1, \mathcal{X}^1)}\right.$$

$$\left.\cdot\rho(s)\|\underline{H}_1(\mu, W_0, K_1(\mu,\cdot)\underline{W}_1^0 + \underline{M}_1\|_{\mathcal{X}^1}(s)ds\right\} \;\leq$$

$$\leq\; \delta^*(2 + k_1^*)k^*c^*\left(\mu^2 + 3\delta^* + 9\delta^{*^2}\right) \leq \delta^* \quad.$$

Using (3.24d) one can show similarly that $\mathcal{F}(\mu, W_0, \underline{W}_1^0, \cdot)$ is a strict contraction with respect to $\underline{M}_1 \in B(0, \delta^*/\mu^2) \subset \mathcal{Y}_{1\rho}^1$ uniformly in W_0 and $\underline{W}_1^0$, since

$$|\,\mathcal{F}(\mu, W_0, \underline{W}_1^0, \underline{M}_1) - \mathcal{F}(\mu, W_0, \underline{W}_1^0, \widetilde{\underline{M}}_1)\,|_{\rho,1} \leq$$

$$\leq\; |\,\underline{M}_1 - \widetilde{\underline{M}}_1\,|_{\rho,1}k^*c^*\left(\mu^2 + 12\delta^* + 36\delta^{*^2}\right) \leq \qquad\qquad (3.34)$$

$$\leq\; \frac{3}{4}|\,\underline{M}_1 - \widetilde{\underline{M}}_1\,|_{\rho,1} \quad.$$

Likewise, we obtain the uniform Lipschitz–continuity of $\mathcal{F}$ with respect to W_0 and $\underline{W}_1^0$, uniformly in $\underline{M}_1$, using again (3.24d)

$$|\,\mathcal{F}(\mu, W_0, \underline{W}_1^0, \underline{M}_1) - \mathcal{F}(\mu, \widetilde{W}_0, \widetilde{W}_1^0, \underline{M}_1)\,|_{\rho,1} \leq \left(|\,W_0 - \widetilde{W}_0\,|_{\rho,1}\right.$$

$$\left.+\|\underline{W}_1^0 - \widetilde{\underline{W}}_1^0\|_{\mathcal{X}^1}\right)k^*c^*(1 + k_1^*)\left(\mu^2 + 12\delta^* + 36\delta^{*2}\right) \qquad (3.35)$$

for all $\mu \in (0, \mu_0^*]$.

Now, the theorem is an immediate consequence of Lemma 3.5, whence it follows also, using (3.35),

$$|\,\underline{M}_1^*(\mu, W_0, \underline{W}_1^0) - \underline{M}_1^*\left(\mu, \widetilde{W}_0, \widetilde{\underline{W}}_1^0\right)\,|_{\rho,1} \leq C_{34}(\mu, \delta^*)\cdot$$

$$\cdot\left(|\,W_0 - \widetilde{W}_0\,|_{\rho,1} + \|\underline{W}_1^0 - \widetilde{\underline{W}}_1^0\|_{\mathcal{X}^1}\right) \qquad\qquad (3.36)$$

with

$$C_{34}(\mu, \delta^*) \;=\; 4k^*c^*(1 + k_1^*)(\mu^2 + 12\delta^* + 36\delta^{*2})$$

$$C_{34}^* \;=\; C_{34}(\mu_0^*, \delta^*) \geq C_{34}(\mu, \delta^*) \quad.$$

Invoking also Lemma 3.5, one shows that $\underline{M}^*$ is a C^{p-2} function of its arguments. Since we shall not use this property, we suppress its proof.

In order to obtain the final local and global stability results of $\underline{W} = \underline{0}$ we have to improve the estimates of $\underline{M}_1^*$, in particular with respect to $\underline{W}_1^0$. We do this in the subsequent two corollaries.

176

Corollary 3.7 *Assume* $|W_0|_{\rho,1} \leq d_0\mu^{-p_0}$, $\|\underline{W}_1^0\|_{\mathcal{X}^1} \leq d_1\mu^{-p_1}$ *for* $0 \leq p_j < 2$, *then* $\underline{M}_1^*(\mu, W_0, \underline{W}_1^0)$ *satisfies the estimates*

$$|\underline{M}_1^*(\mu, W_0, \underline{W}_1^0)|_{\rho,1} \leq \widehat{C}_{35}(\mu, d_0\mu^{-p_0}) + \widehat{C}_{35}(\mu, d_1\mu^{-p_1}) \tag{3.37}$$

where $\widehat{C}_{35}(\mu, d) = 4k^*c^*(1 + k_1^*)(\mu^2 d + 2\mu^2 d^2 + 4\mu^4 d^3)$.

In particular, we have for $p_0 = p_1 = 1$

$$|\underline{M}_1^*(\mu, W_0, \underline{W}_1^0)|_{\rho,1} \leq \widehat{C}_{36}(\mu, d) = \widehat{C}_{35}\left(\mu, \frac{d_0}{\mu}\right) + \widehat{C}_{35}\left(\mu, \frac{d_1}{\mu}\right) \quad . \tag{3.38}$$

We set $\widehat{C}_{35}^*(d) = \widehat{C}_{35}(\mu_0^*, d)$ *and* $\widehat{C}_{36}^*(d) = \widehat{C}_{36}(\mu_0^*, d)$.

The proof is an immediate consequence of Lemma 3.1 and the inequalities (3.23d), (3.29), (3.33a), (3.33b) and (3.34). The latter implies

$$|\underline{M}_1^*(\mu, W_0, \underline{W}_1^0)|_{\rho,1} \leq 4|\mathcal{F}(\mu, W_0, \underline{W}_1^0, 0)|_{\rho,1} \tag{3.39}$$

whence (3.37) follows.

However, this estimate is still not strong enough, as can be seen by setting $p_0 \leq \frac{2}{3}$, $p_1 = \frac{4}{3}$. One obtains then

$$|\underline{M}_1^*(\mu, W_0, \underline{W}_1^0)|_{\rho,1} \leq c\mu^{-2/3} \tag{3.40}$$

which is not sufficient for controling the μ–behavior of subsequent estimates. To achieve this purpose we prove refined estimates taking into account the special form of the dependence on $\underline{W}_1^0$.

Corollary 3.8 *Assume that the hypotheses of the previous Corollary hold. Then* $\underline{M}_1^*$ *satisfies the following estimate for* $t \geq 0$

$$\rho(t)\|\underline{M}_1^*(\mu, W_0, \underline{W}_1^0)\|_{\mathcal{X}^1} \leq C_{35}\mu^2 \Big\{ F_1(W_0, \underline{M}_1^*)\left(|W_0|_{\rho,1} + |\underline{M}_1^*|_{\rho,1}\right) \tag{3.41}$$

$$+ e^{-\frac{7\alpha}{8\mu^2}t}\|\underline{W}_1^0\|_{\mathcal{X}^1}\left(F_1(W_0, \underline{M}_1^*) + F_2(W_0, \underline{W}_1^0, \underline{M}_1^*)\cdot\|\underline{W}_1^0\|_{\mathcal{X}^1}\right) \Big\}$$

where C_{35} *is a positive constant depending on* μ_0^* *and* R *only. Moreover* F_1 *and* F_2 *are defined as*

$$F_1(W_0, \underline{M}_1^*) = 1 + 2|W_0|_{\rho,1} + 2|\underline{M}_1^*|_{\rho,1} + 3\mu^2 \cdot \left(|W_0|_{\rho,1}^2 + |\underline{M}_1^*|_{\rho,1}^2\right)$$

$$F_2(W_0, \underline{W}_1^0, \underline{M}_1^*) = 1 + 3\mu^2\left(|W_0|_{\rho,1} + |\underline{M}_1^*|_{\rho,1} + \|\underline{W}_1^0\|_{\mathcal{X}^1}\right) \quad .$$

Proof. Remark that $\underline{M}_1^* = \mathcal{F}(\mu, W_0, \underline{W}_1^0, \underline{M}_1^*)$ and use (3.23d) in (3.27) to obtain the first term on the right hand side of (3.41) as in the proof of Theorem 3.6 for $\underline{W}_1^0 = 0$. When estimating the terms in (3.27) depending on $\underline{W}_1^0$ we collect the terms in $K_1 \underline{W}_1^0$ in (3.23d) and obtain (3.41).

We are now going to prove the second reduction result which gives different weights (measured in powers of μ) to $\|\cdot\|_{\mathcal{X}^0}$ and $\|\cdot\|_{\mathcal{X}^1}$. Let $\delta > 0$, $l \geq 0$, and introduce the closed sets

$$
\begin{aligned}
D_{0,l}^\delta &= \left\{ W_0 \in Y_{0\rho}^1 \,\big|\, |W_0|_{\rho,0} \leq \delta\mu^{l-1}, \ |W_0|_{\rho,1} \leq \delta\mu^{-l-1} \right\} \\
D_{1,l}^\delta &= \left\{ \underline{W}_1 \in \mathcal{Y}_{1\rho}^1 \,\big|\, |\underline{W}_1|_{\rho,0} \leq \delta\mu^{l-1}, \ |\underline{W}_1|_{\rho,1} \leq \delta\mu^{-l-1} \right\} \\
E_{1,l}^\delta &= \left\{ \underline{W}_1^0 \in \mathcal{X}_1^1 \,\big|\, \|\underline{W}_1^0\|_{\mathcal{X}^0} \leq \delta\mu^{l-1}, \ \|\underline{W}_1^0\|_{\mathcal{X}^1} \leq \delta\mu^{-l-1} \right\} \\
P_l^\delta &= D_{0,l}^\delta \times E_{1,l}^\delta
\end{aligned}
\tag{3.42}
$$

Theorem 3.9 *Let δ^* be as in (3.33a), (3.33b), $\mu \in (0, \mu_0^*]$, $(W_0, \underline{W}_1^0) \in P_l^{\delta^*}$. Then there exists a unique family of fixed–points $\underline{M}_1^*(\mu, W_0, \underline{W}_1^0)$ of (3.28) in $D_{1,l}^{\delta^*}$. This family is Lipschitz–continuous in $(W_0, \underline{W}_1^0)$ with constant $C_{36}(\mu_0^*, R)$ given by*

$$
C_{36}(\mu, \delta) = 4k^*c^*(1 + k_1^*)\left(\mu + 6\sqrt{2}\delta + 36\delta^2\mu\right) \quad . \tag{3.43}
$$

Proof. As in the proof of Theorem 3.6 we apply Lemma 3.5 to $\mathcal{F}(\mu, W_0, \underline{W}_1^0, \cdot)$ for $\underline{M}_1$ in $D_{1,l}^{\delta^*}$. The mapping $\mathcal{F}$ maps the set $D_{1,l}^{\delta^*}$ into itself.

Indeed, according to Lemma 3.1, (3.20), (3.23d), (3.29), (3.33a) and (3.33b), we have

$$
|\mathcal{F}(\mu, W_0, \underline{W}_1^0, \underline{M}_1)|_{\rho,0} \leq \sup_{t \geq 0} \rho(t) \int_0^t \|K_1(\mu, t-s)\|_{\mathcal{L}(\mathcal{X}^0, \mathcal{X}^0)}
$$

$$
\cdot \|\underline{H}_1(\mu, W_0, K_1(\mu, \cdot)\underline{W}_1^0 + \underline{M}_1)(s)\|_{\mathcal{X}^0} ds \leq
$$

$$
\leq \delta^*\mu^{l-1}(2 + k_1^*)k^*c^*\left(\mu^2 + 3\sqrt{2}\delta^*\mu + 18\delta^{*^2}\mu^2\right)
$$

$$
\leq \delta^*\mu^l \leq \frac{\delta^*}{2}\mu^{l-1} \quad .
$$

Similarly we obtain either

$$
|\mathcal{F}(\mu, W_0, \underline{W}_1^0, \underline{M}_1)|_{\rho,1} \leq \delta^*\mu^{-l-1}(2+k_1^*)k^*c^*\left(\mu^2 + 3\sqrt{2}\delta^*\mu + 18\delta^{*^2}\mu^2\right) \leq \frac{\delta^*}{2}\mu^{-l-1}
$$

or

$$
\begin{aligned}
|\mathcal{F}(\mu, W_0, \underline{W}_1^0, \underline{M}_1)|_{\rho,1} &\leq \delta^*\mu^{l-1}(2 + k_1^*)k^*c^*\left(\mu + 3\sqrt{2}\delta^* + 18\delta^{*^2}\mu\right) \\
&\leq \delta^*\mu^{l-1} \quad .
\end{aligned}
$$

by using either the estimates for

$$\|K_1(\mu, t-s)\|_{\mathcal{L}(\mathcal{X}^1, \mathcal{X}^1)}\|\underline{H}_1\|_{\mathcal{X}^1}$$

or for

$$\|K_1(\mu, t-s)\|_{\mathcal{L}(\mathcal{X}^0, \mathcal{X}^1)}\|\underline{H}_1\|_{\mathcal{X}^0}$$

which follow from Lemmas 3.1 and 3.3. Due to Lemma 3.1, (3.24d), (3.29), (3.33a) and (3.33b), we can also show that $\mathcal{F}$ is a strict contraction in $\underline{M}_1 \in D_{1,l}^{\delta^*}$, since

$$\mid \mathcal{F}\left(\mu, W_0, \underline{W}_1^0, \underline{M}_1\right) - \mathcal{F}\left(\mu, W_0, \underline{W}_1^0, \widetilde{\underline{M}}_1\right) \mid_{\rho, 1} \le$$

$$\le \sup_{t \ge 0} \int_0^t \|K_1(\mu, t-s)\|_{\mathcal{L}(\mathcal{X}^0, \mathcal{X}^1)} \|\underline{H}_1\left(\mu, W_0, K_1(\mu, \cdot)\underline{W}_1^0 + \underline{M}_1\right)(s)$$

$$-\underline{H}_1\left(\mu, W_0, K_1(\mu, \cdot)\underline{W}_1^0 + \widetilde{\underline{M}}_1\right)(s)\|_{\mathcal{X}^0} ds$$

$$\le \mid \underline{M}_1 - \widetilde{\underline{M}}_1 \mid_{\rho, 0} k^* c^* \left(\mu + 6\sqrt{2}\delta^* + 36{\delta^*}^2 \mu\right) \le \frac{3}{4} \mid \underline{M}_1 - \widetilde{\underline{M}}_1 \mid_{\rho, 0} \quad .$$

Likewise, we have for $(W_0, \underline{W}_1^0)$ and $(\widetilde{W}_0, \widetilde{\underline{W}}_1^0)$ in $P_l^{\delta^*}$ and $\underline{M}_1 \in D_{1,l}^{\delta^*}$

$$\mid \mathcal{F}\left(\mu, W_0, \underline{W}_1^0, \underline{M}_1\right) - \mathcal{F}\left(\mu, \widetilde{W}_0, \widetilde{\underline{W}}_1^0, \underline{M}_1\right) \mid_{\rho, 1} \le \left(\mid W_0 - \widetilde{W}_0 \mid_{\rho, 0} + \right.$$
$$\left. +\|\underline{W}_1^0 - \widetilde{\underline{W}}_1^0\|_{\mathcal{X}^0}\right) k^* c^* (1 + k_1^*) \left(\mu + 6\sqrt{2}\delta^* + 36{\delta^*}^2 \mu\right) \tag{3.44}$$

which implies that $\mathcal{F}$ is uniformly Lipschitz–continuous in $(W_0, \underline{W}_1^0) \in P_l^{\delta^*}$, $\underline{M}_1 \in D_{1,l}^{\delta^*}$ and $\mu \in (0, \mu_0^*]$.

Now, Theorem 3.9 follows from Lemma 3.5. From (3.44) we also deduce

$$\mid \underline{M}_1^*\left(\mu, W_0, \underline{W}_1^0\right) - \underline{M}_1^*\left(\mu, \widetilde{W}_0, \widetilde{\underline{W}}_1^0\right) \mid_{\rho, 1} \le C_{36}(\mu, \delta^*) \cdot$$
$$\cdot \left(\mid W_0 - \widetilde{W}_0 \mid_{\rho, 0} + \|\underline{W}_1^0 - \widetilde{\underline{W}}_1^0\|_{\mathcal{X}^0}\right) \quad . \tag{3.45}$$

Corollary 3.10 *Under the assumptions of Theorem 3.9 and for*

$$\widehat{C}_{37}(\mu, \delta) = 8\sqrt{2}k^* c^* (1 + k_1^*) \left(\mu + 2\sqrt{2}\delta + 8\delta^2 \mu\right) \tag{3.46}$$

the following estimates hold for $0 < \mu \le \mu_0^*$:

$$\mid \underline{M}_1^*\left(\mu, W_0, \underline{W}_1^0\right) \mid_{\rho, 0} \le \delta^* \mu^l \widehat{C}_{37}\left(\mu, \delta^*\right)$$

$$\mid \underline{M}_1^*\left(\mu, W_0, \underline{W}_1^0\right) \mid_{\rho, 1} \le \delta^* \widehat{C}_{37}\left(\mu, \delta^*\right) \inf\left(\mu^{l-1}, \mu^{-l}\right) \tag{3.47}$$

$$\sup_{t \ge 0} \rho(t) \left\|(1 + a)\underline{M}_1^*\left(\mu, W_0, \underline{W}_1^0\right)\right\|_\infty \le \delta^* \widehat{C}_{37}\left(\mu, \delta^*\right) \inf\left(1, \mu^{l-\frac{1}{2}}\right)$$

179

Proof. Corollary 3.10 follows immediately from (3.20), the proof of Theorem 3.9 and the subsequent remark. According to that proof we have

$$| \underline{M}_1^* \left(\mu, W_0, \underline{W}_1^0 \right) |_{\rho, j} \leq \frac{3}{4} | \underline{M}_1^* \left(\mu, W_0, \underline{W}_1^0 \right) |_{\rho, 0} +$$

$$+ | \mathcal{F} \left(\mu, W_0, \underline{W}_1^0, 0 \right) |_{\rho, j}$$

which implies

$$| M_1^* \left(\mu, W_0, \underline{W}_1^0 \right) |_{\rho, j} \leq 4 | \mathcal{F} \left(\mu, W_0, \underline{W}_1^0, 0 \right) |_{\rho, j} \quad . \tag{3.48}$$

Remark. We could derive stronger reduction results by treating 'nongeneric' cases like e.g. $D_{uu}^2 g(0, 0, 0) = 0$. In that case μ^{l-1} (resp. μ^{-l-1}) could be replaced by $\mu^{l-4/3}$ (resp. $\mu^{-l-4/3}$) in the definition of $D_{0,l}^\delta$, $D_{1,l}^\delta$ and $E_{1,l}^\delta$. However, we shall not consider these exceptional cases.

Corollary 3.11 *Assume that the hypotheses of Corollary 3.10 hold. Then $\underline{M}_1^*$ satisfies the following estimates for $t \geq 0$ and $j = 0, 1$:*

$$\rho(t) \| \underline{M}_1^* \left(\mu, W_0, \underline{W}_1^0 \right) (t) \|_{\mathcal{X}^j}$$

$$\leq C_{37} \mu^2 \Big\{ \widetilde{F}_1 \left(W_0, \underline{M}_1^* \right) \left(| W_0 |_{\rho, j} + | \underline{M}_1^* |_{\rho, j} \right) +$$

$$+ e^{-\frac{7\alpha}{8\mu^2} t} \Big[\| \underline{W}_1^0 \|_{\mathcal{X}^j} \left(\widetilde{F}_1 \left(W_0, \underline{M}_1^* \right) + 6\mu^2 \| \underline{W}_1^0 \|_{\mathcal{X}^0} \| \underline{W}_1^0 \|_{\mathcal{X}^1} \right) +$$

$$+ \| \underline{W}_1^0 \|_{\mathcal{X}^0}^{1/2} \| \underline{W}_1^0 \|_{\mathcal{X}^1}^{1/2} \left(| W_0 |_{\rho, j} + | \underline{M}_1^* |_{\rho, j} + \| \underline{W}_1^0 \|_{\mathcal{X}^j} \right) \Big] \Big\} \tag{3.49}$$

where

$$\widetilde{F}_1 \left(W_0, \underline{M}_1^* \right) = 1 + \sup_{t \geq 0} \| a W_0 \|_\infty + \sup_{t \geq 0} \| a \underline{M}_1^* \|_\infty +$$

$$+ 3\mu^2 \left(\sup_{t \geq 0} \| a W_0 \|_\infty^2 + \sup_{t \geq 0} \| a \underline{M}_1^* \|_\infty^2 \right)$$

The proof uses Lemma 3.1, (3.23d), (3.29) and an estimate of $\int_0^t K_1(\mu, t-s) K_1(\mu, s) \underline{W}_1^0$ in the L^∞–norm of $\underline{W}_1^0$. Otherwise it proceeds similarly to that of Corollary 3.8.

In Theorems 3.6 and 3.9 we have found a Lipschitz–continuous mapping $\underline{M}_1^*(\mu, W_0, \underline{W}_1^0)$ under different hypotheses which solves the 'hyperbolic' part of system (3.5) via

$$\underline{W}_1(t) = e^{-L_{\mu 1} t} \underline{W}_1^0 + \underline{M}_1^* \left(\mu, W_0, \underline{W}_1^0 \right) (t) \equiv$$

$$\equiv \underline{W}_1^* \left(\mu, W_0, \underline{W}_1^0 \right) (t) \quad . \tag{3.50}$$

Thus we have reduced the task to solve (3.5) to the scalar equation

$$\partial_t W_0 + L_{\mu_0} W_0 = Q(W_0) + H_0\left(\mu,\, W_0 + \underline{W}_1^*\left(\mu,\, W_0,\, \underline{W}_1^0\right)\right)$$
$$W_0|_{t=0} = W_0^0 \tag{3.51}$$

under the conditions of Theorem 3.9.

As a next step we prove local stability for the front solution $\underline{U}_\mu^*$ of the original system which corresponds to $\underline{W} = \underline{0}$ of (3.5). We shall be able to use the reduction result and thus treat the reduced equation (3.51) and show stability of $W_0 = 0$ for small initial data W_0^0 and small positive μ. The component $\underline{W}_1^0$ need not to be small, as we shall see below. Instead of (3.51) we use its integral form

$$W_0(t) = K_0(\mu,\, t)W_0^0 + \int_0^t K_0(\mu,\, t - s)\Big\{Q(W_0)(s) +$$
$$+ H_0\left(\mu,\, W_0 + K_1(\mu,\, \cdot)\underline{W}_1^0 + \underline{M}_1^*\left(\mu,\, W_0,\, \underline{W}_1^0\right)\right)(s)\Big\}ds \tag{3.52}$$

where

$$K_0(\mu,\, t) = \exp(-L_{\mu_0} t)\quad,\quad t \geq 0\quad.$$

We treat only the case $\gamma < -2$. For $\rho(t)$ we choose

$$\rho(t) = \exp\left(\frac{b_1}{2} t\right)\quad,\quad \text{where } b_1 < \min(1,\, b_0)\quad. \tag{3.53}$$

According to Lemma 3.1 there exists a positive constant C_{38} such that K_0 satisfies

$$\|K_0(\mu,\, t)\|_{\mathcal{L}(X^m,\, X^j)} \leq C_{38}\, t^{\frac{m-j}{2}}\, e^{-b_1 t}\quad,\quad t > 0 \tag{3.54}$$

where $0 \leq m \leq j \leq 1$. C_{38} depends on μ_0^* alone. We choose $\mu_0 > 0$ small enough so that $16\mu_0^2 < \alpha$. Thus, the function ρ satisfies (3.29).

Introduce the positive constants δ_0^0, δ_0, δ_1^0 which satisfy

$$\max\left(\delta_0^0,\, \delta_1^0\right) \leq \delta_0 \leq \delta^*\quad,\quad \delta_0^0 \leq \frac{1}{2}\delta_0 C_{38}^{-1} \tag{3.55}$$

and define the following closed sets for $l \geq 0$, $j = 0, 1$:

$$E_{1,l}^{\delta_1^0} = \left\{\underline{W}_1^0 \in \mathcal{X}_1^1 \,\big|\, \|\underline{W}_1^0\|_{\mathcal{X}^j} \leq \delta_1^0 \mu^{l(1-2j)-1}\right\}$$
$$F_{0,l}^0 = \left\{W_0^0 \in X_0^1 \,\big|\, \|W_0^0\|_{X_0^j} \leq \delta_0^0 \mu^{l(1-2j)}\right\} \tag{3.56}$$
$$G_{0,l}^{\delta_0} = \left\{W_0 \in Y_{0\rho}^1 \,\big|\, |W_0|_{\rho,j} \leq \delta_0 \mu^{l(1-2j)}\right\}$$

The following theorem gives local exponential stability for the case $\gamma < -2$. The slight generalization towards a global result in $\underline{W}_1^0$ by the factor μ^{-1} stems from the spectral properties of the linearized system. In the following sections a globalisation in W_0 is proved having its origin in a global stability property of the scalar front, which is proved in Section 4.

Theorem 3.12 *The constants δ^*, μ_0^* are taken from Theorem 3.9 and $\gamma < -2$ is assumed. Then, there are positive constants δ_0^0, δ_1^0, δ_0 satisfying (3.55), and a constant $\overline{\mu}_0 \in (0, \mu_0^*]$ so that, if $(W_0^0, \underline{W}_1^0) \in F_{0,l}^0 \times E_{1,l}^{\delta_1^0}$, there exists a unique solution $W_0(t)$ of (3.52) in $G_{0,l}^{\delta_0}$. Furthermore the inequality*

$$\|W_0(t)\|_{X^j} \leq \delta_0 \mu^{l(1-2j)} e^{-\frac{b_1}{2}t} \tag{3.57a}$$

holds, for $t \geq 0$.

This implies that, for $(W_0^0, \underline{W}_1^0) \in F_{0,l}^0 \times E_{1,l}^{\delta_1^0}$, the equation (3.5) has a unique solution $\underline{W}(t)$ in $G_{0,l}^{\delta_0} \times D_{1,l}^{\delta^*+\delta_1^0}$ with $\underline{W}(0) = (W_0^0, \underline{W}_1^0)$ and, for $t \geq 0$,

$$\|\underline{W}(t)\|_{X^1} \leq |\underline{W}|_{\rho,1} e^{-\frac{b_1}{2}t} \quad . \tag{3.57b}$$

Proof. Let $W_0^0 \in F_{0,l}^0$ be fixed. We show that $\mathcal{K}_0(\mu, W_0, \underline{W}_1^0)$ is a strict contraction in $G_{0,l}^{\delta_0}$, where $\mathcal{K}_0$ stands for the right–hand side of (3.52). Since $W_0 \in G_{0,l}^{\delta_0}$ and (3.55) holds, Theorem 3.9 can be applied (with δ^* replaced by δ_0) and thus, (3.5) and (3.52) are completely equivalent. We consider all constants depending only on δ^*, μ_0^* as universal. They are denoted by letters like c, $c_1, \dots$. Moreover, $j = 0, 1$ always holds.

The following estimates are essential

$$|Q(W_0)|_{\rho,j} \leq c_1 \delta_0^2 \mu^{l(1-2j)} \tag{3.58a}$$

$$|M_1^*(\mu, W_0, \underline{W}_1^0)|_{\rho,j} \leq c_2 \delta_0 \mu^{l(1-2j)} \tag{3.58b}$$

$$\rho(t)\|M_1^*(\mu, W_0, \underline{W}_1^0)(t)\|_{X^j} \leq c_3 \mu^{l(1-2j)} \mu^2 m_1^*(t) \tag{3.58c}$$

where

$$m_1^*(t) = 1 + \frac{1}{\mu^2}\left(\delta_1^{0^2} + \mu\delta_1^0\right) \exp\left(-7\alpha/8\mu^2\right) t$$

for all $t \geq 0$. Estimate (3.20) has been used in each of the inequalities. Lemma 3.3, Corollary 3.10 (δ_0 replacing δ^*) and Corollary 3.11 have been applied in (3.58a), (3.58b) and (3.58c) respectively.

182

$\mathcal{K}_0$ maps $G_{0,l}^{\delta_0}$ into itself, since Lemma 3.1 (see (3.54)), Lemma 3.3 and (3.20), (3.55), (3.58a)–(3.58c) imply

$$\rho(t)\|\mathcal{K}_0\left(\mu,\,W_0,\,\underline{W}_1^0\right)(t)\|_{X^j} \leq c_4 \mu^{l(1-2j)}\left(\delta_0^2 + {\delta_1^0}^2 + \mu^2 + \mu\delta_1^0\right) + \frac{\delta_0}{2}\ . \tag{3.59}$$

Similarly we obtain from Lemma 3.4 and by using (3.11c), (3.43) and (3.45)

$$\begin{aligned}\rho(t)\|\mathcal{K}_0\left(\mu,\,W_0,\,\underline{W}_1^0\right)(t) - \mathcal{K}_0\left(\mu,\,\widetilde{W}_0,\,\underline{W}_1^0\right)(t)\|_{X^j} \leq\\ c_5\left(\mu + \delta_1^0 + \delta_0\right)\,|\,W_0 - \widetilde{W}_0\,|_{\rho,0}\ .\end{aligned} \tag{3.60}$$

Here we have used that

$$\frac{\delta_1^0}{\mu}\int_0^t e^{-\frac{b_1}{2}(t-s)}(t-s)^{-1/2}e^{-\frac{\alpha}{\mu^2}s}ds \leq c_6\delta_1^0$$

for all $t \geq 0$. Inequalities (3.59), (3.60) show that, when $0 < \overline{\mu}_0 \leq \mu_0^*$, δ_0 and δ_1^0 are chosen sufficiently small, $\mathcal{K}_0$ defines a strict contraction in $G_{0,l}^{\delta_0}$ and thus the theorem is proved.

In the local stability result of Theorem 3.12, we did not make any additional assumption about the coefficients of $g_0(\cdot,\,\cdot)$. If we look more closely at the proofs of Theorem 3.12 and of Lemmas 3.3, 3.4, we see that the worst term, that has to be estimated, is the term $D_{\underline{u}_1^2}^2 g_0(0,\,0)(\underline{W}_1,\,a\underline{W}_1)$. Below we give an improved local stability result when $D_{\underline{u}_1^2}^2 g_0(0,\,0) = 0$. In this case, we can allow $\|\underline{W}_1^0\|_{\mathcal{X}^1}$ (and therefore $\|W_1\|_{\infty,\,a}$) to be of order $\delta_1^0 \mu^{-4/3}$. To prove this result, we use the first Reduction–Theorem 3.6 as well as Corollary 3.8. Then we can state the following stability result.

Theorem 3.13 *Assume that $D_{\underline{u}_1^2}^2 g_0(0,\,0) = 0$ and $\gamma < -2$. Then, there are positive constants δ_0^0, δ_1^0, δ_0 satisfying (3.55) and a constant $\overline{\mu}_0 \in (0,\,\mu_0^*]$ such that, for every $\mu \in (0,\,\overline{\mu}_0]$ and every pair $(W_0^0,\,\underline{W}_1^0) \in B_{X_0^1}(0,\,\delta_0^0) \times B_{\mathcal{X}^1}(0,\,\delta_1^0\mu^{-4/3})$, there exists a unique solution $W_0(t)$ of (3.52) in $B_{Y_{0\rho}^1}(0,\,\delta_0)$ with $W_0(0) = W_0^0$. Furthermore, the inequality*

$$\|W_0(t)\|_{X^1} \leq \delta_0 e^{-\frac{b_1}{2}t} \tag{3.61a}$$

holds, for $t \geq 0$.

This implies that, for $(W_0^0,\,\underline{W}_1^0) \in B_{X_0^1}(0,\,\delta_0^0) \times B_{\mathcal{X}_1^1}(0,\,\delta_1^0\mu^{-4/3})$, the equation (3.5) has a unique solution $\underline{W}(t)$ in $B_{Y_{0\rho}^1}(0,\,\delta_0) \times B_{\mathcal{Y}_{1\rho}^1}(0,\,\delta^* + \delta_1^0)$ with $\underline{W}(0) = (W_0^0,\,\underline{W}_1^0)$ and, for $t \geq 0$,

$$\|\underline{W}(t)\|_{\mathcal{X}^1} \leq |\underline{W}|_{\rho,1}e^{-\frac{b_1}{2}t}\ . \tag{3.61b}$$

"Sketch of the proof". Here we use the Reduction-Theorem 3.5 in order to obtain the equations (3.52) or (3.53). To prove Theorem 3.13, we apply the strict contraction fixed point theorem to the mapping $W_0 \mapsto \mathcal{K}_0(\mu, W_0, \underline{W}_1^0)$. Arguing as in the proof of Theorem 3.12, but, using here Corollaries 3.7 and 3.8 and the estimates (3.24b) and c) of Lemma 3.4, one proves that $\mathcal{K}_0(\mu, \cdot, \underline{W}_1^0)$ is a strict contraction from $B_{Y_{0\rho}^1}(0, \delta_0)$ into itself, if $(W_0^0, \underline{W}_1^0)$ belongs to $B_{X_0^1}(0, \delta_0^0) \times B_{\mathcal{X}_1^1}(0, \delta_1^0 \mu^{-4/3})$.

4   Global stability of fronts for the central equation

In the previous section we have proved the local stability of the front solution U_{00}^* for the central equation. After having set $U_0 = U_{00}^* + e^{\frac{\gamma}{2}x}W_0$, we showed, in the case $\gamma < -2$, that, if the initial data W_0^0 are small enough in the norm of X^1, then the solution $W_0(t)$ of the equation

$$W_{0t} - W_{0xx} + \left(2U_{00}^* + \frac{\gamma^2}{4} - 1\right)W_0 = -aW_0^2 \quad , \tag{4.1}$$

satisfying $W_0(0) = W_0^0$, exists for all time, belongs to X^1 and converges exponentially fast to 0 in X^1. In this section, we address the question of the global stability of the front U_{00}^* in X^1 for $\gamma \le -2$; that is, we want to obtain stability results when $\|W_0^0\|_{X^1}$ is not small. If $\gamma < -2$, we show that $W_0(t)$ converges exponentially fast to 0 in X^1. In the case $\gamma = -2$, we shall show in [RK95] that $W_0(t)$ converges to 0 like $t^{-1/4}$ when W_0^0 belongs to $X^1 \cap L^1(\mathbb{R})$. The hardest part of the study below is when W_0^0 changes sign. To handle this case, we need first to perform the analysis for $W_0^0 \le 0$ and $W_0^0 \ge U^0$, where U^0 is defined in (4.5) below. We study this case as well as the case when W_0^0 changes sign by introducing an auxiliary unknown function $B(t, x)$ and by applying the maximum principle.

In this section we develop and improve the ideas from [Ki92]. We express the stability results for U_{00}^*, so that they can be directly used in the global stability analysis of the front $\underline{U}_\mu^*$ for the system.

Later on, we shall also use the linear operators L_0^d, for $-1 \le d \le 1$, which coincide with $L_{\mu0}^d$ in (3.9) for $\mu = 0$.

$$L_0^d = -\partial_{xx}^2 + \left(\frac{\gamma^2}{4} - 1 + (1 + d)U_{00}^*\right) \quad . \tag{4.2}$$

184

With this notation, the equation (4.1) can be rewritten as

$$W_{0t} + L_0^1 W_0 = -aW_0^2 \quad . \tag{4.3}$$

4.1 An auxiliary problem

We introduce the auxiliary unknown $B(t, x)$ given by

$$W_0(t) = B(t, x)U^0 \tag{4.4}$$

where

$$U^0(x) = -U_{00}^*(x)e^{-\frac{\gamma}{2}x} \quad . \tag{4.5}$$

Due to (3.4), U^0 corresponds to the trivial solution of the original central equation (1.5). We remark that $W_0(t)$ is a solution of the equation (4.1) with $W_0(0) = W_0^0$ if and only if $B(t) = B(t, \cdot)$ satisfies

$$B_t - \frac{1}{f^2} \left(f^2 B_x \right)_x + U_{00}^* B - U_{00}^* B^2 = 0 \tag{4.6}$$

with $B(0) = \dfrac{W_0^0}{U^0}$, where

$$f = e^{-\frac{\gamma}{2}x} U_{00}^* = -U^0 \quad . \tag{4.7}$$

Since $W_0 = -Bf$ and $aW_0 = -BU_{00}^*$ holds, $W_0 \in X^k$, $k = 0, 1, 2$, is equivalent to $B \in X_f^k := H^k(\mathbb{R}, f) \cap H^k(\mathbb{R}, U_{00}^*)$. The linear operator $L_{f,d}$ given in (4.8) corresponds to L_0^d in (4.2) and is defined by

$$L_{f,d} = -\frac{1}{f^2}\partial_x \left(f^2 \partial_x \right) + dU_{00}^* \quad , \quad L_{f,1} = L_f \quad , \tag{4.8}$$

so that (4.6) reads

$$B_t + L_{f,d}B = U_{00}^* B(B - (1 - d)) \quad .$$

Moreover

$$\frac{1}{f} e^{-L_0^d t} \left(f B_0 \right) = e^{-L_{f,d}t} B_0 \tag{4.9}$$

holds for $B_0 \in X_f^0$.

The operator $L_f := L_{f,1}$ is sectorial in X_f^1. The nonlinearity in (4.6) defines a C^k-mapping from X_f^1 into X_f^1 for every integer $k \geq 0$. Hence, there exists, for every $B_0 \in X_f^1$, a maximal time $T(B_0) > 0$, such that the initial value problem $B(0) = B_0$

185

for Equation (4.6) has a unique solution $B \in C^0([0, T_{B_0}), X_f^1) \cap C^1((0, T_{B_0}), X_f^0)$. Either $T_{B_0} = +\infty$ or there exists a sequence $t_n \to T_{B_0} - 0$ with $\|B(t_n)\|_{X_f^1} \to +\infty$ (cf. [He80], Thms.3.33, 3.34 and Section III). Moreover, according to ([He80], Corollary 3.4.6), $B \in C^2((0, T_{B_0}), X_f^1)$ holds as well. Via a "bootstrap argument" one concludes that $B_x \in C^0((0, T_{B_0}), X_f^2)$ and $B_{xt} \in C^2((0, T_{B_0}), X_f^1)$. This implies finally that $B \in C^2((0, T_{B_0}) \times I\!R, I\!R)$. Since $B \in C^0([0, T_{B_0}), X_f^1)$ and that X_f^1 is continuously embedded into $L^\infty(I\!R, f) \cap L^\infty(I\!R, U_{00}^*)$, for any τ, $0 < \tau < T_{B_0}$, there exists a positive constant $K_\tau = K(B_0, \tau)$ such that

$$\max\left(\sup_{0 \le t \le \tau} \|B(t)\|_{X_f^1}, \sup_{0 \le t \le \tau} \|B(t)f\|_{L^\infty(I\!R)}, \sup_{0 \le t \le \tau} \|B(t)U_{00}^*\|_{L^\infty(I\!R)}\right) \le K_\tau .$$
(4.10)

And also $B(t)$ belongs to $C^0([0, \tau] \times I\!R, I\!R)$. From (4.10), we deduce that, for any $\delta > 0$,

$$\lim_{R \to +\infty} e^{-\delta R^2}\left(\max_{\substack{0 \le |x| \le R \\ 0 \le t \le \tau}} |B(t, x)|\right) = 0 .$$
(4.11)

All these properties will allow us later to apply to $B(t)$ the maximum principle as stated in ([PW67] p.183).

Let us point out that there exists a non-decreasing positive function $\tilde{K}(\cdot)$ such that, if, for some $t_0 > 0$, we have, for all $t \in [0, t_0]$,

$$\max\left(\|B(t)U_{00}^*\|_{L^\infty(I\!R)}, \|B(t)\|_{X_f^0}\right) \le R_0 ,$$
(4.12)

then

$$\sup_{0 \le t \le t_0} \|B(t)\|_{X_f^1} \le \tilde{K}(R_0) .$$
(4.13)

We do not give the proof of this result, because it is very similar to the proof of the estimates given in Proposition 4.6.

In the process of our analysis we shall frequently encounter an equation of the form

$$C_t + L_{f,\delta} C = D$$
(4.14)

where C and D stand for different functions of B. We shall use (4.14) to simplify the notation. The following arguments will give invariant regions of $B(t)$ of increasing precision for $t \in [0, T_{B_0})$.

Let $B_0 \ge 0$ and $B(t)$ be the corresponding solution of (4.6) with $B(0) = B_0$. Then, $-B = C$ solves (4.14) for $\delta = 1$, $D = -U_{00}^*(-B)^2 \le 0$. Thus, applying the maximum principle in $[0, \tau]$, $\tau < T_{B_0}$, leads to

$$B(t) \geq 0 \quad \text{for} \quad t \in [0, T_{B_0}) \quad . \tag{4.15}$$

Consider $C = (B - 1) \exp(-K_\tau t)$, where K_τ is given in (4.10), then C satisfies

$$C_t + L_{f,0} C + (K_\tau - U_{00}^* B) C = 0$$

where $K_\tau - U_{00}^* B \geq 0$. Therefore, $B_0 \leq 1$ implies

$$B(t) \leq 1 \quad \text{for} \quad t \in [0, T_{B_0}) \quad . \tag{4.16}$$

Finally, let $d \in [0, 1]$. Assume $0 \leq B_0 \leq 1 - d$. Since $C = B - (1 - d)$ solves (4.14) for $\delta = 0$, $D = U_{00}^* B(B - 1)$, we conclude via (4.15) and (4.16) that $D \leq 0$ holds, and by the maximum principle

$$0 \leq B(t) \leq 1 - d \quad \text{for} \quad t \in [0, T_{B_0}) \quad . \tag{4.17}$$

The relation (4.17) can still be improved by considering $C = B - \exp(-L_{f,d} t) B_0$. It solves (4.14) for $\delta = d$, $D = U_{00}^* B(B - (1 - d)) \leq 0$. One concludes from the maximum principle

$$0 \leq B(t) \leq \exp\left(-L_{f,d} t\right) B_0 \quad \text{for} \quad t \in [0, T_{B_0}) \quad . \tag{4.18}$$

The validity of the inequality

$$\| \exp\left(-L_{f,0} t\right) B_0 \|_{X_f^0} \leq C \left(\|B_0\|_{X_f^1} \right) T \tag{4.19}$$

for $0 \leq t < T < \infty$ and $0 \leq B_0 \leq 1$ is essential for the global existence in time of $B(t)$. The estimate (4.19) will be proved in Lemma 4.2.

If $0 \leq B_0 \leq 1$, we derive from (4.16), (4.18) and (4.19), for all $t \in [0, T_{B_0})$,

$$\|B(t)\|_{X_f^0} + \|B(t) U_{00}^*\|_{L^\infty(\mathbb{R})} \leq T_{B_0} C \left(\|B_0\|_{X_f^1} \right) + 1$$

which, by (4.13), implies that

$$\|B(t)\|_{X_f^1} \leq \tilde{K} \left(T_{B_0} C \left(\|B_0\|_{X_f^1} \right) + 1 \right) \quad . \tag{4.20}$$

This inequality shows that $\|B(t)\|_{X_f^1}$ stays bounded when $t \to T_{B_0} - 0$. Therefore, $T_{B_0} = \infty$. We have proved the following result

187

Proposition 4.1 *If $B_0 \in X_f^1$ and $0 \leq B_0 \leq 1 - d$ hold, where $d \in [0,1]$, then the solution $B(t)$ of (4.6) with $B(0) = B_0$ exists on $[0,\infty)$ and satisfies (4.17) and (4.18).*

It remains to bound $e^{-L_{f,0}t}B_0$ in X_f^0. From the Lemma 3.1 and the equality (4.9), it follows that for $t \geq 0$

$$\|e^{-L_{f,0}t}B_0\|_{j,f} \leq C_{41}e^{-b_0 t}\|B_0\|_{j,f}, \quad j = 0,1, \tag{4.21}$$

holds, where $b_0 = \frac{\gamma^2}{4} - 1$ and $\|\cdot\|_{j,f}$ denotes the norm in $H^j(\mathbb{R}, f)$. We also denote by $\|\cdot\|_{j,U_{00}^*}$ the norm in $H^j(\mathbb{R}, U_{00}^*)$. The estimate of $e^{-L_{f,0}t}B_0$ in X_f^0 will be a direct consequence of (4.21) and of the Lemma 4.2 below.

Lemma 4.2 *Let $B_0 \in H^1(f) \cap H^0(U_{00}^*)$, $0 \leq B_0 \leq 1$. Then, for $t \geq 0$,*

$$0 \leq e^{-L_{f,0}t}B_0 \leq 1 \quad.$$

Moreover, for any positive constant α_0 satisfying $\frac{\gamma}{2}(1 + \alpha_0) + \sqrt{\frac{\gamma^2}{4} + 1} > 0$, there exists a positive constant $C_{42}(\alpha_0, \gamma)$ such that, for $t \geq 0$,

$$\|e^{-L_{f,0}t}B_0\|_{0,U_{00}^*}^2 \leq \|B_0\|_{0,U_{00}^*}^2 + C_{42}(\alpha_0, \gamma)l(t)\|B_0\|_{1,f}^{\alpha_0}, \tag{4.22}$$

where

$$l(t) = \begin{cases} t & , \text{ if } b_0 = 0, \text{ i.e. } \gamma = -2 \\ \dfrac{1}{\alpha_0 b_0}\left(1 - e^{-\alpha_0 b_0 t}\right) & , \text{ if } b_0 > 0. \end{cases} \tag{4.23}$$

Proof. Note that $\widetilde{B} = e^{-L_{f,0}t}B_0$ is a solution of the equation

$$\widetilde{B}_t - \frac{1}{U_{00}^{*2}}\partial_x\left(U_{00}^{*2}\widetilde{B}_x\right) + \gamma\widetilde{B}_x = 0 \quad. \tag{4.24}$$

Applying the maximum principle to (4.24), we show, that $0 \leq \widetilde{B}(t) \leq 1$ holds. Taking the inner product in $L^2(\mathbb{R})$ of (4.24) with $\widetilde{B}U_{00}^{*2}$, we obtain, for $t \geq 0$,

$$\frac{d}{dt}\|\widetilde{B}\|_{0,U_{00}^*}^2 + 2\|\widetilde{B}_x\|_{0,U_{00}^*}^2 = 2\gamma\int_{\mathbb{R}} \widetilde{B}^2 U_{00}^* U_{00x}^* \, dx \quad. \tag{4.25}$$

From (4.21) and (4.25), we infer

$$\begin{aligned}
\frac{d}{dt}\|\widetilde{B}\|_{0,U_{00}^*}^2 &\leq 2|\gamma|\,\|\widetilde{B}e^{-\frac{\gamma}{2}x}U_{00}^*\|_\infty^{\alpha_0} \times |\int_{\mathbb{R}} \widetilde{B}^{2-\alpha_0}U_{00}^{*1-\alpha_0}e^{\frac{\alpha_0\gamma x}{2}}U_{00x}^* \, dx| \\
&\leq \widetilde{C}(\gamma)e^{-\alpha_0 b_0 t}\|B_0\|_{1,f}^{\alpha_0}|\int_{\mathbb{R}} e^{\frac{\alpha_0\gamma x}{2}}U_{00x}^* \, dx| \quad.
\end{aligned} \tag{4.26}$$

188

Since $U_{00x}^{*} e^{-(\frac{\gamma}{2}+\sqrt{\frac{\gamma^2}{4}+1})x}$ is bounded as x goes to $-\infty$ (see Section 2), the integral $|\int_{I\!R} e^{\frac{\alpha_0 \gamma x}{2}} U_{00x}^{*} \, dx|$ ist bounded by a constant $C(\alpha_0)$, and (4.26) becomes

$$\frac{d}{dt} \|\widetilde{B}\|_{0,U_{00}^{*}}^{2} \leq C_{42} (\alpha_0, \gamma) \, e^{-\alpha_0 b_0 t} \, \|B_0\|_{1,f}^{\alpha_0} \quad ,$$

where $C_{42}(\alpha_0, \gamma) = C(\alpha_0)\widetilde{C}(\gamma)$,

and (4.22) follows, by integration from 0 to t. ∎

4.2 The case of initial data $W_0^0 \leq 0$, bounded below by U^0

We recall that, as in Section 4.1, for any $W_0^0 \in X^1$, there is a positive maximal time $T_{W_0^0} \leq +\infty$, such that the equation (4.1) has a unique solution $W_0(t) \equiv S_0(t)W_0^0$ in $C^0([0, T_{W_0^0}), X^1)$. Either $T_{W_0^0} = +\infty$, or $\|W_0(t_n)\|_{X^1} \to +\infty$ for some sequence t_n tending to $T_{W_0^0} - 0$. The function $W_0(t, x)$ belongs to $C^2((0, T_{W_0^0}) \times I\!R, I\!R) \cap C^0([0, T_{W_0^0}) \times I\!R, I\!R)$ and, for any $\tau \in (0, T_{W_0^0})$ we have, for some $K_\tau = K(\tau, W_0^0)$

$$\sup_{0 \leq t \leq \tau} \|W_0(t)\|_{X^1} \leq K_\tau \quad . \tag{4.27}$$

In view of (3.21) the same estimate holds for $\|W_0(t)\|_\infty$ and $\|aW_0(t)\|_\infty$. Therefore, $W_0(t)$ satisfies (4.11) for every positive δ, and the maximum principle may be applied. Moreover, if there are positive numbers t_0 and R_0 such that

$$\max(\|W_0(t)\|_{X^0}, \|aW_0(t)\|_\infty) \leq R_0 \tag{4.28}$$

holds for all $t \in [0, t_0]$, then we have also

$$\sup_{0 \leq t \leq t_0} \|W_0(t)\|_{X^1} \leq \widetilde{K}(R_0) \tag{4.29}$$

for some positive number $\widetilde{K}(R_0)$, which can be assumed to be

a nondecreasing function in R_0. The proof of (4.29) is similar to the proof of the estimates in Proposition 4.6. For this reason it is omitted.

For $W_0^0 \in X^1$ the function $e^{-L_0^1 t} W_0^0$ exists for all positive times, belongs to $C^2((0, +\infty) \times I\!R, I\!R) \cap C^0([0, +\infty) \times I\!R, I\!R)$, satisfies the estimates of Lemma 3.1, and has the property (4.11). Since

$$\left(W_0 - e^{-L_0^1 t} W_0^0\right)_t - \left(W_0 - e^{-L_0^1 t} W_0^0\right)_{xx} + (2U_{00}^{*} + b_0)\left(W_0 - e^{-L_0^1 t} W_0^0\right) = -aW_0^2 \leq 0 \quad ,$$

189

holds, where $b_0 + 2U_{00}^* \geq 0$, we can apply the maximum principle to conclude

$$S_0(t)W_0^0 \leq e^{-L_0^1 t}W_0^0 \quad \text{on} \quad \left[0, T_{W_0^0}\right) \; . \tag{4.30}$$

Likewise, the maximum principle implies that, if $W_0^0 \leq 0$,

$$S_0(t)W_0^0 \leq 0 \quad \text{on} \quad \left[0, T_{W_0^0}\right) \; . \tag{4.31}$$

Remember that $S_0(t)W_0^0 = B(t)U^0$, where $B(t)$ is the solution of (4.6) with $B(0) = W_0^0/U^0$. The Proposition 4.3 below is then a direct consequence of the Proposition 4.1, the inequalities (4.17), (4.18), (4.30), and the equality (4.9).

Proposition 4.3 *Let W_0^0 be given in X^1, with $(1-d)U^0 \leq W_0^0 \leq 0$, where $0 \leq d \leq 1$. Then $S_0(t)W_0^0$ exists on $[0,\infty)$ and we have, for $t \geq 0$,*

$$\begin{aligned} (1-d)U^0 &\leq S_0(t)W_0^0 \leq 0 \\ e^{-L_0^d t}W_0^0 &\leq S_0(t)W_0^0 \leq e^{-L_0^1 t}W_0^0 \; . \end{aligned} \tag{4.32}$$

4.3 The case of general initial data W_0^0, bounded below by U^0

It remains to treat the case $W_0^0 \geq U^0$, but otherwise arbitrary. We shall use the same type of argument as in [Ki92].

Theorem 4.4 *Let $W_0^0 \in X^1$, with $W_0^0 \geq (1-d)U^0$, where $0 \leq d \leq 1$. Then $S_0(t)W_0^0$ exists on $[0,\infty)$ and we have, for $t \geq 0$,*

$$\sup\left((1-d)U^0, e^{-L_0^d t}\widetilde{W_0^0}\right) \leq S_0(t)W_0^0 \leq e^{-L_0^1 t}W_0^0 \tag{4.33}$$

where $\widetilde{W_0^0} = \mathrm{Inf}\,(0, W_0^0)$.

Proof. Consider the solution $\widetilde{W}_0(t) = S_0(t)\widetilde{W_0^0}$ of (4.1). Since $\widetilde{W_0^0} \in X^1$ satisfies the hypotheses of Proposition 4.3, $\widetilde{W}_0(t)$ exists for all $t \geq 0$ and satisfies (4.32). Moreover, according to (4.5) and (4.32), $\widetilde{W}_0(t)$ fulfills

$$\|a\widetilde{W}_0(t)\|_\infty \leq 1 - d \; , \quad \text{for} \quad t \geq 0 \; . \tag{4.34}$$

Since $(1-d)U^0 \leq \widetilde{W_0^0} \leq 0$ one concludes, using (4.4), (4.5) and (4.20), that for every $T_0 > 0$

$$\|\widetilde{W}_0(t)\|_{X^1} \leq c_1 \tilde{K}\left(T_0 C\left(c_0 \|W_0^0\|_{X^1}\right) + 1\right), \quad \text{for} \quad 0 \leq t \leq T_0 \tag{4.35}$$

190

holds, where c_0, c_1 are positive constants depending only on γ and U^*_{00}. The function $W_0(t) = S_0(t)W^0_0$ satisfies (4.27), (4.30) and (4.11) on $[0, T_{W^0_0})$. Moreover

$$C_t - C_{xx} + \left[(2U^*_{00} + b_0) + K_\tau(1-d) + a(\widetilde{W}_0 + W_0) \right] C = 0 \quad , \tag{4.36}$$

holds, where

$$C(t) = e^{-(K_\tau + (1-d))t} \left(\widetilde{W}_0(t) - W_0(t) \right)$$

and K_τ is taken from (4.27).

Since $\widetilde{W}_0(0) - W_0(0) \le 0$, and by (4.27) and (4.34), $K_\tau + (1-d) + a(\widetilde{W}_0 + W_0) \ge 0$, we can apply the maximum principle on every interval $[0, \tau]$, $0 < \tau < T_{W^0_0}$, which implies

$$\widetilde{W}_0(t) \le W_0(t) \quad \text{on} \quad \left[0, T_{W^0_0} \right)$$

and, with (4.30),

$$\widetilde{W}_0(t) \le W_0(t) \le e^{-L^1_0 t} W^0_0 \quad \text{on} \quad \left[0, T_{W^0_0} \right) \quad . \tag{4.37}$$

By Lemma 3.1 and (4.35), $\|e^{-L^1_0 t} W^0_0\|_{X^1}$ and $\|\widetilde{W}_0(t)\|_{X^1}$ are bounded by a positive constant $C_0(T_{W^0_0})$ on $[0, T_{W^0_0}]$. Thus, by (3.20) and (4.37), $\|W_0(t)\|_{X^0}$ and $\|aW_0(t)\|_\infty$ are also bounded by $C_0(T_{W^0_0})$ there. The properties (4.28), (4.29) then imply that $\|W_0(t)\|_{X^1}$ is bounded by $\widetilde{K}(C_0(T_{W^0_0}))$ on $[0, T_{W^0_0})$. It follows that $W_0(t)$ exists for all $t \ge 0$ in X^1. The estimate (4.33) is a direct consequence of (4.37) and (4.32).

It remains to study the decay of $\|S_0(t)W^0_0\|$ as $t \to \infty$. In this part we shall treat the case $\gamma < -2$ alone and postpone the case $\gamma = -2$ to part II [RK95]. ■

4.4   Decay of $S_0(t)W^0_0$ in X^1 when $W^0_0 \ge (1-d)U^0$ and $\gamma < -2$

The first five estimates of the next proposition are a direct consequence of Lemma 3.1 and Theorem 4.4, while the last estimate follows at once from Lemma 4.2 and Theorem 4.4. If $d > 0$ and $0 < \delta < d$, we set

$$b_{d,\delta} = \inf (b_0, d - \delta) \quad .$$

Whenever $\gamma < -2$, we have $b_{d,\delta} > 0$.

Proposition 4.5 *Let* $W_0^0 \in X^1$, $W_0^0 \geq (1-d)U^0$, $0 \leq d \leq 1$, *and* $0 < \delta < d$ *if* $d > 0$. *Then* $S_0(t)W_0^0$ *satisfies the following estimates for* $t \geq 0$:

$$\|S_0(t)W_0^0\|_0 \leq e^{-b_0 t}\|W_0^0\|_0 \tag{4.38a}$$

$$\|S_0(t)W_0^0\|_\infty \leq C_{43}e^{-b_0 t}\left(\|W_0^0\|_0\|W_0^0\|_1\right)^{1/2} \tag{4.38b}$$

$$\|aS_0(t)W_0^0\|_\infty \leq C_{44}(\gamma,\delta)\max\left(1-d,\ e^{-b_1,\delta t}\left(\|W_0^0\|_{X^0}\|W_0^0\|_{X^1}\right)^{1/2}\right) \tag{4.38c}$$

If moreover $d > 0$, *then*

$$\|S_0(t)W_0^0\|_{0,a} \leq C_{45}(\gamma,\delta,d)e^{-b_{d,\delta}t}\|W_0^0\|_{X^0} \tag{4.39a}$$

$$\|aS_0(t)W_0^0\|_\infty \leq C_{45}(\gamma,\delta,d)e^{-b_{d,\delta}t}(\|W_0^0\|_{X^0}\|W_0^0\|_{X^1})^{1/2} \qquad . \tag{4.39b}$$

For $d = 0$

$$\|S_0(t)W_0^0\|_{0,a} \leq \left\{\|W_0^0\|_{X^0}^2 + C_{42}\left(\alpha_0,\gamma\right)\frac{1-e^{-\alpha_0 b_0 t}}{\alpha_0 b_0}\|W_0^0\|_1^{\alpha_0}\right\}^{1/2} \tag{4.39c}$$

where α_0 *is taken from Lemma 4.2.*

The estimates of $S_0(t)W_0^0$ in $H^1(I\!R)$ and in X^1 are given in the following proposition.

Proposition 4.6 *Under the hypotheses of Proposition 4.5,* $S_0(t)W_0^0$ *satisfies the following estimates for* $t \geq 0$:

$$\|S_0(t)W_0^0\|_1 \leq C_{46}(\gamma,\delta)e^{-b_0 t}\|W_0^0\|_{X^1}\max\left(1-d,\ e^{-b_1,\delta t}\|W_0^0\|_{X^1}\right) \qquad . \tag{4.40}$$

Whenever $d > 0$, *we have*

$$\|S_0(t)W_0^0\|_1 \ \leq \ C_{47}(\gamma)e^{-b_0 t}\|W_0^0\|_1 + \tag{4.40a}$$
$$+C_{48}(\gamma,\delta,d)e^{-(b_0+b_{d,\delta})t}\|W_0^0\|_0\left(\|W_0^0\|_{X^0}\|W_0^0\|_{X^1}\right)^{1/2}$$

$$\|S_0(t)W_0^0\|_{1,a} \ \leq \ C_{48}(\gamma,\delta,d)\left[e^{-b_{d,\delta}t}\|W_0^0\|_{X^1} + \right. \tag{4.40b}$$
$$\left. +e^{-2b_{d,\delta}t}\|W_0^0\|_{X^1}^{1/2}\|W_0^0\|_{X^0}^{3/2}\right]$$

If $d = 0$, *then*

$$\|S_0(t)W_0^0\|_{1,a} \ \leq \ C_{49}(\gamma)\left[\|W_0^0\|_{X^1}^2 + \right. \tag{4.40c}$$
$$\left. +\frac{C_{42}(\alpha_0,\gamma)}{\alpha_0 b_0}\|W_0^0\|_1^{\alpha_0}\right]^{1/2}\max\left(1,\ e^{-b_1,\delta t}\|W_0^0\|_{X^1}\right) \qquad .$$

The constant α_0 *is again taken from Lemma 4.2.*

We remark that $\|S_0(t)W_0^0\|_{X^1}$ decays to 0 exponentially fast, if $\gamma < -2$, and $0 < d \le 1$.

Proof. Taking the inner product in $L^2(\mathbb{R})$ of the equality (4.1) with $W_0(t) \equiv S_0(t)W_0^0$ yields

$$\frac{1}{2}\frac{d}{dt}\|W_0\|_0^2 + \|W_{0x}\|_0^2 + b_0\|W_0\|_0^2 + \int_{\mathbb{R}} 2U_{00}^* W_0^2 \, dx \le \|aW_0\|_\infty \|W_0\|_0^2$$

which gives, after an integration,

$$\int_t^{t+1} \|W_{0x}\|_0^2 \, ds \le \frac{1}{2}\|W_0(t)\|_0^2 + \int_t^{t+1} \|aW_0\|_\infty \|W_0\|_0^2 \, ds \quad . \tag{4.41}$$

If we take the inner product in $L^2(\mathbb{R})$ of (4.1) with $-W_{0xx}$, we obtain, for $t \ge 0$,

$$\frac{d}{dt}\|W_{0x}\|_0^2 + 2\|W_{0xx}\|_0^2 + 2b_0\|W_{0x}\|_0^2 \le 4|\int_{\mathbb{R}} U_{00}^* W_0 W_{0xx} \, dx| + 2|\int_{\mathbb{R}} aW_0^2 W_{0xx} \, dx|$$

or also

$$\frac{d}{dt}\|W_{0x}\|_0^2 \le 4\|W_0(t)\|_0^2 + \|aW_0\|_\infty^2 \|W_0\|_0^2 \quad . \tag{4.42}$$

If we apply the uniform Gronwall Lemma to (4.42) and take into account the estimate (4.41), we obtain for $t \ge 1$

$$\|W_{0x}(t)\|_0^2 \le \frac{1}{2}\|W_0(t-1)\|_0^2 + \int_{t-1}^t \left(4 + 2\|aW_0\|_\infty^2\right)\|W_0\|_0^2 \, ds \quad . \tag{4.43}$$

For $0 \le t \le 1$, we simply deduce from (4.41) and (4.42),

$$\|W_{0x}(t)\|_0^2 \le \|W_0(0)\|_1^2 + \int_0^t \left(4 + \|aW_0\|_\infty^2\right)\|W_0\|_0^2 ds \quad . \tag{4.44}$$

The estimates (4.40) and (4.40a) follow at once from (4.43), (4.44), (4.38a), (4.38c) and (4.39b). We remind that the equality (4.1) also writes

$$\partial_t W_0 - \frac{1}{a^2}\left(a^2 W_{0x}\right)_x + \gamma W_{0x} + (2U_{00}^* + b_0)W_0 + aW_0^2 = 0 \tag{4.45}$$

Taking the inner product in $L^2(\mathbb{R})$ of (4.45) with $a^2 W_0(t)$ yields

$$\frac{1}{2}\frac{d}{dt}\|W_0\|_{0,a}^2 + \frac{1}{2}\|W_{0x}\|_{0,a}^2 + \int_{\mathbb{R}} a^2\left(2U_{00}^* - 1 - \frac{\gamma^2}{4}\right)W_0^2 \, dx \le \|aW_0\|_\infty \|W_0\|_{0,a}^2$$

193

which implies after an integration

$$\int_t^{t+1} \|W_{0x}\|_{0,a}^2 \, ds \leq \|W_0(t)\|_{0,a}^2 + \int_t^{t+1} \left(2 + \frac{\gamma^2}{2} + 2\|aW_0\|_\infty\right) \|W_0\|_{0,a}^2 \, ds \quad . \tag{4.46}$$

If we take now the inner product in $L^2(\mathbb{R})$ of (4.45) with $-(a^2 W_{0x})_x$, we obtain, after a short computation,

$$\frac{d}{dt}\|W_{0x}\|_{0,a}^2 \leq \gamma^2 \|W_{0x}\|_{0,a}^2 + 4\left(\frac{\gamma^2}{4} + 1\right)^2 \|W_0\|_{0,a}^2 + 4\|aW_0\|_\infty^2 \|W_0\|_{0,a}^2 \quad . \tag{4.47}$$

Applying the uniform Gronwall Lemma to (4.47) and taking into account the estimate (4.46) yield, for $t \geq 1$,

$$\begin{aligned}
\|W_{0x}(t)\|_{0,a}^2 \leq & (\gamma^2 + 1) \, \|W_0(t-1)\|_{0,a}^2 + \int_{t-1}^t \left[\left(\frac{\gamma^2}{4} + 1\right)(2\gamma^2 + 6) + \right. \\
& \left. +2\left(\gamma^2 + 1\right)\|aW_0\|_\infty + 4\|aW_0\|_\infty^2\right]\|W_0\|_{0,a}^2 \, ds \quad .
\end{aligned} \tag{4.48}$$

For $0 \leq t \leq 1$, we derive from (4.46) and (4.47) a similar estimate, where $\|W_0(t-1)\|_{0,a}^2$ is replaced by $\|W_0(0)\|_{1,a}^2$. The estimates (4.40b) and (4.40c) are direct consequences of (4.48), (4.39a), (4.39b) and (4.38c), (4.39c). ∎

5 Global stability of fronts for the system

The global stability results for the front solution of the scalar equation have implications on the stability of the front for the full system (3.5). In this section we shall analyse some of these consequences for the case that γ satisfies the strict inequality $\gamma < -2$. We show that, if

$$\begin{aligned}
\|W_0^0\|_{X^j} & \leq \frac{\delta_0}{2\widetilde{C}_{30}} g_0(\mu)\mu^{l(1-2j)} \\
\|\underline{W}_1^0\|_{X^j} & \leq \delta_1^0 \mu^{l(1-2j)-1} \\
W_0^0 & \geq (1-d_0)U^0
\end{aligned} \tag{5.1}$$

holds for some positive constants δ_0, δ_1^0, $d_0 \in (0,1]$, $l \geq 0$ and $\widetilde{C}_{30}$, then the front solution of (3.5) is exponentially asymptotically stable. $\widetilde{C}_{30}$ will be defined in (5.4).

194

In comparison with Theorem 4.4, the lower bound of W_0^0 has a stricter form. The upper bound of $\|W_0^0\|_{X^j}$ contains the function $g_0(\mu)$ which will satisfy $g_0(\mu) \to \infty$ as $\mu \to 0$. This function reflects the effect of the global result in Section 4. Various forms of admissible $g_0(\mu)$ will be given. Let us point out that we always have $j = 0$ or 1, a fact, which is used throughout the section without explicit mentioning.

First we prove a so–called short time or fast flow result. It shows that, within a time $T_0(\mu) = O(\mu^2|\log\mu|)$, the solution of (3.5), with initial conditions satisfying (5.1), reaches a set where $\underline{W}_1^0$ is small. This step will be achieved under the following conditions:

$$0 < \delta^* < 1 \quad , \quad \delta_1^0 < \delta^* \quad ,$$
$$1 \le \delta_0 g_0(\mu) \le \frac{\delta^*}{2\mu} \quad , \quad 1 \le g_0(\mu) \le \mu^{-p_0} \quad , \quad \mu \in (0, \mu_0^*]$$

(5.2)

where $p_0 \in (0, 1)$ and δ^*, μ_0^* are given in (3.33a) and (3.33b). Moreover we have:
$$\lim_{\mu \to 0+} g_0(\mu) = +\infty.$$

In the second part we consider the long–time or slow flow of (3.51), the reduced form of (3.5), which starts at $\underline{W}(T_0(\mu))$. Then the solution $\widetilde{W}_0^*(t)$ of the scalar equation (4.1) with $\widetilde{W}_0^*(0) = W_0(T_0(\mu))$ will be a suitable guiding function to estimate the decay of $\widetilde{W}_0(t)$, the solution of (3.51) with the same initial condition $W_0(T_0(\mu))$. Further restrictions have to be imposed on $g_0(\mu)$ to prove finally the stability. They will depend strongly on the size of d_0.

A few words on the role of μ_0^*. It has been defined in (3.33a), (3.33b). In the course of our analysis it will be redefined a finite number of times to satisfy certain additional inequalities. This will be done frequently without any further comments, like, for instance, in $\mu^{-p_0} + \mu^{-1} \le c\mu^{-1}$, $p_0 \le 1$. Constants within a proof, which depend on (δ^*, μ_0^*) only, are considered as universal and denoted by small $c's$: $c_1, \widetilde{c}\ldots$

For the *weight function* $\rho(t)$ we choose

$$\rho(t) = \exp\left(\frac{b_1}{2}t\right) \quad , \quad 0 < b_1 < \min(b_0, 1 - \delta)$$

(5.3)

and define $\widetilde{C}_{30} > 1$ such that

$$\|K_0(t)\|_{\mathcal{L}(X^j, X^j)} \le \widetilde{C}_{30}\rho^{-2}(t)$$
$$\|K_0(t)\|_{\mathcal{L}(X^0, X^1)} \le \widetilde{C}_{30}t^{-1/2}\rho^{-2}(t)$$

(5.4)

for $t > 0$, where $K_0(t) = \exp(-L_{\mu 0}t)$ (cf. Lemma 3.1). Moreover we assume μ_0^* to

195

be so small that

$$b_1 < \frac{\alpha}{8\mu_0^{*2}} \quad .$$

5.1   A short–time or fast flow result

We begin with a technical lemma

Lemma 5.1 *Let η and d be any positive numbers. Then there exists C_{51} positive such that $W_{0k} \in X^1$, $k = 0, 1$, and*

$$\|W_{01} - W_{02}\|_{\infty, a^j} \leq C_{51}\eta, \quad j = 0, 1, \quad W_{01} \geq (1 - d)U^0$$

implies

$$W_{02} \geq (1 - (d - \eta))\, U^0 \quad . \tag{5.5}$$

Proof. Choose $x_0 \in \mathbb{R}$ so that $U_{00}^*(x) \geq 1/2$ holds on $(-\infty, x_0]$. Since $(a^{-1}U_{00}^*)(x) \to \infty$ as $x \to \infty$, we have $a^{-1}U_{00}^*(x) \geq c_0 > 0$ for $x \in [x_0, \infty)$. In $(-\infty, x_0]$ we can write

$$\frac{|(W_{01} - W_{02})(x)|}{|U^0(x)|} = \frac{a(x)|(W_{01} - W_{02})(x)|}{U_{00}^*(x)} \leq 2C_{51}\eta$$

while in $[x_0, \infty)$ we have

$$\frac{|(W_{01} - W_{02})(x)|}{|U^0(x)|} = \frac{|(W_{01} - W_{02})(x)|}{(U_{00}^*a^{-1})(x)} \leq \frac{C_{51}}{c_0}\eta \quad .$$

Choose $C_{51} = \min(\frac{1}{2}, c_0)$ and the lemma is proved.

Now let $\underline{W}(t) = (W_0(t), \underline{W}_1(t))$ be the solution of (3.5) with initial data $(W_0^0, \underline{W}_1^0) = \underline{W}^0$ satisfying (5.1), (5.2). $\underline{W}(t)$ is continuous in t as a function in $\mathcal{X}^j$. Therefore, there exists a time $T^*(\mu, \underline{W}^0)$, simply denoted by $T^*(\mu)$, such that

$$|W_0|_{\rho, j, t} < 2\delta_0 g_0(\mu)\mu^{l(1-2j)} \quad \text{for } 0 \leq t < T^*(\mu) \tag{5.6a}$$

and

$$(\rho(t)\|W_0(t)\|_{X^j})_{|t = T^*(\mu)} = 2\delta_0 g_0(\mu)\mu^{l(1-2j)} \tag{5.6b}$$

for $j = 0$ or $j = 1$. Here we have used the notation

$$|W_0|_{\rho, j, t} = \sup_{0 \leq \tau \leq t} \rho(\tau)\|W_0(\tau)\|_{X^j} \quad . \tag{5.7}$$

196

Similarly, we set

$$|\underline{W}|_{\rho, j, t} = \sup_{0 \leq \tau \leq t} \rho(\tau) \|\underline{W}(\tau)\|_{\mathcal{X}^j} \quad .$$

We remark that $|\underline{W}|_{\rho, j} = |\underline{W}|_{\rho, j, \infty}$.

Now we choose the time

$$T_0(\mu) = \frac{2\mu^2}{\alpha} \log \frac{1}{\mu^3 \delta_0^2 g_0^2} \tag{5.8}$$

and prove that $T_0(\mu) < T^*(\mu)$. The argument is as follows: For $t \in [0, T^*(\mu)]$ the Theorem 3.9 can be applied, and thus (3.52) is equivalent to (3.5). If we can prove that (3.52) is solvable for $t \in [0, T_0(\mu)]$ and that (5.6a) is valid there, then the definition of $T^*(\mu)$ implies $T_0(\mu) < T^*(\mu)$. In the following theorem $\underline{W}(t)$ denotes the solution of (3.5) with initial condition $\underline{W}^0 = (W_0^0, W_1^0)$ at $t = 0$, whereas $W_0^*(t)$ denotes the solution of (4.1) with initial condition W_0^0.

Theorem 5.2 *Let g_0, δ_0 be fixed satisfying (5.2) with $p_0 \in (0, 1)$, and let $l \geq 0$, $d_0 \in (0, 1]$, $\eta \in (0, 1]$ be given. Then there are positive constants $\delta_1^0 \leq \delta^*$ and $\mu_1^* \leq \mu_0^*$ such that, for every $\mu \in (0, \mu_1^*]$ and every $\underline{W}^0$ satisfying (5.1), a unique solution $\underline{W}(t)$ of (3.5) with $\underline{W}(0) = \underline{W}^0$, exists on the time interval $[0, T_0(\mu)]$. Moreover, $W_0(t)$ and $W_0^*(t)$ both obey the inequality*

$$|W_0|_{\rho, j, T_0(\mu)} \leq \delta_0 g_0(\mu) \mu^{l(1-2j)} \quad . \tag{5.9a}$$

The function $\underline{W}_1(t)$ satisfies

$$\|\underline{W}_1(T_0(\mu))\|_{\mathcal{X}^j} \leq C_{52} \mu^{l(1-2j)} \mu^2 \delta_0^2 (g_0(\mu))^2 \quad . \tag{5.9b}$$

Furthermore, there exists $\mu_1^*(\eta) \leq \mu_1^*$ so that, if $\|\underline{W}_1^0\|_{\mathcal{X}^j} \leq \delta_1^0 \eta \mu^{l(1-2j)-1}$, we have, for $t \in [0, T_0(\mu)]$,

$$\|W_0(t) - W_0^*(t)\|_{\mathcal{X}^j} \leq \frac{C_{51}}{\sqrt{2}} \eta \mu^{l(1-2j)} \quad , \tag{5.10}$$

which implies, via Lemma 5.1,

$$W_0(t) \geq (1 - (d_0 - \eta)) U^0 \quad . \tag{5.11}$$

Proof. According to the preliminary arguments it suffices to prove the solvability of (3.52) in $I(\mu) = [0, T_0(\mu)]$. This will be done via a fixed point argument in the closed set $D \subset Y_{0, \rho, T_0}^1$, where

$$Y_{0, \rho, T_0}^j = \{W_0 \in C_b^0([0, T_0(\mu)], X^j) \mid |W_0|_{\rho, j, T_0(\mu)} < \infty\}$$

$$D = \{W_0 \in Y_{0, \rho, T_0}^1 \mid |W_0|_{\rho, j, T_0(\mu)} \leq \delta_0 g_0 \mu^{l(1-2j)}\} \cap \{W_0(0) = W_0^0\} \quad .$$

197

We denote the right–hand side of (3.52) by $\mathcal{K}_0 = \mathcal{K}_0(\mu, W_0, \underline{W}_1^0)$. To estimate it we need the Lemmata 3.1, 3.3 and 3.4. We can apply Theorem 3.9 and its Corollaries 3.10 and 3.11, since $W_0 \in D$ and (5.1–2) holds. Also we need (3.20) to estimate the L_∞–bounds. We summarize the essential inequalities

$$|W_0|_{\rho, j, T_0} \le g_0 \delta_0 \mu^{l(1-2j)} \quad , \quad \sup_{0 \le t \le T_0} \|W_0(t)\|_{\infty, a^j} \le \sqrt{2} g_0 \delta_0$$

$$|M_1^*|_{\rho, j, T_0} \le \widehat{C}_{37} \delta^* \mu^{l(1-2j)}$$

$$\rho(t)\|\underline{M}_1^*(t)\|_{\mathcal{X}^j} \le c_1 \mu^2 \mu^{l(1-2j)} \left(\delta_0^2 g_0^2 + \mu^{-2} \left(\delta_1^{0^2} + \mu \delta_0 g_0 \delta_1^0 \right) \exp \left(-\frac{7\alpha}{8\mu^2} t \right) \right)$$

$$\|K_1(\mu, t)\underline{W}_1^0\|_{\mathcal{X}^j} \le c_2 \delta_1^0 \mu^{-1} \exp \left(-\frac{\alpha}{\mu^2} t \right) \mu^{l(1-2j)} \quad ,$$

$$(5.12)$$

where $0 \le t \le T_0(\mu)$.

Now using (3.20), (5.4) and applying Lemma 3.3, we estimate $\mathcal{K}_0$ as follows

$$\rho(t)\|\mathcal{K}_0(t)\|_{X^j} \le \mu^{l(1-2j)} \left\{ \frac{1}{2} g_0 \delta_0 + c_3 \int_0^t e^{-\frac{b_1}{2}(t-s)} \left[g_0^2 \delta_0^2 + \delta_1^0 \mu^{-1} \exp \left(-\frac{\alpha}{2\mu^2} s \right) + \right. \right.$$

$$\left. \left. + \left(g_0 \delta_0 \delta_1^0 \mu^{-1} + (\delta_1^0)^2 \mu^{-2} \right) \exp \left(-\frac{\alpha}{\mu^2} s \right) \right] ds \right\} \quad .$$

$$(5.13)$$

The first term under the integral has to be controlled by $T_0(\mu)$. All other terms under the integral gain a factor μ^2 by integration. Referring to (5.8) we finally obtain

$$\rho(t)\|\mathcal{K}_0(t)\|_{X^j} \le \mu^{l(1-2j)} \left\{ \frac{g_0 \delta_0}{2} + c_4 \left(T_0(\mu) g_0^2 \delta_0^2 + \mu g_0 \delta_0 \delta_1^0 + (\delta_1^0)^2 \right) \right\}$$

$$\le \mu^{l(1-2j)} g_0 \delta_0 \quad ,$$

$$(5.14)$$

for μ_1^* and δ_1^0 small enough. Thus $\mathcal{K}_0$ maps D into itself.

To prove that $\mathcal{K}_0$ is a strict contraction we have to indicate the dependence of $\mathcal{K}_0$ on W_0. From (5.4), we obtain for $W_0, \widetilde{W}_0 \in D$

$$\rho(t)\|K_0(\mu, t) \left(W_0 - \widetilde{W}_0 \right) \|_{X^1} \le \widetilde{C}_{30} t^{-1/2} e^{-b_1 t} \|W_0 - \widetilde{W}_0\|_{X^0} \quad .$$

Moreover, we apply Lemma 3.4 as well as (3.45) to conclude

$$\rho(t)\| \left(\mathcal{K}_0(W_0) - \mathcal{K}_0(\widetilde{W}_0) \right)(t)\|_{X^1} \le \qquad\qquad\qquad (5.15)$$

$$\le c_5 \int_0^t \frac{e^{-\frac{b_1}{2}(t-s)}}{\sqrt{t-s}} \left\{ g_0 \delta_0 + \frac{\delta_1^0}{\mu} \exp \left(-\frac{\alpha}{2\mu^2} s \right) \right\} ds \, |W_0 - \widetilde{W}_0|_{\rho, 0, T_0}$$

198

$$\leq \ \tilde{c}_5 \, | \, W_0 - \widetilde{W}_0 \, |_{\rho, 0, T_0} \left\{ T_0(\mu)^{1/2}(g_0 \delta_0) + \delta_1^0 \right\} \quad .$$

Again, the term in the bracket can be made small by choosing μ_1^*, δ_1^0 appropriately. Thus, $\mathcal{K}_0$ is a strict contraction in D. Therefore, there exists a unique solution of (3.52) in D for $t \in [0, T_0(\mu)]$. Hence, the first part of the theorem is proved. A similar argument works for $W_0^*(t)$. Therefore the above inequalities hold for W_0^* as well.

One could show that $W_0(t)$ is a Lipschitz–continuous function of $\underline{W}_1^0$ by applying Lemma 3.5. Since we do not need this result, we omit the proof.

The estimate (5.9b) is a direct consequence of (5.8) and (5.12). For the second part of the assertion we choose $\eta \in (0, 1]$ and δ_0, δ_1^0 as given in the first part of the theorem. We require $W_0 \in D$ and $\|\underline{W}_1^0\|_{\mathcal{X}^j} \leq \eta \delta_1^0 \mu^{l(1-2j)-1}$. Let $W_0(t)$ be the solution of (3.52) for $0 \leq t \leq T_0(\mu)$; and $W_0^*(t)$ be the solution of (4.1) with $W_0^*(0) = W_0(0)$. Then we obtain

$$(W_0 - W_0^*)(t) = \int_0^t K_0(\mu, \, t - s) \Big\{ 2 \left(U_{00}^* - U_{\mu 0}^* \right) W_0^*(s) +$$

$$+ Q(W_0)(s) - Q(W_0^*)(s) + H_0(\mu, \, W_0, \, \underline{W}_1^0)(s) \Big\} ds \quad .$$

To estimate the right–hand side we use Lemma 2.1 for $|U_{\mu 0}^* - U_{00}^*| = O(\mu^2)$, Lemma 3.3 for Q and (5.13) for the H_0–term. It follows

$$| \, W_0 - W_0^* \, |_{\rho, j, T_0} \leq c_6 \mu^{l(1-2j)} \left(T_0(\mu) \delta_0^2 g_0^2 + \delta_1^0 \eta \right) \quad . \tag{5.16}$$

We see from (5.16) and from (5.8) that, for δ_1^0 small enough, we can choose $\mu_1^*(\eta) \in (0, \mu_1^*]$ so that, for $\mu \leq \mu_1^*(\eta)$, (5.10) holds. ∎

5.2 Long–time or slow flow result

We know now that the solution $\underline{W}(t)$ which has been constructed in Theorem 5.2, exists in $[0, T_0(\mu)]$ and satisfies (5.11). Set $\tilde{d}_0 = d_0 - \eta$ and choose $\eta > 0$ so small that $\tilde{d}_0 > 0$. With the help of the solution $\widetilde{W}_0^*$ of (4.1), satisfying the initial condition $\widetilde{W}_0^*(0) = W_0(T_0(\mu))$, we shall prove global existence and exponential asymptotic decay under the one sided restriction (5.11). A peculiar situation arises: we have to distinguish the cases $\tilde{d}_0 \in (0, \frac{1}{2}]$, $\tilde{d}_0 > \frac{1}{2}$. Our program can be achieved in the first case if $g_0(\mu)$ is further severely restricted, essentially to logarithmic growth in μ,

199

whereas in the second case $g_0(\mu)$ can grow like μ^{-p_0}, where $0 < p_0 < \min(\frac{1}{2}, \frac{2(3l+2)}{11})$. The decisive point is the control of a part of the quadratic term, which, in the second case, can be achieved by the linear part.

We set $\tilde{\mu}_1 = \mu_1^*(\eta)$, where $\mu_1^*(\eta)$ is given in Theorem 5.2. The solution $\widetilde{W}(t)$ of (3.5), which satisfies $\widetilde{W}(0) = \underline{W}(T_0(\mu)) \equiv \widetilde{W}^0$, exists locally in time. Moreover, we have $W_0(T_0(\mu)) \geq (1 - \tilde{d}_0)U_1^0$, where $\tilde{d}_0 \in (0, 1)$. We are treating now the first case $d_0 \in (0, \frac{1}{2}]$. Subsequently, we make the following assumption:

$$n_1 \left(\delta_0 g_0(\mu) + \mu^{2l}(\delta_0 g_0(\mu))^2 \right) < \delta^* \mu^{-1} , \tag{5.17}$$

for $0 < \mu \leq \tilde{\mu}_1$, where the constant n_1 satisfies $n_1 \geq 4$. It is defined below. This assumption further restricts $g_0(\mu)$, if $0 < l < \frac{1}{2}$ holds. However, the forthcoming analysis will show that it is obsolete.

$\widetilde{W}(t)$ is a continuous function with values in $\mathcal{X}^1$. Therefore, there is a positive time $T_1(\mu)$ such that for $t \in [0, T_1(\mu))$

$$\begin{aligned} \rho(t)\|\widetilde{W}_0(t)\|_{\mathcal{X}^j} &< n_1 \delta_0 g_0(\mu) \mu^{l(1-2j)} \left(1 + j\delta_0 g_0 \mu^{2l} \right) \\ \rho(t)\|\widetilde{W}_0(t)\|_{\infty, a^k} &< n_1 \delta_0 g_0(\mu) , \quad k = 0, 1 \end{aligned} \tag{5.18}$$

holds. If $T_1(\mu) < \infty$, then at least one of the four inequalities is an equality. Our aim is to show $T_1(\mu) = +\infty$. In $[0, T_1(\mu)]$ we can apply Theorem 3.9 and write $\widetilde{W}(t)$ as $(\widetilde{W}_0(t), K_1(\mu, t)\widetilde{W}_1^0 + \underline{M}_1^*(\mu, \widetilde{W}_0, \widetilde{W}_1^0))$, where $\widetilde{W}_0(t)$ solves (3.52) and $\widetilde{W}_1^0$ is bounded by (5.9b).

We shall use $\widetilde{W}_0^*$ as a guiding function, i.e. we shall compare $\widetilde{W}_0$ with $\widetilde{W}_0^*$. Recall that $\widetilde{W}_0^*(t)$ solves (4.1) with initial condition $\widetilde{W}_0^*(0) = W_0(T_0(\mu))$. Due to Theorem 4.4, Propositions 4.5, 4.6 and to the inequality $W_0(T_0(\mu)) \geq (1 - \tilde{d}_0)U^0$, $\widetilde{W}_0^*$ exists for all time and satisfies

$$\begin{aligned} \|\widetilde{W}_0^*(t)\|_{\mathcal{X}^j} &\leq C_{53} e^{-\tilde{b}_1 t} \delta_0 g_0 \mu^{l(1-2j)} \left(1 + j e^{-\tilde{b}_1 t} \delta_0 g_0 \mu^{2l} \right) \\ \|\widetilde{W}_0^*(t)\|_{\infty, a^k} &\leq C_{53} e^{-\tilde{b}_1 t} \delta_0 g_0 , \quad k = 0, 1 \end{aligned} \tag{5.19}$$

where $\tilde{b}_1 < \min(\tilde{d}_0, b_1)$. We choose $\rho(t) = \exp(\frac{\tilde{b}_1}{2}t)$, and $n_1 = 4C_{53}$, $C_{53} \geq 1$. Applying Corollaries 3.10, 3.11 and also (5.9b), we obtain the estimates

$$\begin{aligned} \rho(t)\|\underline{M}_1^*(t)\|_{\mathcal{X}^j} &\leq c_7 \mu^{l(1-2j)} \mu^2 n_1^2 \delta_0^2 g_0^2(\mu) \left(1 + j\mu^{2l} \delta_0 g_0(\mu) \right), \\ \rho(t)\|\underline{M}_1^*(t)\|_{\infty, a^k} &\leq c_7 \mu^2 n_1^2 \delta_0^2 g_0^2(\mu) \left(1 + \mu^l (\delta_0 g_0(\mu))^{1/2} \right) , \quad k = 0, 1. \end{aligned} \tag{5.20}$$

200

Now $\widetilde{\mu}_1$ is chosen so that $\underline{\widetilde{W}}_1 = K_1 \widetilde{W}_1^0 + \underline{M}_1^*(\mu, \widetilde{W}_0, \widetilde{W}_1^0)$ satisfies

$$
\begin{aligned}
\|\underline{\widetilde{W}}_1(t)\|_{\mathcal{X}^j} &\leq c_8 \mu^{l(1-2j)} \mu^2 \delta_0^2 g_0^2(\mu) \left[1 + n_1^2 \left(1 + \mu^{2l} \delta_0 g_0(\mu)\right)\right] \\
&< \delta^* \mu^{l(1-2j)}
\end{aligned}
\tag{5.21}
$$

for $\mu \in (0, \widetilde{\mu}_1]$.

Theorem 5.3 *There exist positive constants p and $\widetilde{\mu} \leq \min(\widetilde{\mu}_1, \mu_1^*)$ such that $g_0(\mu) = \log(\mu^{-p})$ satisfies (5.2), (5.17) for $\mu \in (0, \widetilde{\mu})$ and such that, with this choice of g_0, the equation (3.5) has a unique global solution $\underline{W}(t)$, $t \in [0, \infty)$, for initial values satisfying (5.1). And $\underline{W}(t)$ belongs to $\mathcal{Y}_\rho^1$ with $\rho(t) = \exp(\frac{b_1}{2} t)$.*

Proof. It suffices to show that $T_1(\mu) = +\infty$. We prove this by contradiction. Assume $\sup_n T_1(\mu_n) < +\infty$ for a sequence $\mu_n \to +0$. Set $\widetilde{W}_0 = \widetilde{W}_0^* + \widetilde{V}_0$. Using (3.52) we obtain

$$
\begin{aligned}
\widetilde{V}_0(t) = \int_0^t K_0(\mu, t-s) \Big\{ &2\left(U_{00}^*(s) - U_{\mu 0}^*(s)\right) \widetilde{W}_0^*(s) + \\
&+ Q\left(\widetilde{W}_0\right)(s) - Q\left(\widetilde{W}_0^*\right)(s) + H_0\left(\mu, \widetilde{W}_0 + \underline{\widetilde{W}}_1\right)(s) \Big\} ds
\end{aligned}
\tag{5.22}
$$

where $\widetilde{W}_0$, $\widetilde{W}_0^*$, $\underline{\widetilde{W}}_1$ are estimated in (5.18), (5.19) and (5.21). We conclude that

$$
\begin{aligned}
\rho(t)\|\widetilde{V}_0(t)\|_{X^0} \leq c_1 \int_0^t e^{-\frac{\bar{b}_1}{2}(t-s)} \Big\{ &\rho(s)\|\widetilde{V}_0(s)\|_{X^0} \left[3\|\widetilde{W}_0^*(s)\|_{\infty, a} + \right. \\
&\left. + \|\widetilde{W}_0(s)\|_{\infty, a}\right] + 2C_0^* \mu^2 \rho(s)\|\widetilde{W}_0^*(s)\|_{X^0} + \rho(s)\|H_0(\mu, \widetilde{W}_0 + \underline{\widetilde{W}}_1)(s)\|_{X^0} \Big\} ds
\end{aligned}
$$

and therefore, by (5.18), (5.19), (5.21) and Lemma 3.3,

$$
\rho(t)\|\widetilde{V}_0(t)\|_{X^0} \leq c_2 C_{53} \delta_0 g_0 \int_0^t e^{-\frac{\bar{b}_1}{2} s} \rho(s)\|\widetilde{V}_0(s)\|_{X^0} ds + c_3 \mu^{l+2} n_1^3 \delta_0^3 g_0^3 \left(1 + \mu^l (\delta_0 g_0)^{1/2}\right) \quad .
$$

Applying the Gronwall Lemma yields

$$
\rho(t)\|\widetilde{V}_0(t)\|_{X^0} \leq c_3 \mu^{l+2} n_1^3 \delta_0^3 g_0^3 \left(1 + \mu^l (\delta_0 g_0)^{1/2}\right) \exp\left(\frac{2}{b_1} c_2 C_{53} \delta_0 g_0\right) \quad .
\tag{5.23}
$$

Similarly we estimate

$$\rho(t)\|\widetilde{V}_0(t)\|_{X^1} \le c_1 \int_0^t e^{-\frac{\widetilde{b}_1}{2}(t-s)}\rho(s)\|\widetilde{V}_0(s)\|_{X^1}\rho^{-1}(s)\Big[\rho(s)\|\widetilde{W}_0^*(s)\|_{\infty,a}+$$

$$+\ \rho(s)\|\widetilde{W}_0(s)\|_{\infty,a}\Big]ds + c_1 \int_0^t e^{-\frac{\widetilde{b}_1}{2}(t-s)}(t-s)^{-1/2}\Big[2C_0^*\mu^2\rho(s)\|\widetilde{W}_0^*(s)\|_{X^0}+$$

$$+\ 2\|\widetilde{W}_0^*\|_{\infty,a}\rho(s)\|\widetilde{V}_0(s)\|_0 + \rho(s)\|H_0\left(\mu,\widetilde{W}_0+\underline{\widetilde{W}}_1\right)(s)\|_{X^0}\Big]ds \quad .$$

Using (5.23) yields

$$\rho(t)\|\widetilde{V}_0(t)\|_{X^1} \ \le \ c_4C_{53}\delta_0 g_0 \int_0^t e^{-\frac{\widetilde{b}_1}{2}s}\rho(s)\|\widetilde{V}_0(s)\|_{X^1}ds+$$

$$+c_3\mu^{l+2}n_1^3\delta_0^3 g_0^3\left(1+\mu^l(\delta_0 g_0)^{1/2}\right)+$$

$$+c_5\mu^{l+2}n_1^3\delta_0^4 g_0^4\left(1+\mu^l(\delta_0 g_0)^{1/2}\right)\exp\left(c_7^0\delta_0 g_0\right)$$

where $c_7^0 = 2c_2 C_{53}/\widetilde{b}_1$. Finally applying the Gronwall Lemma again gives

$$\rho(t)\|\widetilde{V}_0(t)\|_{X^1} \ \le \ c_6\mu^{l+2}n_1^3\delta_0^4 g_0^4\left(1+\mu^l(\delta_0 g_0)^{1/2}\right)\exp\left(c_7\delta_0 g_0\right) \tag{5.24}$$

where $c_7 = c_7^0 + 2c_4 C_{53}/\widetilde{b}_1$.

The estimates hold for all $\mu \in (0, \widetilde{\mu}]$. Set $\Gamma = c_7\delta_0$, $\Gamma^0 = c_7^0\delta_0$, then we have

$$\rho(t)\|\widetilde{V}_0(t)\|_{X^0} \ \le \ c_3 n_1^3\delta_0^3\left(\log(\mu^{-p})\right)^3\left(1+\mu^l\delta_0^{1/2}\left(\log\mu^{-p}\right)^{1/2}\right)\mu^{l+2-\Gamma^0 p}$$

$$\rho(t)\|\widetilde{V}_0(t)\|_{X^1} \ \le \ c_6 n_1^3\delta_0^4\left(\log(\mu^{-p})\right)^4\left(1+\mu^l\delta_0^{1/2}(\log\mu^{-p})^{1/2}\right)\mu^{l+2-\Gamma p} \quad . \tag{5.25}$$

We choose $p < \min(2(\Gamma^0)^{-1}, (2l+2)\Gamma^{-1})$ and reach a contradiction if $\sup_n T_1(\mu_n) < +\infty$.

5.3   Improved global stability for $d_0 > \frac{1}{2}$

If we require $d_0 > 1/2$ in (5.1), then we may take $\eta > 0$ so that $d_0 - \eta = \widetilde{d}_0 > \frac{1}{2}$ in (5.11) of Theorem 5.2. For proving global existence we proceed like in the preceding section; however the weight function $\widetilde{\rho}(t)$ is chosen more appropriately than before. As long as the solution $\underline{\widetilde{W}}(t)$ of (3.5) with $\underline{\widetilde{W}}(0) = \underline{W}(T_0(\mu))$ exists and satisfies (5.18), we can apply Theorem 3.9 and write $\widetilde{W}(t) = (\widetilde{W}_0(t), K_1(\mu, t)\widetilde{W}_1^0 + \underline{M}_1^*(\mu, \widetilde{W}_0, \widetilde{W}_1^0))$. As before, we set $\widetilde{W}(t) = \widetilde{W}_0^*(t) + \widetilde{V}_0(t)$. In contrast to (5.22) we incorporate the term $2a\widetilde{W}_0^*(t)$ into the linear operator. Thus we have

$$
\begin{aligned}
\partial_t \widetilde{V}_0 + \widetilde{L}_{\mu 0}(t)\widetilde{V}_0 \;=\; & 2\left(U_{00}^* - U_{\mu 0}^*\right)\widetilde{W}_0^* - a\widetilde{V}_0^2 + \\
& + H_0\left(\mu, \widetilde{W}_0 + K_1(\mu, t)\widetilde{W}_1^0 + \underline{M}_1^*\left(\mu, \widetilde{W}_0, \widetilde{W}_1^0\right)\right)
\end{aligned}
\tag{5.26}
$$

where

$$
\widetilde{L}_{\mu 0} \equiv -\partial_{xx}^2 + 2U_{\mu 0}^* + b_0 + 2a\widetilde{W}_0^*(t) \quad .
$$

Remark that $\|U_{\mu 0}^* - U_{00}^*\|_{C_b^2(\mathbb{R})} \le C_0^* \mu^2$ holds, and that, since $U^0 = -U_{00}^*/a$,

$$
2\left(U_{00}^*(x) + a\widetilde{W}_0^*(x, t)\right) \ge 2\widetilde{d}_0 U_{00}^* > U_{00}^*(x) \quad .
\tag{5.27}
$$

Let $\widetilde{K}_0 = \widetilde{K}_0(\mu, t, s)$ be the linear evolution operator associated with the equation

$$
\frac{dV_0}{dt} + \widetilde{L}_{\mu 0}(t)V_0 = 0 \quad , \quad V_0(s) = V_0^0 \quad .
\tag{5.28}
$$

This operator is well defined on X^0 and X^1 (see [He80], chapter 7 for instance) and the solution $\widetilde{V}_0$ of (5.26) solves the integral equation

$$
\widetilde{V}_0(t) = \int_0^t \widetilde{K}_0(\mu, t, s)\left[2\left(U_{00}^* - U_{\mu 0}^*\right)\widetilde{W}_0^* - a\widetilde{V}_0^2 + \right.
\tag{5.29}
$$

$$
\left. + H_0\left(\mu, \widetilde{W}_0 + K_1(\mu, \cdot)\widetilde{W}_1^0 + \underline{M}_1^*\left(\mu, \widetilde{W}_0, \widetilde{W}_1^0\right)\right)(s)\right]ds
$$

since $\widetilde{V}_0(0) = 0$. We estimate $\widetilde{K}_0$ as in Lemma 3.1.

Lemma 5.4 *Let* $\epsilon > 0$ *satisfy* $\epsilon < \min(2\widetilde{d}_0 - 1, b_0)$ *and* $\widetilde{b} = \min(2\widetilde{d}_0 - 1 - \epsilon, b_0 - \epsilon)$. *Then there are positive constants* $\widetilde{\mu}_2 \le \widetilde{\mu}_1$ *and* C_{54} *such that*

$$
\|\widetilde{K}_0(\mu, t, s)V_0^0\|_{X^j} \le C_{54}(t - s)^{\frac{(m-j)}{2}}(\delta_0 g_0(\mu))^j e^{-\widetilde{b}(t-s)}\|V_0^0\|_{X^m}
\tag{5.30}
$$

holds for $\mu \in (0, \widetilde{\mu}_2]$, $0 \le m \le j \le 1$. *The constants* $\widetilde{\mu}_2$ *and* C_{54} *depend on* ε.

Proof. It suffices to prove the lemma for $s = 0$. We proceed similarly as in Lemma 3.1. Taking the inner product of (5.28) with V_0 one obtains, for $\mu > 0$ small enough,

$$
\frac{d}{dt}\|V_0\|_0^2 + 2\|V_{0x}\|_0^2 + 2\left(b_0 - \frac{\epsilon}{4}\right)\|V_0\|_0^2 \le 0
\tag{5.31}
$$

203

and

$$2 \int_t^{t+1} \|V_{0x}\|_0^2(s)ds \le \|V_0(t)\|_0^2 \quad .$$

Applying the Gronwall Lemma we get

$$\|V_0(t)\|_0^2 \le e^{-2(b_0 - \frac{\epsilon}{4})t}\|V_0^0\|_0^2 \quad . \tag{5.32}$$

Take now the inner product of (5.28) with $(-V_{0xx})$ and use (5.19)

$$\frac{d}{dt}\|V_{0x}\|_0^2 + \|V_{0xx}\|_0^2 + 2b_0\|V_{0x}\|_0^2 \le 4\left(1 + \frac{\epsilon}{4} + C_{53}\delta_0 g_0(\mu)\right)^2 \|V_0\|_0^2 \quad .$$

Applying the uniform Gronwall Lemma and using (5.31) and (5.32) yields for $t > 0$

$$\|V_{0x}(t)\|_0 \le t^{-\frac{(1-j)}{2}} c_1 e^{-(b_0 - \frac{\epsilon}{2})t} \delta_0 g_0(\mu)\|V_0^0\|_j \quad . \tag{5.33}$$

Equation (5.28) can also be written as

$$\frac{d}{dt}V_0 - \frac{1}{a^2}\left(a^2 V_{0x}\right)_x + \gamma V_{0x} + \left(b_0 + 2\left(U_{\mu 0}^* - U_{00}^*\right) + 2U_{00}^* + 2a\widetilde{W}_0^*(t)\right)V_0 = 0 \quad . \tag{5.34}$$

Take the inner product with $a^2 V_0$ to obtain

$$\frac{d}{dt}\|V_0\|_{0,a}^2 + 2\int_{I\!R} a^2 \left(2\tilde{d}_0 U_{00}^* - 1 - 2C_0^*\mu^2\right)V_0^2 \le 0 \quad . \tag{5.35}$$

Since $U_{00}^*(-\infty) = 1$, we can find $\tilde{\beta} = \tilde{\beta}(\tilde{d}_0, \epsilon)$ such that

$$2\tilde{d}_0 U_{00}^*(\tilde{\beta}) - 1 \ge 2\tilde{d}_0 - 1 - \frac{\epsilon}{4} \quad .$$

Thus, for μ small enough and $x \in (-\infty, \tilde{\beta}]$, we have

$$2\tilde{d}_0 U_{00}^*(x) - 1 - 2C_0^*\mu^2 \ge 2\tilde{d}_0 - 1 - \frac{\epsilon}{2} \quad . \tag{5.36}$$

From (5.35) and (5.36), we derive that

$$\frac{d}{dt}\|V_0\|_{0,a}^2 + 2\left(2\tilde{d}_0 - 1 - \frac{\epsilon}{2}\right)\|V_0\|_{0,a}^2 \le$$
$$\le 2\left(1 + \frac{\epsilon}{4} + 2\tilde{d}_0\right)e^{\tilde{\beta}\gamma}\|V_0\|_0^2 \le 6e^{-2(b_0 - \frac{\epsilon}{4})t}e^{\tilde{\beta}\gamma}\|V_0^0\|_0^2 \quad ,$$

and, due to the Gronwall Lemma, we infer

$$\|V_0\|_{0,a}^2 \le c_2 \left(e^{-2(b_0 - \frac{\epsilon}{2})t} + e^{-2(2\tilde{d}_0 - 1 - \frac{\epsilon}{2})t}\right)\|V_0^0\|_{X^0}^2 \quad . \tag{5.37}$$

204

If we take the inner product of (5.34) with $a^2 V_0$, use the inequality

$$2|\gamma \int_{I\!R} a^2 V_0 V_{0x} \, dx| \leq \frac{2\gamma^2}{2\tilde{d}_0 - 1 + \gamma^2} \|V_{0x}\|_{0,a}^2 + \frac{2\tilde{d}_0 - 1 + \gamma^2}{2} \|V_0\|_{0,a}^2$$

and take into account the properties (5.36) and (5.37), we obtain

$$\int_t^{t+1} \|V_{0x}\|_{0,a}^2 \, ds \leq c_3 \left(e^{-2(b_0 - \frac{\xi}{2})t} + e^{-2(2\tilde{d}_0 - 1 - \frac{\xi}{2})t} \right) \|V_0(0)\|_{X^0}^2 \quad . \tag{5.38}$$

Now take the scalar product of (5.34) with $-(a^2 V_{0x})_x$ and obtain

$$\frac{d}{dt} \|V_{0x}\|_{0,a}^2 + 2 \int_{I\!R} \frac{1}{a^2} \left((a^2 V_{0x})_x \right)^2 dx - 2\gamma \int_{I\!R} (a^2 V_{0x})_x V_{0x} \, dx + 2b_0 \|V_{0x}\|_{0,a}^2 \leq$$

$$\leq \|2U_{\mu 0}^* + 2a\widetilde{W}_0^*\|_\infty^2 \|V_0\|_{0,a}^2 \quad .$$

Since

$$|2\gamma \int_{I\!R} (a^2 V_{0x})_x V_{0x} \, dx| \leq \gamma^2 \|V_{0x}\|_{0,a}^2 + \int_{I\!R} \frac{1}{a^2} \left((a^2 V_{0x})_x \right)^2 dx \quad ,$$

we deduce from the above inequality that,

$$\frac{d}{dt} \|V_{0x}\|_{0,a}^2 \leq c_4 \left(\delta_0 g_0(\mu) \right)^2 \|V_0\|_{0,a}^2 + \gamma^2 \|V_{0x}\|_{0,a}^2 \quad .$$

Applying the uniform Gronwall Lemma and taking into account the estimates (5.32), (5.33), (5.37) and (5.38), we finally obtain the estimate (5.30) for $j = 1$.

Now we choose a new weight function $\tilde{\rho}$

$$\tilde{\rho}(t) = \exp \left(\tilde{b} - \tilde{\epsilon} \right) t \quad , \quad 0 < \tilde{\epsilon} \leq \tilde{b}/4 \quad . \tag{5.39}$$

For each solution $\widetilde{W}(t)$ of (3.5), with $\widetilde{W}(0) = W(T_0(\mu))$, there exists a unique $T_1 = T_1(\mu)$ so that, for $0 \leq t \leq T_1(\mu)$, (5.18) holds, where ρ is replaced by $\tilde{\rho}$. On $[0, T_1(\mu)]$, we may apply Theorem 3.9; thus $\widetilde{W}_0(t)$ solves (3.51) and the equation (5.26) holds. Since $\tilde{d}_0 > 1/2$, we are able to apply Lemma 5.4 and show that $T_1(\mu) = \infty$, if

$$g_0(\mu) = \mu^{-p_0} \quad , \quad 0 < p_0 < \min \left(\frac{1}{2}, \frac{2(3l + 2)}{11} \right) \quad . \tag{5.40}$$

The estimates (5.20) and (5.21) are still valid, when $\rho(t)$ is replaced by $\tilde{\rho}(t)$.

Theorem 5.5 *Assume that $g_0(\mu)$ satisfies (5.40) and $\underline{\widetilde{W}}^0 = (\widetilde{W}_0^0, \widetilde{W}_0^1)$ the inequalities (5.9a), (5.9b) and (5.11). Then there is a positive number $\widetilde{\mu} \leq \min(\widetilde{\mu}_2, \mu_1^*)$ such that, for each $\mu \in (0, \widetilde{\mu}]$, a unique solution $\underline{\widetilde{W}}(t)$ of (3.5), with the initial value $\underline{\widetilde{W}}(0) = \underline{\widetilde{W}}^0$, exists on $[0, \infty)$, and $\underline{\widetilde{W}} \in \mathcal{Y}_{\widetilde{\rho}}^1$ holds.*

Remark. Theorem 5.5 implies immediately, that the initial-value problem (3.5) for $\underline{W}(t)$, with $\underline{W}(0) = \underline{W}^0$, where $\underline{W}^0$ satisfies (5.1), has a unique solution on $[0, \infty)$ and belongs to $\mathcal{Y}_{\widetilde{\rho}}^1$. The result follows from Theorems 5.2 and 5.5.

Proof of Theorem 5.5. We prove that $T_1(\mu) = \infty$. Assume the contrary. Then there exists a $T_2(\mu) < T_1(\mu)$, such that

$$\widetilde{\rho}(t)\|\widetilde{V}_0(t)\|_{X^j} < \frac{1}{\sqrt{2}}\delta_0\mu^{p_0}\mu^{l(1-2j)} \tag{5.41}$$

for all t, $0 \leq t < T_2(\mu)$ and at least one of the inequalities is an equality for $t = T_2(\mu)$. Applying the Lemmatas 3.3, 5.4 and Theorem 2.1, and using the properties (5.9), (5.20), (5.39), (5.40), we obtain for $0 \leq t < T_2(\mu)$

$$\widetilde{\rho}(t)\|\widetilde{V}_0(t)\|_{X^0} \leq c_5 \int_0^t e^{-\widetilde{\epsilon}(t-s)}e^{-(\widetilde{b}-\widetilde{\epsilon})s}\|a\widetilde{\rho}(s)\widetilde{V}_0\|_\infty \widetilde{\rho}(s)\|\widetilde{V}_0(s)\|_{X^0}\,ds +$$

$$+c_5 \int_0^t e^{-\widetilde{\epsilon}(t-s)}\left[2C_0^*\mu^2\widetilde{\rho}(s)\|\widetilde{W}_0^*(s)\|_{X^0} + \right.$$

$$\left. +\widetilde{\rho}(s)\|H_0(\mu, \widetilde{W}_0 + K_1(\mu,\cdot)\underline{\widetilde{W}}_1^0 + \underline{M}_1^*)(s)\|_{X^0}\right]\,ds$$

and hence

$$\widetilde{\rho}(t)\|\widetilde{V}_0(t)\|_{X^0} \leq c_5\delta_0\mu^{p_0} \int_0^t e^{-(\widetilde{b}-\widetilde{\epsilon})s}\widetilde{\rho}(s)\|\widetilde{V}_0(s)\|_{X^0}\,ds +$$

$$+c_6 n_1^3 \mu^{l+2}\delta_0^3 g_0^3\left(1 + \mu^l(\delta_0 g_0)^{1/2}\right) \quad.$$

Applying the Gronwall Lemma yields

$$\widetilde{\rho}(t)\|\widetilde{V}_0(t)\|_{X^0} \leq c_6 n_1^3 \mu^{l+2}\delta_0^3 g_0^3\left(1 + \mu^l(\delta_0 g_0)^{1/2}\right)\exp\left(\frac{4}{3\widetilde{b}}c_5\delta_0\mu^{p_0}\right) \quad. \tag{5.42}$$

Likewise, using (5.30) with $j = 1$, we obtain as in (5.24)

$$\widetilde{\rho}(t)\|\widetilde{V}_0(t)\|_{X^1} \leq c_7 n_1^3 \mu^{l+2}\delta_0^4 g_0^4\left(1 + \mu^l(\delta_0 g_0)^{1/2}\right)\exp\left(\frac{8}{3\widetilde{b}}c_5\delta_0^2\right) \tag{5.43}$$

206

for $0 \leq t \leq T_2(\mu)$. From (5.42), (5.43) and (5.40) we conclude that

$$\widetilde{\rho}(t) \|\widetilde{V}_0(t)\|_{X^j} \leq \mu^{l(1-2j)} \frac{1}{2\sqrt{2}} \delta_0 \mu^{p_0} < \mu^{l(1-2j)} \frac{1}{2} \delta_0 \mu^{p_0} \quad , \qquad \text{for } 0 \leq t \leq T_2(\mu) \quad .$$

This is in contradiction to the fact that one of the two strict inequalities in (5.41) should be an equality for $t = T_2(\mu)$. Hence $T_1(\mu) = T_2(\mu) = +\infty$, and the theorem is proved.

5.4   Remarks on the non–generic case $D^2_{\underline{u}_1,\underline{u}_1} g_0(0,0) = 0$.

As we have already remarked it in Section 3 (see Theorem 3.13), we can improve the global stability results by allowing $\|W^0_1\|_{X^1}$ to be of order $\delta\mu^{-4/3}$ if $D^2_{\underline{u}_1,\underline{u}_1} g_0(0,0) = 0$. The only changes in the proof occur in the short–time or fast flow argument.

More precisely, we assume here that

$$D^2_{\underline{u}_1 \underline{u}_1} g_0(0,0) = 0 \tag{5.44}$$

holds and that we are given initial data $\underline{W}^0 = (W^0_0, \underline{W}^0_1)$ such that

$$\|W^0_0\|_{X^1} \leq \frac{\delta_0}{2\widetilde{C}_{30}} g_0(\mu)$$

$$\|\underline{W}^0_1\|_{X^1} \leq \delta^0_1 \mu^{-p_1} \tag{5.45}$$

$$W^0_0 \geq (1 - d_0)U^0$$

where $0 < p_1 \leq 2$ and $d_0 \in (0,1]$, holds. We assume also that (5.2) is true.

Arguing as in the proof of Theorem 5.2, but using here the Reduction–Theorem 3.6 and its Corollaries 3.7 and 3.8, we prove the following result:

Let

$$T^*_0(\mu) = \frac{2\mu^2}{\alpha} |\log \frac{1}{\mu^{2+p_1} g^2_0 \delta^2_0}| \quad .$$

Theorem 5.6 *Let* $p_0 \in (0,1)$, $p_1 \in (0, \frac{4}{3}]$ *with* $p_0 + p_1 \leq 2$, $d_0 \in (0,1]$, $\eta > 0$ *be given. Let* g_0, δ_0 *be fixed satisfying* (5.2). *Then there are positive constants* $\delta^0_1 \leq \delta^*$ *and* $\mu^*_1 \leq \mu^*_0$ *such that, for every* $\mu \in (0, \mu^*_1]$ *and every* $\underline{W}^0$ *satisfying* (5.45), *a unique solution* $\underline{W}(t)$ *of* (3.5) *exists on the time interval* $[0, T_0(\mu)]$ *with* $\underline{W}(0) = \underline{W}^0$. *Moreover* $W_0(t)$ *and the solution* $W^*_0(t)$ *of* (4.1) *with initial condition* W^0_0 *obey the inequality*

$$|W_0|_{\rho,1,T^*_0(\mu)} \leq \delta_0 g_0(\mu) \quad . \tag{5.46a}$$

207

The function $\underline{W}_1(t)$ satisfies

$$\|\underline{W}_1(T_0^*(\mu))\|_{\mathcal{X}^1} \le C_{52}\mu^2 \left[\delta_0^2 g_0^2 + \mu^{4-4p_1} \left(\delta_1^0\right)^4\right] \quad. \tag{5.46b}$$

Furthermore, for every $\eta \in (0,1]$, there exists $\mu_1^*(\eta) \le \mu_1^*$ such that, if $\|\underline{W}_1^0\|_{\mathcal{X}_1^1} \le \delta_1^0 \eta \mu^{-p_1}$, we have, for all $t \in [0, T_0^*(\mu)]$,

$$\|W_0(t) - W_0^*(t)\|_{X^1} \le \frac{C_{51}}{\sqrt{2}}\eta \quad. \tag{5.47}$$

This implies, via Lemma 5.1, for $t \in [0, T_0^*(\mu)]$,

$$W_0(t) \ge (1 - (d_0 - \eta))U^0 \quad. \tag{5.48}$$

Using Theorem 5.6 and arguing like in the Sections 5.2 and 5.3, one proves the following global stability results

Theorem 5.7 *Assume that (5.44) holds. Then the statement of Theorem 5.3 is still true, for initial data satisfying (5.45), provided that* $0 < p_1 \le \frac{4}{3}$.

Likewise, if $d_0 > \frac{1}{2}$, *the statement of Theorem 5.5 still holds, for initial data satisfying (5.45), provided that* $0 < p_1 \le \frac{4}{3}$, $g_0(\mu) = \mu^{-p_0}$, *with* $0 < p_0 \le \frac{4}{11}$.

Chapter 4
A Spatial Center Manifold Approach to Steady Bifurcations from Spatially Periodic Patterns

by **A MIELKE**

Contents

1   Introduction

The appearance of spatially periodic patterns is a common feature of many physical, chemical, or biological systems, in particular when the underlying physical domain is sufficiently large compared to the wave length which is prefered by the system.

Here, we consider only one aspect, namely steady bifurcations associated to spatially periodic solutions of problems in infinite cylinders. The infinitely long cylinder is an idealization often done in physics or mathematics: it simplifies the analysis considerably but still matches experiments on large finite domains very well, e.g., in the Taylor–Couette problem. Some mathematical justification for this idealization is given in [Mi90] in the context of small bifurcating solutions and in [Mi94a, Mi96b] for large solutions. In the latter work the attractors for problems on large and unbounded domains are compared.

Typical applications comprise the flow of viscid or inviscid fluids in channels or pipes and periodic travelling waves in reaction–diffusion systems (so–called wave trains). The theory for bifurcations in the classes of periodic solutions (period doubling, subharmonic branching) is well developed using the Liapunov–Schmidt reduction, see e.g. [Va90, Va91]. The appearance of nonperiodic solutions was studied much less, even so some instability mechanisms exactly lead to such solutions, for example the Eckhaus instability which will be treated in Section 7.2. The present work establishes methods which help to fill exactly this gap: we show that center–manifold theory can be generalized in a suitable way when the axial variable plays the role of the time (spatial dynamics). Thus, we are able to characterize all patterns lying orbitally close to the original periodic one. This is done by reducing the elliptic problem to a finite dimensional ODE. Using normal form theory near closed orbits and perturbation arguments it is then possible to predict spatially periodic, quasiperiodic, or chaotic solutions as well as solutions which are asymptotic to periodic solutions at both infinities (defects or fronts).

The work concentrates on the infinite dimensional functional analytic problems concerning the reduction of the elliptic PDE to the finite dimensional ODE on the center manifold, in particular for applications in reaction–diffusion systems and in hydrodynamics. The discussion of the reduced system (normal forms, bifurcations, etc.) is only sketched in oder to connect the results from center manifold theory to previous results in finite dimensional dynamical systems theory.

The general approach developed here is the use of spatial center manifolds for elliptic systems in cylindrical domains. This tool was put forward by Kirchgässner [Ki82]; and since then it proved very effective in many areas, such as elasticity theory [Mi90], Navier–Stokes problems [IoMD89], and inviscid fluid flow including surface waves [IoK92, BrM95]. Up to now all applications of this type were restricted to autonomous systems, that is, the axial variable x does not appear explicitly. Small nonautonomous perturbations where allowed in [Mi86], but the linear part was autonomous.

To be more precise, let us consider the spatial dynamical system (i.e., a differential equation with respect to the spatial varaible $x \in \mathbb{R}$)

$$\frac{d}{dx}w = \mathcal{F}(w),$$

with w in some Hilbert space H, which has a periodic solution p with $p(x+T) = p(x)$. To study bifurcations from the orbit $\Gamma_{per} = \{\, p(x) \in H \,:\, x \in [0,T]\,\}$ we have to deal with the linearization around p:

$$Lu = \frac{d}{dx}u - \widetilde{A}(x)u = 0, \quad \text{where } \widetilde{A}(x) = D\mathcal{F}(p(x)).$$

Often we will write $\widetilde{A}(x)$ in the form $A + B(x)$, then A is the dominant autonomous part and $B(x)$ are lower order perturbations. Bifurcations from Γ_{per} are to be expected only if $Lu = 0$ allows for bounded solutions over $\mathbb{R}$. All necessary information on L can be obtained by studying the associated Floquet operator L_{per} acting on the space of periodic rather than bounded functions u. An eigenvalue λ of L_{per} is called Floquet exponent, and the Floquet multipliers are given by $\rho = e^{\lambda T}$. From $L_{per}u = \lambda u$ we find $L(e^{-\lambda x}u) = 0$, and hence, Floquet exponents on the imaginary axis are closely related to bifurcations.

In Section 2 we extensively study the properties of L_{per} and L in different functional spaces. A remarkable difficulty in elliptic problems is that L_{per} may have the whole complex plane as its spectrum. Fortunately, the case with discrete spectrum (and compact resolvent) is generic if the nonautonomous term $B(x)$ is lower order compared to the autonomous part A. For simple problems like the Schrödinger operator with periodic potential this exotic case can be excluded in general, see Section 6. Also for the Stokes problem of periodic surface waves we give a special formulation showing the discreteness of the spectrum, see Sect. 7.1.

Yet, for general applications in hydrodynamics or nonlinear elasticity we are not able to show that the spectrum is always discrete. Thus, we have to impose this as an assumption (which generically is true anyhow). In the appendix we provide an easy example showing that the non–generic behavior really exists even in semilinear systems.

Starting from a linear nonautonomous system $u' - \widetilde{A}(x)u = 0$ ($' = \frac{d}{dx}$) with discrete Floquet exponents we have to separate the critical exponents on the imaginary axis from the other ones. This means we are looking for a periodic projection

211

$P \in C^1([0,T], \mathcal{L}(H))$ commuting with L_{per} such that $L_{per}|_{\text{Range }P}$ has spectrum on $i\mathbb{R}$ while $L_{per}|_{\text{Range}(I-P)}$ has spectrum in $\{\lambda \in \mathbb{C} : |\text{Re }\lambda| > \delta\}$ for some $\delta > 0$. Note that $L_{per}P = PL_{per}$ is equivalent to $P' - \widetilde{A}P + P\widetilde{A} = 0$. Since the spectral part to be separated is unbounded the construction of P is not immediate. However, the generalized eigenspace at $i\omega$ is related to the eigenspace at $i(\omega + 2\pi k/T)$ simply by multiplying the eigenfunctions by $e^{i2\pi kx/T}$. Thus, the Dunford projector for the spectral set in $R_{\delta,n+1/2} = \{\lambda \in \mathbb{C} : |\text{Re }\lambda| < \delta, |\text{Im }\lambda| < (2n+1)\pi/T\}$, $n \in \mathbb{N}$, can be expressed in terms of that on $R_{\delta,1/2}$. Letting $n \to \infty$ the limit exists and equals the desired $P = P(x)$.

Further we derive several results for the linear operator L in spaces of functions which are p–integrable or bounded after multiplication with the exponential weight $e^{-\alpha|x|}$, $|\alpha| < \delta$. Since we have to deal with quasilinear problems later on, we need maximal regularity results as introduced in [Mi87]. Consider

$$u' - \widetilde{A}(x)u = g, \quad \text{with } e^{-\alpha|\cdot|}g(\cdot) \in L_p(\mathbb{R}, H),$$

where $p \in (1, \infty)$. Then, we prove $e^{-\alpha|\cdot|}u(\cdot) \in L_p(\mathbb{R}, D(A)) \cap W^{1,p}(\mathbb{R}, H)$ (see Lemma 2.5). This result relies on the resolvent estimate $\|(A-i\xi)^{-1}\|_{\mathcal{L}(H)} \leq C/(1+|\xi|)$, $\xi \in \mathbb{R}$, for the dominant linear part.

In Section 3 we apply the linear theory to nonlinear problems in a neighborhood of a periodic orbit $\Gamma_{per} = \{p(x) : x \in [0,T]\}$, where $p(x+T) = p(x)$. Studying bifurcations from Γ_{per} the standard ansatz $w(x) = p(x) + u(x)$ is not appropriate for finding orbitally close solutions, e.g., if w is periodic with a slightly different period T_1 then u would be quasiperiodic and not small. It is more convenient to use the local coordinate change $w = p(\tau) + v$ with $\langle q(\tau), v \rangle = 0$, where q is also T–periodic and $\langle q(\tau), p'(\tau) \rangle = 1$ for all τ. Now, τ is the coordinate along the orbit Γ_{per} and v is transversal to it. Hence, "orbitally close" is equivalent to v being small. We use τ as new independent (time–like) variable instead of x and obtain

$$\frac{d}{d\tau}x = 1 + a(\tau, v) = \frac{1 - \langle q'(\tau), v \rangle}{\langle q(\tau), \mathcal{F}(p(\tau) + v) \rangle},$$

$$\frac{d}{d\tau}v = \widehat{\mathcal{F}}(\tau, v) = (1 + a(\tau, v))\mathcal{F}(p(\tau) + v) - \mathcal{F}(p(\tau)), \quad \langle q(\tau), v(\tau) \rangle = 0.$$

By construction, the functions a and $\widehat{\mathcal{F}}$ are T–periodic in τ. Note that $\widehat{\mathcal{F}}(\tau, v)$ is quasilinear in v even for semilinear $\mathcal{F}(w)$.

Using the spectral splitting constructed above ($v_0(\tau) = P(\tau)v(\tau)$, $v_1(\tau) = v(\tau) -$

$v_0(\tau)$), the v–equation can be linearly decoupled:

$$\frac{d}{d\tau}v_0 - \widehat{A}_0(\tau)v_0 = \mathcal{N}_0(\tau, v_0 + v_1),$$

$$\frac{d}{d\tau}v_1 - \widehat{A}_1(\tau)v_1 = \mathcal{N}_1(\tau, v_0 + v_1).$$

After some manipulations we are able to apply the center manifold theorem to the equation for v. We obtain a compact center manifold

$$\widehat{\mathcal{M}}_C = \{\, p(\tau) + v_0 + h(\tau, v_0) \in D(A) \,:\, P(\tau)v_0 = v_0,\ \langle q(\tau), v_0(\tau)\rangle = 0,\ \|v_0\| \leq \varepsilon \,\}$$

where $h(\tau, v_0) = h(\tau + T, v_0) = \mathcal{O}(\|v_0\|^2)$ and $\tau \in S^1 = \mathbb{R}/T\mathbb{Z}$. This manifold contains the original periodic orbit Γ_{per} and all solutions $w : \mathbb{R} \to D(A)$ which are orbitally close to Γ_{per}. All bifurcating solutions are now found by studying the reduced problem

$$\frac{d}{d\tau}x = 1 + a(\tau, v_0 + h(\tau, v_0)), \quad \frac{d}{d\tau}v_0 - \widehat{A}_0(\tau)v_0 = \mathcal{N}_0(\tau, v_0 + h(\tau, v_0)),$$

which is an ODE since Range $P(x)$ is finite–dimensional.

In Section 4 we concentrate on reversible and Hamiltonian systems. In such systems periodic orbits generically appear in families of periodics with a smoothly varying period. If the elliptic problem on the cylinder has a reflection symmetry with respect to $x \to -x$, then the associated differential equation is reversible. This means that there is a linear operator $R \in \mathcal{L}(H)$ with $R^2 = I$ such that $\mathcal{F}(Rw) = -R\mathcal{F}(w)$. A periodic orbit p is called *reversible* if there is an x_0 such that $Rp(x_0 + x) = p(x_0 - x)$ for all x. We show that a spatial center manifold for a reversible orbit can be chosen reversible also, i.e., $R\widehat{\mathcal{M}}_C = \widehat{\mathcal{M}}_C$. Moreover, on $\widehat{\mathcal{M}}_C$ coordinates exist such that the reduced reversibility operation is diagonal with entries ± 1. A well–known example of this kind are the Taylor vortices to be studied in Section 7.2.

If the elliptic problem was derived from a variational problem, then we can associate to it a Hamiltonian structure. This idea was introduced in [Mi91] and exploited in the case of center manifolds at fixed points or group orbits if a symmetry action is present. We generalize the method to the case of periodic orbits here. Thus, we are able to recover the Hamiltonian structure of the reduced ODE. This is useful in many respects. First of all we have the energy of Hamiltonian function which is a conserved quantity. Moreover, for Hamiltonian systems there is a powerful machinery for studying bifurcations from periodic orbits. For an interesting application we refer to [BM92, BrM95] and Section 7.1 where bifurcations from the Stokes waves are investigated.

An important question related to the bifurcation analysis of steady patterns is the question of stability of these patterns. Assuming that the basic periodic pattern is neutrally stable for parameters below a certain threshold, one often sees an exchange of stability from the basic pattern to one of the bifurcating branches. For problems on unbounded cylinders recently a new approach was developed in [BrM96, BrM95], which is based on the principle of reduced instability. The idea is to study the spectrum of the linearization around the new steady state again as a (linear) spatial dynamical system which is then again amenable to the spatial center manifold reduction. In Section 5 we generalize these ideas to the present situation where the basic state is the periodic orbit Γ_{per}. We are able to study the linearization of all steady solutions which are orbitally close to Γ_{per}.

In Section 6 we study spatially periodic travelling waves in reaction–diffusion systems. For constant and diagonal diffusion matrices we show that the Floquet spectrum around any periodic traveling wave is discrete. As a consequence the existence of center manifolds is guaranteed as long as the nonlinearities do not depend on derivatives of the concentrations. Furthermore, we give a specific example where the Floquet operator decouples and can be studied via Hill's theory. The example is reversible and Hamiltonian such that $\rho_{1,2} = 1$ is always a double Floquet multiplier. Interesting bifurcations can arise when an additional pair of multipliers lies on the unit circle, i.e., $\rho_{3,4} = e^{\pm i\theta}$. Particularly the case $\rho_{3,4} = -1$ is studied which leads to period doubling.

The last section is devoted to hydrodynamic problems. First we are concerned with the Stokes problem of periodic surface waves in an inviscid heavy fluid of finite depth. Using a special formulation the system appears as a semilinear problem with linear Cauchy–Riemann equations in the transformed flow domain and nonlinear boundary conditions on the free surface. From this special structure we are able to conclude the discreteness of the Floquet spectrum for the linearization around any periodic wave, e.g. the Stokes waves. Thus, the center manifold theory applies here and gives a rigorous mathematical foundation to the analysis and numerical study given in [BM92].

Second we consider viscous incompressible flows which are governed by the Navier–Stokes equations, in particular the Bénard problem for a fluid layer heated from below or the Taylor–Couette experiment of a fluid between to rotating cylinders. In both cases we are concerned with a reversible system exhibiting roll–like solutions

214

in certain regimes of the parameters. We provide the general structure appearing in spatial period doubling bifurcation which was numerically investigated for the Bénard problem in [BO86]. Moreover, the Eckhaus instability [Ec65] is interpreted in our new setting. In this case the steady problem has again a family of periodic states for each fixed control parameter. However, only the solutions in a limited band of wave lengths are stable while the other periodic patterns are unstable under perturbations with a slightly different wave length (sideband instability). For such instability results we refer to [BrM96, Mi95, Mi96a], here we use our stationary approach to show that these instabilities also lead to stationary bifurcations. In [IoP93] it is shown that such a situation corresponds to the case of a quadruple Floquet exponent 0 in the spatial setting. Thus, we have provided the functional analytic background in order to show that the analysis in [IoP93] is applicable to the Taylor–Couette problem in full mathematical rigor.

2   Floquet theory for elliptic systems

We consider linear elliptic problems in an infinite cylinder $\mathbb{R} \times \Omega$, which are periodic in the axial direction x with period T. Without loss of generality we can assume $T = 2\pi$ and write S^1 for the interval $[0, 2\pi]$ when both ends are identified. This automatically accounts for periodicity. We restrict ourself to the case of systems where the x–dependence only appears in the lower order terms. This is true whenever a semilinear problem is linearized about a periodic solution. We write the system in abstract form

$$Lu = \frac{d}{dx}u - [A + B(x)]u = g(x), \tag{2.1}$$

where u lies in a Hilbert space H. The operator $A : D(A) \to H$ is assumed to be closed and densely defined and to have compact inverse. Our aim is to find conditions such that for each bounded g (2.1) has a unique bounded solution. Therefore we study the associated Floquet operator $L_{per} : D(L_{per}) \subset \mathcal{H}_{0,0} \to \mathcal{H}_{0,0}$ given by

$$L_{per}u = \frac{d}{dx}u - [A + B(x)]u, \quad D(L_{per}) = \mathcal{H}_{1,0} \cap \mathcal{H}_{0,1}. \tag{2.2}$$

Here we have used the abbreviation $\mathcal{H}_{k,\beta} = H^k(S^1, D(A^\beta))$, where $D(A^\beta)$, $\beta \in [0, 1]$, is the interpolation space $[D(A), H]_\beta$ in the sense of [LM68], e.g., $[H^k(\Omega), L_2(\Omega)]_\beta = H^{\beta k}(\Omega)$.

The eigenvalues $\lambda \in \mathbb{C}$ of L_{per} are called Floquet exponents of (2.1), since $L_{per}U_\lambda = \lambda U_\lambda$ implies that $u(x) = e^{-\lambda x}U_\lambda(x)$ is a solution of (2.1) with $g = 0$. Note that λ is a Floquet exponent if and only if $\lambda + ik$, $k \in \mathbb{Z}$, is one (take $U_{\lambda+ik} = e^{-ikx}U_\lambda$). The values $\rho = e^{2\pi\lambda}$ are called the Floquet multipliers. To each ρ there exists a whole family $\lambda + ik$, $k \in \mathbb{Z}$, of Floquet exponents.

As a relation between L in (2.1) and L_{per} we immediately see that uniqueness of bounded solutions for (2.1) can only hold, if no Floquet multipliers are on the imaginary axis. This condition will also prove to be sufficient.

To be precise we make the following assumptions.

(A1) H is a Hilbert space and $A : D(A) \subset H \to H$ is a closed, densely defined operator, such that $A - i\xi$ is invertible for all $\xi \in \mathbb{R}$ with $\|(A - i\xi)^{-1}\|_{H\to H} = \mathcal{O}(1/|\xi|)$ for $|\xi| \to \infty$. $A^{-1} : H \to H$ is a compact operator.

(A2) $\exists \beta \in [0,1)\, \exists r \geq 2 : B = B(x) \in C^r(S^1, \mathcal{L}(D(A^\beta), H))$.

(A3) There exists a $\lambda \in \mathbb{C}$ which is not a Floquet exponent of L_{per}.

(A3*) The operator L_{per} has no Floquet exponent on the imaginary axis.

2.1 General remarks on the Floquet operator

We let $\widetilde{L}_{per} = d/dx - A$ which is the autonomous Floquet operator with $B = 0$. Its Floquet exponents are given by $\sigma_n + ik$, where σ_n are the eigenvalues of A. Obviously (A3*) is now a consequence of (A1).

For elliptic problems on cylinders with bounded cross–section Σ the set $\Gamma = \{\,\mathrm{Re}\,\sigma : \sigma \in \mathrm{Spectrum}(A)\,\}$ is discrete and has no large gaps. This means, there is an $\ell > 0$ such that $\Gamma \cap [a, a+\ell] \neq \emptyset$ for all $a \in \mathbb{R}$. This implies that the spectrum of $\widetilde{L}_{per}$ has no large gaps in $\mathbb{C}$, namely $\{\,\sigma_n + ik : n \in \mathbb{N},\ k \in \mathbb{Z}\,\} \cap \{\,z \in \mathbb{C} : |z - z_0| < \ell+1\,\} \neq \emptyset$ for all $z_0 \in \mathbb{C}$. As a consequence, the norm of the resolvent $(\widetilde{L}_{per} - \lambda)^{-1}$ in $\mathcal{L}(\mathcal{H}_{0,0})$ is bounded from below by $1/(\ell+1)$.

Including nonautonomous perturbations B, we want to show that $L_{per} = \widetilde{L}_{per} - B$ also has a compact resolvent. Applying $\widetilde{L}_{per}^{-1}$ or $(\widetilde{L}_{per} - \lambda)^{-1}$ to the resolvent equation results in two equivalent problems:

$$u - \widetilde{L}_{per}^{-1}(B + \lambda)u = \widetilde{L}_{per}^{-1}g, \quad u - (\widetilde{L}_{per} - \lambda)^{-1}Bu = (\widetilde{L}_{per} - \lambda)^{-1}g, \quad g \in \mathcal{H}_{0,0}.$$

Note that $(\widetilde{L}_{per} - i\xi)^{-1}$ is a bounded operator from $\mathcal{H}_{0,0}$ to $D(\widetilde{L}_{per}) = \mathcal{H}_{1,0} \cap \mathcal{H}_{0,1}$, and hence compact from $\mathcal{H}_{0,0}$ into $\mathcal{H}_{0,\beta}$ for $\beta < 1$. This is easily seen by expanding u into a Fourier series $\sum e^{ikx} u_n$, see the next subsection. The compactness of $(\widetilde{L}_{per} - \lambda)^{-1}B$ allows us to use Fredholm's theory, such that it suffices to show that the homogeneous problem only has the trivial solution. For parabolic problems the decay of $(\widetilde{L}_{per} - \lambda)^{-1}$ for $\lambda \to \infty$ yields a λ such that $\|(\widetilde{L}_{per} - \lambda)^{-1}B\| < 1$ and thus $(L_{per} - \lambda)^{-1}$ exists. For elliptic systems the same argument fails for large B, since the resolvent is bounded from below. In fact, it is possible to construct examples, where the spectrum of L_{per} is the whole complex plane. Assumption (A3) is stated exactly for the purpose to exclude this exotic case. We have the following result.

Theorem 2.1

Let (A1) and (A2) be satisfied.

a) Then, either (i) or (ii) holds:

    (i) *For all $\lambda \in \mathbb{C}$ there exists $u \in D(L_{per}) \setminus \{0\}$ with $L_{per} u = \lambda u$.*

    (ii) *The spectrum of L_{per} is discrete and the resolvent is compact.*

b) Genericity of case (ii):

Let $\beta \in [0,1)$ and $\mathcal{B} = C(S^1, \mathcal{L}(D(A^\beta), H))$ with norm $|B|_\infty = \sup\{ \|B(x)u\|_H : x \in S^1, \|u\|_{D(A^\beta)} = 1 \}$. Then, the set $\mathcal{B}_{(ii)} = \{ B \in \mathcal{B} : \widetilde{L}_{per} - B \text{ satisfies (ii)} \}$ is open and dense. Moreover, in any one–parameter family $L_{per}(\alpha) = \widetilde{L}_{per} - \alpha B$, $\alpha \in \mathbb{C}$, there is a discrete set $\mathcal{C}_{(i)} \subset \mathbb{C}$ without accumulation points such that $L_{per}(\alpha)$ satisfies (ii) if $\alpha \notin \mathcal{C}_{(i)}$.

Proof: Assume that (i) does not hold. Then there is a $\lambda \in \mathbb{C}$ such that $L_{per} u = \lambda u$, or equivalently $u = M(\lambda)u = \widetilde{L}_{per}^{-1}(B + \lambda)u$, has only the trivial solution. By Fredholm's theory $I - M(\lambda)$ is invertible and hence $(L_{per} - \lambda)^{-1} = (I - M(\lambda))^{-1}\widetilde{L}_{per}^{-1}$ is the desired compact resolvent. The discreteness of the spectrum of L_{per} is a consequence of the compactness of the resolvent. This proves part a).

For showing that $\mathcal{B}_{(ii)}$ is open, take any $B \in \mathcal{B}_{(ii)}$. Then there is a λ such that the compact operator $\widetilde{L}_{per}^{-1}(B + \lambda)$ does not have the eigenvalue 1. However, this remains true when B is replaced by $\widetilde{B}$ and $|\widetilde{B} - B|_\infty$ is small enough. To prove the density of $\mathcal{B}_{(ii)}$ it is sufficient to show the last assertion.

Assume $L_{per}(\alpha_0)$ satisfies (i) and let $K(\alpha) = \widetilde{L}_{per}^{-1}\alpha B$. Then $K(\alpha_0)$ has the eigenvalue 1 which is isolated by compactness of $K(\alpha_0)$. That is, there is $\varepsilon > 0$ such that $\mathrm{Spectrum}(K(\alpha_0)) \cap \{ z \in \mathbb{C} : |z - 1| < \varepsilon \} = \{1\}$. Since $K(\alpha) = (\alpha/\alpha_0)K(\alpha_0)$ we find

217

Spectrum$(K(\alpha))\cap\{z\in\mathbb{C}:|z-1|<\varepsilon/2\}=\{\alpha/\alpha_0\}$ whenever $|\alpha_0/\alpha-1|<\varepsilon/(2+\varepsilon)$. Hence, (ii) holds in a whole neighborhood of α_0 except for α_0 itself. $\qquad\square$

Of course, for small enough B case (ii) always holds true. For some elliptic problems the appearance of case (i) can be excluded for arbitrarily large B. This is the case if B has smoothing properties. Assume there is a Banach space $X\supset\mathcal{H}_{0,0}$ (e.g., $X=\mathcal{H}_{\alpha,\beta}$ with $\alpha,\beta<0$) such that $B\in\mathcal{L}(X,\mathcal{H}_{0,0})$. Moreover, assume that $\inf\{\,\|(\widetilde{L}_{per}-\lambda)^{-1}\|_{\mathcal{L}(\mathcal{H}_{0,0},X)}:\lambda\notin\text{Spectrum}(\widetilde{L}_{per})\}=0$. Then λ can be chosen such that

$$\|(\widetilde{L}_{per}-\lambda)^{-1}B\|_{\mathcal{L}(\mathcal{H}_{0,0})}\leq\|(\widetilde{L}_{per}-\lambda)^{-1}\|_{\mathcal{L}(\mathcal{H}_{0,0},X)}\|B\|_{\mathcal{L}(X,\mathcal{H}_{0,0})}<1,$$

and the existence of $(\widetilde{L}_{per}-B-\lambda)^{-1}$ follows as above. This approach will be taken in Lemma 6.1.

An interesting characterization of case (i) is given in [Ku82, Thm.12] for scalar elliptic equations. It relates the Floquet operator L_{per} to the operator $L=d/dx-A-B(x)$ on square integrable functions. Here we give, in our more general context, a short self–contained proof except for one of the implications.

Theorem 2.2
Assume that (A1) and (A2) hold. Then the following four statements are equivalent:
(i) $\quad\forall\lambda\in\mathbb{C}\;\exists u\neq0:L_{per}u=\lambda u.$
(i.1) $\quad\exists u\in L_2(\mathbb{R},D(A))\cap H^1(\mathbb{R},H):u\neq0,\;Lu=0.$
(i.2) $\quad\forall\lambda\in\mathbb{C}\;\exists u\in L_2(\mathbb{R},D(A))\cap H^1(\mathbb{R},H):u\neq0,\;Lu=\lambda u.$
(i.3) $\quad\exists u\in L_2(\mathbb{R},D(A))\cap H^1(\mathbb{R},H):u\neq0,\;Lu=0,\;\lim\limits_{|x|\to\infty}\frac{1}{|x|}\log\|u(x)\|=-\infty.$

Proof: The last statement means that u decays faster than any exponential. In fact, if u is the solution stated in (i.3), then $e^{\lambda x}u(x)$ satisfies $Lu=\lambda u$ and still lies in $L_2(\mathbb{R},D(A))\cap H^1(\mathbb{R},H)$. Thus, the implications (i.3) $\Longrightarrow$ (i.2) $\Longrightarrow$ (i.1) are immediate.

To show that (i.1) is a consequence of (i) let $K(\omega)=\text{Kernel}(L_{per}-i\omega)$ and we consider $n(\omega)=\dim K(\omega)=\dim\text{Kernel}(I-\widetilde{L}_{per}(B+i\omega))$. The function n is bounded from below by 1 and periodic with period 1. We choose $\omega_0\in[0,1)$ such that $n(\omega_0)=\min n$. Then, by upper semi–continuity of the dimension there is $\varepsilon>0$ such that $n(\omega)=n(\omega_0)$ and $K(\omega)$ depends continuously on ω, for $|\omega-\omega_0|<\varepsilon$. Hence, we can choose a continuous 1–periodic function $v:[0,1]\to\mathcal{H}_{0,0}$ such that $v(\omega)\in K(\omega)$ and $\int_0^1\int_{S^1}\|v(\omega,x)\|^2\,dx\,d\omega>0$ (e.g., with $v(\omega,\cdot)=0$ for $|\omega-\omega_0|\geq\varepsilon$).

From $v(\omega, \cdot) \in K(\omega)$ we have $L(e^{-i\omega x}v(\omega, x)) = 0$. We let

$$u(x + 2k\pi) = \int_0^1 e^{-i2k\pi\omega}e^{-i\omega x}v(\omega, x)\, d\omega, \quad \text{for } x \in [0, 2\pi) \text{ and } k \in \mathbb{Z}, \qquad (2.3)$$

which is a direct integral in the sense of [RS80]. Note that u is continuous and bounded, and satisfies $Lu = 0$. Moreover, keeping $x \in [0, 2\pi)$ fixed, the sequence $(u(x + 2k\pi))_{k\in\mathbb{Z}}$ contains exactly the Fourier coefficients of the function $[0, 1) \ni \omega \mapsto e^{-i\omega x}v(\omega, x)$ (assumed to be continued periodically). Hence, Parseval's identity (H and $D(A)$ are Hilbert spaces) gives $\sum_{k\in\mathbb{Z}} \|u(x + 2k\pi)\|^2 = \int_{S^1} \|e^{-i\omega x}v(\omega, x)\|^2 d\omega$. Integration over $x \in [0, 2\pi)$ yields

$$\int_{\mathbb{R}} \|u\|^2 dx = \int_0^{2\pi} \sum_{\mathbb{Z}} \|u(x + 2k\pi)\|^2 dx = \int_0^1 \int_{S^1} \|v(\omega, x)\|^2 d\omega dx > 0. \qquad (2.4)$$

Here the norms can either be taken in H or $D(A)$. Thus, (i.1) is established.

For the opposite implication (i.1) $\Longrightarrow$ (i) we simply reverse the above argument. Given $u \in L_2(\mathbb{R}, D(A))$ we define v by inverting (2.3): $v(\omega, x) = \sum_{k\in\mathbb{Z}} e^{i2k\pi\omega}u(x - 2k\pi)$. As above we have (2.4), $Lv(\omega, \cdot) = 0$ and $u_\omega(x) = e^{-i\omega x}v(\omega, x)$ is 2π–periodic in x with $L_{per}u_\omega = i\omega u_\omega$. It remains to show that $u_\omega \neq 0$. However, (2.4) implies $u_\omega \neq 0$ on a set of positive measure in $[0, 1]$. Thus, the Floquet spectrum is not discrete and Theorem 2.1 tells us that (i) holds.

The implication (i) $\Longrightarrow$ (i.2) follows from the equivalence of (i) and (i.1). If L_{per} satisfies (i) then so does $L_{per} - \lambda$ for any $\lambda \in \mathbb{C}$. Since (i) implies (i.1) we find a nontrivial u such that $(L - \lambda)u = 0$. But this is the assertion of (i.2).

The only implication left over is (i) $\Longrightarrow$ (i.3). For this we refer the reader to [Ku82], since there much deeper methods are needed, e.g., holomorphic bundles over Fréchet spaces. $\qquad\square$

We use the characterization (i.1) in the Appendix in order to give an example where the nongeneric case (i) holds. It consists of a (nonlocal) lower order perturbation of the Laplace operator in the two–dimensional strip with Dirichlet boundary conditions.

2.2 Invertibility on $L_2(\mathbb{R}, H)$

We now are interested in the equation $Lu = g$ with $g \in L_2(\mathbb{R}, H)$. The following result was also established in [Ku82, Thm.14] for scalar elliptic equations. We give a

219

self–contained analytical proof which allows us to proceed further into the direction of applications.

Theorem 2.3

Let A and B satisfy (A1), (A2), and (A3*). Then the following statements hold.

a) The set of Floquet exponents of L_{per} is discrete and there is a positive δ such that all Floquet exponents λ satisfy $|\operatorname{Re}\lambda| > \delta$.

b) For each $g \in L_2(\mathbb{R}, H)$ equation (2.1) has a unique solution u in $W^{1,2}(\mathbb{R}, H) \cap L_2(\mathbb{R}, D(A))$.

Proof: The discreteness of the Floquet spectrum is shown in Theorem 2.1. The existence of $\delta > 0$ follows from the discreteness, (A3*) and the fact that $\lambda \in \Sigma(L_{per}) \Longleftrightarrow \lambda + ik \in \Sigma(L_{per})$. Assume $\lambda_n \in \Sigma(L_{per})$ and $\operatorname{Re}\lambda_n \to 0$, then we may assume $|\operatorname{Im}\lambda_n| \leq 1$ and hence a subsequence converges to λ^* with $\operatorname{Re}\lambda^* = 0$. The discreteness of $\Sigma(L_{per})$ yields $\lambda^* \in \Sigma(L_{per})$ which is in contradiction to (A3*). This proves part a).

In part b) we are interested in functions $g \in L_2(\mathbb{R}, H)$. In contrast to the periodic case the operator $u \to Bu$ is not a relatively compact perturbation of the autonomous part $u \to u' - Au$. Thus, we use a refined method employing Fourier transform:

$$\mathcal{F}^{\pm}u(\xi) = \frac{1}{\sqrt{2\pi}} \int_{\mathbb{R}} e^{\mp i\xi x} u(x)\, dx,$$

then $\mathcal{F}^{\pm} : L_2(\mathbb{R}, D(A^{\gamma})) \to L_2(\mathbb{R}, D(A^{\gamma}))$ are bounded invertible operators with $(\mathcal{F}^{\pm})^{-1} = \mathcal{F}^{\mp}$, where γ takes the values 0, β, and 1.

Applying $\mathcal{F}^+$ to (2.1) we obtain for $w = \mathcal{F}^+ u$ the equation

$$i\xi w(\xi) - Aw - [B^* w](\xi) = \mathcal{F}^+ g(\xi), \quad \text{where } [B^* w](\xi) = \sum_{n \in \mathbb{Z}} B_n w(\xi - n), \qquad (2.5)$$

with $B_n \in \mathcal{L}(D(A^{\beta}), H)$ such that $B(x) = \sum_{n \in \mathbb{Z}} e^{inx} B_n$. From (A2) we conclude that there is a constant $C > 0$ such that

$$\|B_n\|_{D(A^{\beta}) \to H} \leq C/(1 + |n|)^r, \quad n \in \mathbb{Z}. \qquad (2.6)$$

According to (A1) we may invert the autonomous part and are lead to

$$w + Mw = (i\xi - A)^{-1}\mathcal{F}^+ g(\xi), \quad \text{where } M = -(i\xi - A)^{-1}B^*. \qquad (2.7)$$

Here, M is not a compact operator, such that the Fredholm theory is not directly applicable. However, the convolution B^* has a special structure deriving from the periodicity of B. The convolution only mixes ξ–values which differ by an integer. Thus, we can restrict the analysis to such subsets first.

We fix $\theta \in [0,1)$ and let $W_n = w(\theta + n)$, $n \in \mathbb{Z}$, and define

$$\ell_2^\gamma = \{ W = (W_n)_{n \in \mathbb{Z}} : W_n \in D(A^\gamma), \ |W|_\gamma^2 := \sum_{n \in \mathbb{Z}} \|W_n\|_{D(A^\gamma)}^2 < \infty \},$$

$$M_\theta : \ell_2^\gamma \to \ell_2^\gamma;$$
$$(M_\theta W)_n = (i(\theta + n) - A)^{-1} \sum_{m \in Z} B_m W_{n-m}.$$

We use the notation $\|\cdot\|_\gamma$ as a short–hand of $\|\cdot\|_{D(A^\gamma)}$.

Lemma 2.4

Let $(A1)$–$(A3^*)$ hold. Then, M_θ is compact and $I + M_\theta$ is invertible with an inverse bounded independently of θ.

Remark: The elements $W \in \ell_2^\gamma$ correspond via $u(x) = e^{i\theta x} \sum W_n e^{inx}$ to functions u such that $e^{-i\theta x} u(x)$ is 2π–periodic in x.

Proof: The operator M_θ is the composition of the convolution $B^* : \ell_2^\beta \to \ell_2^0; W \mapsto (\sum_{m \in \mathbb{Z}} B_m W_{n-m})_n$ and the multiplication operator

$$R_\theta : \ell_2^0 \to \ell_2^\beta; W \mapsto ((i(\theta + n) - A)^{-1} W_n)_n.$$

We show that B^* is bounded and that R_θ is compact.

Using Cauchy–Schwartz's inequality we obtain

$$|B^* W|_0^2 = \sum_n \left\| \sum_m B_m W_{n-m} \right\|_0^2 \leq \sum_n \left[\sum_m \|B_m\|^{1/2} \left(\|B_m\|^{1/2} \|W_{n-m}\|_\beta \right) \right]^2$$

$$\leq \sum_n \left[\sum_m \|B_m\| \cdot \sum_m \|B_m\| \|W_{n-m}\|_\beta^2 \right] = \left(\sum_m \|B_m\| \right)^2 |W|_\beta^2.$$

Here all sums range over $\mathbb{Z}$ and the norm of B_m is taken in $\mathcal{L}(D(A^\beta), H)$. With (2.6) the boundedness is established.

From the resolvent estimate in (A1) we conclude, that for any $W \in \ell_2^0$ the image $V = R_\theta W$ satisfies $(nV_n)_n$, $(AV_n)_n \in \ell_2^0$. By interpolation we obtain $((1+|n|)^{1-\beta} V_n)_n \in \ell_2^\beta$.

For a given bounded sequence $W^k \in \ell_2^0$ with $|W^k|_0 \leq \rho$ we can choose a subsequence, again called W^k, such that each component V_n^k converges for $k \to \infty$ to V_n^∞ in $D(A^\beta)$. To this end use that $D(A)$ is compactly embedded in $D(A^\beta)$ and that the sequence V_n^k, $k \in \mathbb{N}$, is bounded in $D(A)$. For any N we get

$$|V^k - V^m|_\beta^2 \leq \sum_{|n|<N} \|V_n^k - V_n^m\|_\beta^2 + (1 + |N|)^{-1+\beta} \sum_{|n|\geq N} (1 + |n|)^{1-\beta} \|V_n^k - V_n^m\|_\beta^2.$$

Here the second sum is bounded independently of N, k, and m. Thus, for given $\varepsilon > 0$ we may choose N so large that the second term is less than $\varepsilon/2$. Then for $k, m > M(\varepsilon)$ the first term also is less than $\varepsilon/2$. This proves the compactness.

In order to show the invertibility of the operator $I + M_\theta$ from ℓ_2^β into itself, it is sufficient, according to Fredholm's theory, to establish the injectivity of $I + M_\theta$. However, any solution W of $W + M_\theta W = 0$ gives rise to a Floquet exponent $i\theta$ with eigenfunction $u(x) = e^{i\theta x} \sum_n e^{inx} W_n$. Thus, assumption (A3*) implies injectivity. The uniform bound in θ of the inverse is immediate, since M_θ is continuous and periodic in θ. This proves Lemma 2.4. $\qquad\square$

Proof of Theorem 2.3 continued: Using the decoupling we find the solution in the case $\xi \in \mathbb{R}$ as follows. For $\xi = \theta + n$ we obtain

$$w(\xi) = w(\theta + n) = W_n = ((I + M_\theta)^{-1}(i(\theta + \cdot) - A)^{-1}\mathcal{F}^+ g(\theta + \cdot))_n.$$

From $W + M_\theta W = \widehat{G}(\theta + \cdot) := (i(\theta + \cdot) - A)^{-1}\mathcal{F}^+ g(\theta + \cdot)$ we obtain

$$
\begin{aligned}
\int_{\mathbb{R}} \|w(\xi)\|_\beta^2 d\xi &= \int_0^1 \sum_{n \in \mathbb{Z}} \|w(\theta + n)\|_\beta^2 d\theta = \int_0^1 |(I + M_\theta)^{-1}\widehat{G}(\theta + \cdot)|_\beta^2 d\theta \\
&\leq C \int_0^1 |\widehat{G}(\theta + \cdot)|_\beta^2 d\theta = C \int_{\mathbb{R}} \|\widehat{G}(\xi)\|_\beta^2 d\xi \\
&= C \int_{\mathbb{R}} \|(i\xi - A)^{-1}\mathcal{F}^+ g(\xi)\|_\beta^2 d\xi \leq C^2 \int_{\mathbb{R}} \|\mathcal{F}^+ g(\xi)\|_0^2 d\xi.
\end{aligned}
$$

Thus, we have a unique solution $w \in L_2(\mathbb{R}, D(A^\beta))$.

Moreover, from $w = (i\xi - A)^{-1}(\mathcal{F}^+ g + B^* w)$ we conclude with (A1) that $\xi w(\xi)$ and $Aw(\xi)$ are both in $L_2(\mathbb{R}, H)$. Hence, inverse Fourier transform yields $u = \mathcal{F}^- w \in W^{1,2}(\mathbb{R}, H) \cap L_2(\mathbb{R}, D(A))$, which is the desired result. Now Theorem 2.3 is proved. $\qquad\square$

2.3 L_p- and $C-$spaces and higher regularity

For the center manifold theory it is important to be able to use the function spaces $L_p(I\!R, H)$ or $C_{bdd}(I\!R, H)$ with exponential weight functions:

$$L_{p,\alpha}(I\!R, H) = \{\, u : I\!R \to H \ : \ e^{-\alpha|\cdot|}u(\cdot) \in L_p(I\!R, H)\,\},$$

$$\|u\|_{p,\alpha} = \left(\int_{I\!R} \|e^{-\alpha|x|}u(x)\|^p \, dx \right)^{1/p},$$

$$C_\alpha(I\!R, H) = \{\, u \in C^0(I\!R, H) \ : \ \|u\|_{\infty,\alpha} = \sup_{x \in I\!R} \|e^{-\alpha|x|}u(x)\| < \infty \,\}.$$

Analogously we define $W_\alpha^{n,p}(I\!R, H)$. Of course, H can be replaced by any of the Hilbert spaces $D(A^\gamma)$.

Lemma 2.5

a) There is a $\delta > 0$ such that for each $\alpha \in (-\delta, \delta)$ and each $p \in (1, \infty)$ equation (2.1) has for each $g \in L_{p,\alpha}(I\!R, H)$ a unique solution $u \in W_\alpha^{1,p}(I\!R, H) \cap L_{p,\alpha}(I\!R, D(A))$.

b) For each $g \in C_\alpha(I\!R, H)$ the unique solution u lies in $C_\alpha(I\!R, D(A^{\tilde{\beta}}))$ for each $\tilde{\beta} \in [0, 1)$.

Proof: We follow the analysis in [Mi87] and determine δ by the transformation $u(x) = (\cosh(x))^\alpha v(x)$. We have $u \in L_{2,\alpha}(I\!R, H)$ if and only if $v \in L_2(I\!R, H)$. Moreover, u solves (2.1) if v is a solution of

$$\frac{d}{dx} v - (A + B(x))v = -\alpha \tanh(x)v + g(x)/(\cosh(x))^\alpha.$$

Since the left–hand side is invertible on $L_2(I\!R, H)$, we find a $\delta > 0$ such that this equation is solvable for all $\alpha \in [-\delta, \delta]$.

To handle the case $p \in (2, \infty)$ we increase the possible p values by induction. Assume the result is known for all $p \in [2, q]$, then we show that is also holds for all p with $-\frac{1}{p} \le -\frac{1}{q} + (1 - \beta)$. Altogether only a finite number of steps is necessary to reach every $p \in [2, \infty)$. For such p and $|\alpha| < \gamma < \delta$ the space $L_{p,\alpha}$ embeds continuously into $L_{q,\gamma}$. Hence, for any $g \in L_{p,\alpha}(I\!R, H)$ we find a solution $u = Kg \in W_\gamma^{1,q}(I\!R, H) \cap L_{q,\gamma}(I\!R, D(A))$. By interpolation we have $A^\beta u \in W_\gamma^{1-\beta,q}(I\!R, H)$, and Sobolev's embedding yields $A^\beta u \in L_{p,\gamma}(I\!R, H)$.

To recover the correct decay rate α, we use the shift operator $(T_m g)(x) = g(x - 2\pi m)$ and the cut–off functions $\chi_m = \chi_{[2\pi m, 2\pi(m+1))}$. Now, the function $g \in L_{p,\alpha}$ can

223

be written as $g = \sum_{n \in \mathbb{Z}} g_n$ with $g_n = \chi_n g$. This sum converges in $L_{q,\gamma}$, whence $A^\beta u = \sum_n A^\beta u_n$ with $u_n = K g_n$ converges in $L_{p,\gamma}$. However, each $g_n \in L_{q,-\gamma}$ and hence the inductional assumption implies $A^\beta u_n \in L_{p,-\gamma}$. From the periodicity $B(x + 2\pi) = B(x)$ and uniqueness we conclude $K T_m = T_m K$. Thus, there is a constant $C_0 > 0$ such that

$$\|T_k A^\beta u_n\|_{p,-\gamma} \le C_0 \|T_k g_n\|_{q,-\gamma} \le 2\pi C_0 \|T_k g_n\|_{p,-\gamma} \quad \forall\, k \in \mathbb{Z}.$$

Here, we have used the embedding of L_p into L_q for $p > q$ on intervals of length 2π. Using the above estimate for $k = n$ and restricting u_n to $[2\pi m, 2\pi(m+1))$ we find

$$
\begin{aligned}
\|\chi_m A^\beta u_n\|_{p,\alpha} &\le C_1 e^{\alpha|m|} e^{-\gamma|n-m|} \|\chi_m A^\beta T_n u_n\|_{p,-\gamma} \le C_1 e^{\alpha|m| - \gamma|n-m|} \|A^\beta T_n u_n\|_{p,-\gamma} \\
&\le C_1 C_0 e^{\alpha|m| - \gamma|n-m|} \|T_n g_n\|_{p,-\gamma} \le C_2 e^{\alpha|m| - \gamma|n-m|} \|T_n g_n\|_{p,\alpha} \\
&\le C_3 e^{\alpha|m| - \gamma|n-m|} e^{-\alpha|n|} \|g_n\|_{p,\alpha} \le C_3 e^{-(\gamma - |\alpha|)|n-m|} \|\chi_n g\|_{p,\alpha}.
\end{aligned}
$$

Thus, the influence of g_n on $u = \sum u_n$ decays exponentially with the distance.

Now, we argue as for Young's inequality for convolutions ($\nu \in (0,1)$):

$$
\begin{aligned}
\|A^\beta u\|_{p,\alpha}^p &= \sum_m \|\chi_m A^\beta u\|_{p,\alpha}^p \le \sum_m \|\chi_m A^\beta u\|_{p,\alpha}^{p-1} \left(\sum_n \|\chi_m A^\beta u_n\|_{p,\alpha} \right) \\
&\le C_3 \sum_m \sum_n \|\chi_m A^\beta u\|_{p,\alpha}^{p-1} e^{-(\gamma - |\alpha|)|n-m|} \|\chi_n g\|_{p,\alpha} \\
&\le C_3 \left(\sum_{n,m} [\|\chi_m A^\beta u\|_{p,\alpha}^{p-1} e^{-\nu(\gamma - |\alpha|)|n-m|}]^{\frac{p}{p-1}} \right)^{\frac{p-1}{p}} \\
&\qquad\qquad \times \left(\sum_{n,m} [e^{-(1-\nu)(\gamma - |\alpha|)|n-m|} \|g \chi_n\|_{p,\alpha}]^p \right)^{\frac{1}{p}} \\
&\le C_3 \|A^\beta u\|_{p,\alpha}^{p-1} \left(\frac{2}{1 - e^{\nu(|\alpha| - \gamma)}} \right)^{1-1/p} \left(\frac{2}{1 - e^{(1-\nu)(|\alpha| - \gamma)}} \right)^{1/p} \left(\sum_n \|\chi_n g\|_{p,\alpha}^p \right)^{1/p}.
\end{aligned}
$$

Dividing by $\|A^\beta u\|_{p,\alpha}^{p-1}$ yields $\|A^\beta u\|_{p,\alpha} \le \widehat{C} \|g\|_{p,\alpha}$. Now, the linear problem for u can be rewritten as $u' - Au = g + B(x)u$ with the right–hand side in $L_{p,\alpha}(\mathbb{R}, H)$. The autonomous theory of [Mi87] then provides the desired result. The case $p \in (1,2)$ follows from studying the adjoint problem, since the dual space of $L_{p,\alpha}(\mathbb{R}, H)$ is given by $L_{p',-\alpha}(\mathbb{R}, H)$, where $p' = p/(p-1)$. This proves part a).

It is sufficient to prove part b) for large $\widetilde{\beta} < 1$, hence we may assume $\widetilde{\beta} \ge \beta$ and let $p > 1/(1-\widetilde{\beta})$. Then for all $\gamma > \alpha$ we know that $g \in C_\alpha(\mathbb{R}, H)$ implies $g \in L_{p,\gamma}(\mathbb{R}, H)$.

224

Thus, the solution satisfies $u \in W_\gamma^{1,p}(\mathbb{R}, H) \cap L_{p,\gamma}(\mathbb{R}, D(A))$, and by interpolation $u \in C_\gamma(\mathbb{R}, D(A^{\tilde{\beta}}))$. The correct exponential decay can be found as above. $\qquad \square$

Remark 2.6 In fact, the constant δ in the above Lemma can be chosen as the minimal distance d of all Floquet exponents from the imaginary axis. This follows by considering the auxiliary system (2.1) where B is replaced by $B + \mu I$ with $|\mu| < d$. For each of these μ the assumptions (A1)–(A3*) remain valid. Now, as in [Mi87, Korollar 4.3a] it follows that our Lemma holds for $|\alpha| < d$.

To prove higher regularity in x–direction we use the method of differential quotients. We let

$$(T_h u)(x) = u(x + h) \quad \text{and} \quad u_h(x) = \frac{1}{h}((T_h u)(x) - u(x)).$$

Then, for any solution of $Lu = g$ we have $Lu_h = g_h + B_h T_h u = f_h$. From $B \in C^1(\mathbb{R}, \mathcal{L}(D(A^\beta), H))$ we find $B_h(x) \to B'(x)$ for $h \to 0$. Since $T_h u \to u$ in $L_{p,\alpha}(\mathbb{R}, H)$ the right–hand side f_h converges for each $g \in W_\alpha^{1,p}(\mathbb{R}, H)$ to $f_0 = g' + B'u$. Thus, the invertibility of L gives the convergence of $u_h = L^{-1} f_h$ against $u' = L^{-1} f_0$. Thus, the derivative exists and satisfies

$$(Lu')(x) = g'(x) + B'(x)u(x). \tag{2.8}$$

The same argument works in the case $g \in C_\alpha^1(\mathbb{R}, H)$. Using induction on the order of differentiation the following result is obtained.

Lemma 2.7
Assume that (A1), (A2), and (A3*) holds and let $n \in \mathbb{N}_0$ with $n \leq r$. Then, for each $g \in W_\alpha^{n,p}(\mathbb{R}, H)$ the solution $u = L^{-1}g$ lies in $W_\alpha^{n+1,p}(\mathbb{R}, H) \cap W_\alpha^{n,p}(\mathbb{R}, D(A))$.
Let additionally $\tilde{\beta} \in [0,1)$. Then, for each $g \in C_\alpha^n(\mathbb{R}, H)$ the solution $u = L^{-1}g$ lies in $C_\alpha^n(\mathbb{R}, D(A^{\tilde{\beta}}))$.

Remark 2.8 From (2.8) we see that u' can be expressed in the form $u' = Kg' + K(B'Kg)$, where $K = L^{-1}$ is the solution operator. The same result can be obtained when K can be expressed by a Green's function $G(x, s)$. From $u(x) = \int_\mathbb{R} G(x, s)g(s)ds = \int_\mathbb{R} G(x, x + t)g(x + t)dt$ we obtain by formal differentiation

$$u'(x) = \int_\mathbb{R} \left[G(x, x + t)g'(x + t) + \frac{\partial}{\partial x}[G(x, x + t)]g(x + t) \right] dt.$$

Comparing $K(B'Kg)$ with the last term we find the formula

$$\frac{\partial}{\partial x}[G(x, x+t)] = \int_{I\!R} G(x,s)B'(s)G(s, x+t)\, ds,$$

which was rigorously established in [Sc91].

2.4 Projections onto spectral strips

Spatial center manifold theory is relevant for problems whose linearization has a finite number of Floquet multipliers on the unit circle. This center part has to be separated from the remaining hyperbolic part. Since no time–2π–map is available we have to find another way to extract the associated projections.

We again consider the operator $L_{per} : u \to u' - \widetilde{A}(x)u$ with periodic boundary conditions, where $\widetilde{A}(x) = A + B(x)$. But now we replace (A3*) by a weaker assumption:

(A3) There exists a $\lambda \in C\!\!\!\!/$ which is not a Floquet exponent of L_{per}.

According to Theorem 2.1 we know that the spectrum of L_{per} is discrete and the resolvent $(L_{per} - \lambda)^{-1}$ is compact. Moreover, the set $J_{Re} = \{\, Re\, z \; : \; z \in Spectrum(L_{per})\,\}$ is also discrete, i.e., $[a, b] \cap J_{Re}$ contains only finitely many points, if $b - a < \infty$.

We now give a method to construct the projection P onto the spectral part associated to all Floquet exponents λ in the strip $\alpha < Re\,\lambda < \gamma$, or equivalently, to all Floquet multipliers ρ in the annulus $e^\alpha < \rho < e^\gamma$. (Of course the method is similarly able to separate each Floquet multiplier by itself, but for simplicity we restrict ourself to this case.) We keep now α, γ with $\alpha < \gamma$ fixed, and for each $I \subset I\!R$ we let

$$\Sigma_I = \{\, \lambda \in Spectrum(L_{per}) \; : \; \alpha < Re\,\lambda < \gamma, \; Im\,\lambda \in I \,\}.$$

Since the spectrum is discrete we can define the spectral projection P_I as the Dunford integral $P_I = \frac{1}{2\pi i} \int_{\Gamma_I} (L_{per} - \lambda)^{-1} d\lambda$, where Γ_I is a positively oriented closed C^1–curve, such that Σ_I lies in its interior whereas $\Sigma(L_{per}) \setminus \Sigma_I$ lies outside.

Our aim is to define $P_{I\!R}$ as the spectral projection of the whole strip. Since the Dunford integral does not exist for this unbounded set, we use the special structure connecting the eigenfunctions corresponding to Floquet exponents which only differ by ik, $k \in Z\!\!\!Z$. Since L_{per} has a compact resolvent, the projection $P_{(-1/2, 1/2]}$ can be written as

$$P_{(-1/2, 1/2]}u(x) = \sum_{k=1}^{N} \frac{1}{2\pi} \int_0^{2\pi} \langle u(y), \psi_k(y) \rangle dy\, \phi_k(x),$$

226

where ϕ_k (resp. ψ_k) are (generalized) eigenvectors of L_{per} $(L^* = -\cdot' - \widetilde{A}^*)$ corresponding to eigenvalues λ_k $(\overline{\lambda}_k) \in \Sigma_{(-1/2,1/2]}$. Since $e^{imx}\phi_k(x)$ $(e^{imx}\psi_k(x))$ are eigenvectors to the shifted eigenvalue $\lambda_k + im$ $(\overline{\lambda}_k - im)$ we find

$$
\begin{aligned}
P_{(-n-1/2,n+1/2]}u(x) &= \sum_{m=-n}^{n} \sum_{k=1}^{N} \frac{1}{2\pi} \int_0^{2\pi} \langle u(y), e^{imy}\psi_k(y)\rangle e^{imx}\phi_k(x) \\
&= \sum_{k=1}^{N} Q_n(\langle u, \psi_k\rangle)(x)\phi_k(x).
\end{aligned}
$$

Here Q_n is the orthogonal projection from $L_2(S^1, \mathbb{C})$ onto the span of $e^{-inx}, \ldots, e^{inx}$. Thus, for each $u \in L_2(S^1, H)$ the limit of $P_{(-n-1/2,n+1/2]}u$ exists and is given by $(P_{\mathbb{R}}u)(x) = \sum_{k=1}^{N}\langle u(x), \psi_k(x)\rangle\phi_k(x)$.

This projection is a real projection (i.e. $u(x)$ real $\Longrightarrow$ $(P_{\mathbb{R}}u)(x)$ real), which acts pointwise in x and is 2π–periodic. The first property follows from the fact that $P_{[-n-1/2,n+1/2]}$ is real, since $\Gamma_{[-n-1/2,n+1/2]}$ can be chosen symmetric to the real axis. As $(P_{[a,b]} - P_{(a,b]})u \to 0$ for $a \to -\infty$ we find $P_{[-n-1/2,n+1/2]}u \to P_{\mathbb{R}}u$. Another way to see this is to choose ϕ_k and ψ_k in $P_{(-1/2,1/2)}$ such that $\overline{\phi}_k = \phi_{k'}$ and $\overline{\psi}_k = \psi_{k'}$ for appropriate k'. For $P_{\{1/2\}}$ we may choose ϕ_k and ψ_k such that $\xi_k = e^{-ix/2}\phi_k$ and $\eta_k = e^{-ix/2}\psi_k$ are real. (We have $L\xi_k = 0$ and $\xi_k(0) = -\xi_k(2\pi)$ and the same for η_k.)

Thus, we are able to define the x–dependent projection

$$
P(x) : H \to H; \quad u \mapsto \sum_{k=1}^{N}\langle u, \psi_k(x)\rangle\phi_k(x). \tag{2.9}
$$

Using the fact that the functions ϕ_k and ψ_k are periodic (generalized) eigenfunctions it follows that they are as smooth in x as the function B. Furthermore, P satisfies the differential equation

$$
P'(x) = \widetilde{A}(x)P(x) - P(x)\widetilde{A}(x), \tag{2.10}
$$

which is classical in Floquet theory (cf. [IoA92]). Here this follows since $P = P_{\mathbb{R}}$ commutes with L_{per}, by construction as a limit of Dunford integral. Hence, for any constant u we have $0 = LPu - PLu = (Pu)' - \widetilde{A}Pu - P(u' - \widetilde{A}u) = (P' - \widetilde{A}P + P\widetilde{A})u$. It is not possible to obtain P by solving (2.10) (as usually done in evolutionary problems), since the initial value problem is ill–posed in the elliptic case.

The projection P allows for the separation of the spectrum and hence for decoupling the modes of different exponential decay in the system $u' - \widetilde{A}(x)u = f(x)$. Further on we restrict our attention to the case $-\alpha = \gamma$ small, since this case is the only relevant

227

for center manifold theory. We decompose $u \in H$ into $u_0 + u_1$ with $u_0(x) = P(x)u$ leading to

$$u_0' - \tilde{A}_0(x)u_0 = f_0(x), \quad u_1' - \tilde{A}_1(x)u_1 = f_1(x), \tag{2.11}$$

where $\tilde{A}_0(x) = \tilde{A}(x)|_{\mathrm{Range}(P(x))}$ and $\tilde{A}_1(x) = \tilde{A}(x)|_{\mathrm{Kernel}(P(x))}$. The finite–dimensional part u_0 can easily be solved by choosing an x–dependent basis for $\mathrm{Range}(P(x))$. However, the infinite–dimensional part cannot be treated in a simple manner. Thus, we avoid the splitting and proceed by considering the auxiliary equation

$$L_\mu v := v' - (\tilde{A}(x) + \mu P(x))v = f_1(x) = (I - P(x))f(x). \tag{2.12}$$

Proposition 2.9

The periodic family $\tilde{A}_\mu(x) = A + (B(x) + \mu P(x))$ satisfies the assumptions (A1), (A2), and (A3). The additional term $\mu P(x)$ moves the Floquet exponents from the strip $\alpha < \mathrm{Re}\,\lambda < \gamma$ to $\alpha + \mu < \mathrm{Re}\,\lambda < \gamma + \mu$, while keeping the remaining spectrum fixed. The eigenfunctions of the operator L_{per} are not affected as $P = P_{I\!R}$ and L_{per} commute.

Thus, for $|\mu| > \gamma - \alpha$ the operator L_μ has no Floquet exponents in the strip $\alpha < \mathrm{Re}\,\lambda < \gamma$. In the case $-\alpha = \gamma > 0$ this shows that even (A3*) is satisfied. Hence, Theorem 2.3 and Lemma 2.5 can be applied to yield, for each $f \in L_{p,\tilde{\gamma}}(I\!R, H)$, $|\tilde{\gamma}| < \gamma$, a unique solution $v = K_\mu f$. We claim that this solution is in fact independent of μ and hence solves the desired equation with $\mu = 0$. To see this, we show $v_0(x) = P(x)v(x) = 0$ for all $x \in I\!R$. Using $v_0' = P'v + Pv'$ and $P' = \tilde{A}P - P\tilde{A}$ we are left with $u_0' = \tilde{A}_\mu u_0$. This is a finite dimensional system with no Floquet exponents in the strip $|\mathrm{Re}\,\lambda| < \gamma$. Hence, the only solution in $L_{p,\tilde{\gamma}}(I\!R, H)$ is the trivial one. Consequently we have $P(x)v(x) \equiv 0$ which implies that the additional term $\mu P(x)$ has no effect on the solution at all.

We summarize the results in the following theorem.

Theorem 2.10

Let $Lu = u' - (A - B(x))u$ satisfy the assumptions (A1)–(A3). Then, there is a periodic projection $P \in C_{per}^k(I\!R, \mathcal{L}(H, H))$ and a constant $\delta > 0$, such that for all $s \in I\!R$ and all $\alpha \in (0, \delta)$ the system

$$u' - (A + B(x))u = f(x), \quad P(s)u(s) = \xi,$$

has for each $(\xi, f) \in Range(P(s)) \times L_{p,\alpha}(\mathbb{R}, H)$ a unique solution $u = \widetilde{K}(s, \xi, f) \in W_{\alpha}^{1,p}(\mathbb{R}, H) \cap L_{p,\alpha}(\mathbb{R}, D(A))$. Moreover, we have $Pu = \widetilde{K}(s, \xi, Pf)$ and $(I - P)u = \widetilde{K}(s, 0, (I - P)f)$.

If $m \leq r$ and $f \in W_{\alpha}^{m,p}(\mathbb{R}, H)$, then we have $u = \widetilde{K}(s, \xi, f) \in W_{\alpha}^{m+1,p}(\mathbb{R}, H) \cap W_{\alpha}^{m,p}(\mathbb{R}, D(A))$.

3 Center manifolds

In this section we consider nonlinear systems which arise when studying the solutions in a neighborhood of a periodic solution (here $u \equiv 0$).

$$u' = \widetilde{A}(x)u + \mathcal{N}(x, u), \tag{3.1}$$

where $\widetilde{A}(x) = A + B(x)$ satisfies the assumptions (A1)–(A3) of the previous section. The nonlinearity is assumed to satisfy

(A4): $\exists \delta > 0, r \geq 1 : \mathcal{N} = \mathcal{N}(x, u) \in C^{r+1}(\mathbb{R} \times B_{\delta}^{D(A)}(0), H),$
$\quad \mathcal{N}(x, 0) = 0, \ D_u \mathcal{N}(x, 0) = 0,$ and $\mathcal{N}(x + 2\pi, u) = \mathcal{N}(x, u).$

We continue to use the notation $B_{\delta}^{X}(u_0) = \{ u \in X : \|u - u_0\|_X \leq \delta \}$. Note that we allow for the same order of regularity loss in the nonlinearity as in the linear operator A. This enables us the treatment of quasilinear systems as in [Mi88].

It is the aim of this section to prove that the set of all small bounded solutions lies in a finite dimensional center manifold. For ODEs, parabolic, and semilinear hyperbolic problems the existence of such center manifolds is well–kown, since there the initial value problem is well–posed and the time–2π–map exists and has a local center manifold. Using the flow this manifold gives rise to a time–periodic center manifold in the extended phase space, see [MS86, IoA92].

For elliptic systems we have to deal with the problem that the spectrum of the linearization is unbounded on both sides of the imaginary axis leading to ill–posedness of the initial value problem. Nevertheless, a center manifold can be constructed by using appropriate Green's functions for the infinite–dimensional part off the imaginary axis. This method was developed in [Ki82, Fi84] for autonomous systems and in [Mi86, Mi88] for nonautonomous systems with autonomous linear part.

In the linearly nonautonomous case the center manifold will be a locally invariant manifold modelled over the manifold of the x–dependent linear subspaces:

$$E(x) = \text{Range } P(x), \quad \mathcal{E} = \bigcup_{x \in \mathbb{R}} (x, E(x)) \subset \mathbb{R} \times D(A) \subset \mathbb{R} \times H.$$

Thus, $\mathcal{E}$ is the $(N+1)$–dimensional submanifold $(N = \dim(\text{Range } P(x)))$ containing all those solutions of the linear problem which grow less than any exponential. For later use we let

$$\mathcal{E}_\varepsilon = \{ (x, u) \in \mathcal{E} : \|Au\|^2 + \|u\|^2 \le \varepsilon^2 \} = \mathcal{E} \cap (\mathbb{R} \times B_\varepsilon^{D(A)}(0)).$$

Of course $\mathcal{E}$ is periodic in x, viz., $(x, u_0) \in \mathcal{E} \iff (x + 2\pi, u_0) \in \mathcal{E}$. Since the system is 2π–periodic it is natural to identify the x–values modulo 2π. In this sense $\mathcal{E}_\varepsilon$ is a compact manifold with boundary. Its topological structure is either a cross–product $S^1 \times B_\varepsilon^{\mathbb{R}^N}(0)$ or a Möbius–manifold $\mathcal{M} \times B_\varepsilon^{\mathbb{R}^{N-1}}(0)$, where $\mathcal{M}$ is the Möbius strip. This identification is particularly useful, when (3.1) was obtained by linearizing around a nontrivial periodic solution p of a periodic system $w' = \mathcal{F}(x, w)$, $(x, w) \in S^1 \times H$. Then, the ansatz $w = p + u$ leads to (3.1), where $\tilde{A}(x) = D\mathcal{F}(x, p(x))$, and $\mathcal{E}_\varepsilon$ (and hence the center manifold) can be interpreted as a submanifold in $S^1 \times H$ containing p, which in u–coordinates is identified with $u \equiv 0$. The case of a periodic solution in an autonomous system needs a different approach which is developed below.

Theorem 3.1
Let the assumptions (A1)–(A4) be satisfied. Then there is an $\varepsilon > 0$ and a reduction function $h = h(x, u_0) : \mathcal{E}_\varepsilon \to D(A)$, such that the following assertions hold:

a) $h \in C^{r-1}(\mathcal{E}_\varepsilon, D(A))$ with $h(x, u_0) \in \text{Kernel } P(x) \cap D(A)$, $h(x, 0) = 0$, $D_u h(x, 0) = 0$, and $h(x + 2\pi, u_0) = h(x, u_0)$ for all $(x, u_0) \in \mathcal{E}_\varepsilon$.

b) The center manifold $\mathcal{M}_C = \{ (x, u_0 + h(x, u_0)) \in \mathbb{R} \times D(A) : (x, u_0) \in \mathcal{E}_\varepsilon \}$ is locally invariant with respect to (3.1). Every solution $u \in C^1(\mathbb{R}, D(A))$ with $\|Au(x)\|^2 + \|u(x)\|^2 \le \varepsilon^2$, for all $x \in \mathbb{R}$, is contained in $\mathcal{M}_C$.

c) Every solution $u_0 \in C^1((x_1, x_2), \mathcal{E}_\varepsilon)$ of the reduced problem

$$u_0' - \tilde{A}_0(x) u_0 = P(x) \mathcal{N}(x, u_0 + h(x, u_0)) \qquad (\tilde{A}_0(x) = \tilde{A}(x)|_{E(x)}) \qquad (3.2)$$

leads via $u(x) = u_0(x) + h(x, u_0(x))$ to a solution of the full problem.

d) If (3.1) is equivariant with respect to the action of a symmetry group, then so is (3.2) with respect to the reduced action.

Proof: The proof of the theorem follows exactly the lines of the corresponding case with autonomous linear part. We repeat the main steps and leave it to the reader to transfer the details form [Mi88] to the present case.

First we modify the nonlinearity by multiplication with a cut–off function in order to render the Lipschitz constant of the modified nonlinearity globally small. Let $\chi \in C^r([0,\infty),[0,1])$ with $\chi(t) = 1$ and 0 for $t \leq 1$ and $t \geq 2$, respectively. We define $F(x,u) = \mathcal{N}(x,u)\chi(\|Au\|/\varepsilon)$ for sufficiently small $\varepsilon > 0$.

For this modified system (i.e., (3.1) with $\mathcal{N}$ replaced by F) we look for all, not necessarily small, bounded solutions, or more precisely for all solutions in $W_\alpha^{1,p}(\mathbb{R}, H) \cap L_{p,\alpha}(\mathbb{R}, D(A))$. Here p and α are chosen such that $p > r + 1$ and $0 < \alpha < (r+1)\delta$. As in Theorem 2.10 we have to prescribe the initial value $u_0(s) = P(s)u(s) = \xi$. For notational convenience we introduce the shifted variable $v(x) = u(x+s)$, then v satisfies the integral equation

$$v = S(s,\xi,v) := K(s,\xi,F(\cdot + s, v(\cdot))).$$

Here $K(s,\cdot,\cdot) : E(s) \times L_{p,\alpha}(\mathbb{R}, H) \to L_{p,\alpha}(\mathbb{R}, D(A))$ is the solution operator of the linear problem

$$v'(x) = (A + B(x+s))v(x) + f(x), \quad P(s)v(0) = \xi, \quad (\xi,f) \in E(s) \times L_{p,\alpha}(\mathbb{R}, H).$$

Using the solution operator $\widetilde{K}$ of Theorem 2.10 and the shift operator $T_s : u \mapsto u(\cdot + s)$ we find the relation $K(s,\xi,f) = T_s \widetilde{K}(s,\xi,T_{-s}f)$. Hence, we conclude that K depends smoothly and periodically on $s \in \mathbb{R}$.

Since F has a small Lipschitz constant in the v–variable, the mapping $S(s,\xi,\cdot)$ is a contraction on $L_{p,\alpha}(\mathbb{R}, D(A))$, uniformly in $(s,\xi) \in \mathcal{E}$. Hence, we find a unique solution $v = V(s,\xi)$ where $V : \mathcal{E} \to W_\alpha^{1,p}(\mathbb{R}, H) \cap L_{p,\alpha}(\mathbb{R}, D(A))$ is continuous. Using the fiber contraction principle (cf. [VaV87]) on the scale of the Banach spaces $L_{p/m,\alpha m}(\mathbb{R}, D(A))$, $m = 1,\ldots,r$ it follows as in [Mi88] that $V : \mathcal{E} \to W_{\alpha r}^{1,p/r}(\mathbb{R}, D(A))$ is a C^{r-1}–mapping.

As all the solution $V(s,\xi)$ lie on the (global) center manifold $\widetilde{\mathcal{M}}_C$ we let $h(s,\xi) = (I - P(s))[V(s,\xi)(0)]$ (recall $P(s)[V(s,\xi)(0)] = \xi$ by definition). Now all the remaining properties of h follow as in the classical case, see e.g., [Mi88]. $\square$

Now we want to study center manifolds for periodic solutions p in x–independent problems

$$\frac{d}{dx}w = \mathcal{F}(w). \tag{3.3}$$

Without loss of generality we again assume $p(x + 2\pi) = p(x)$. Linearizing around p we have to be more careful: the ansatz $w = p(x) + u$ leads to an equation of the form (3.1), yet small u does not allow us to capture solutions w which are orbitally close to $\Gamma_{per} = \{ p(x) : x \in \mathbb{R} \}$, like closeby periodic solutions with a slightly different period.

To handle this case we introduce new local coordinates (τ, v) measuring the phase along the orbit Γ_{per} and the distance perpendicular. We choose a 2π–periodic function $q : \mathbb{R} \to H$ with $\langle q(x), p'(x) \rangle = 1$ (e.g., $q = p'/\|p'\|^2$). We let $w = p(\tau) + v$ with $\tau \in [0, 2\pi] = S^1$ and $\langle q(\tau), v \rangle = 0$. The implicit function theorem supplies a local diffeomorphism between w from a neighborhood of Γ_{per} and $(\tau, v) \in \{ (\tau, v) \in S^1 \times H : \|v\| < \varepsilon, \ \langle q(\tau), v \rangle = 0 \}$. We will use τ as a new time–like variable replacing the variable x. Consequently, x, p, and v will be considered as functions of $\tau \in \mathbb{R}$. Further on, the dot ($\cdot$) denotes the derivative with respect to τ. The transformed system (3.3) takes the form

$$\frac{1}{x}(\dot{p} + \dot{v}) = \mathcal{F}(p + v). \tag{3.4}$$

From the orthogonality of v we obtain $\langle \dot{q}, v \rangle + \langle q, \dot{v} \rangle = 0$. Applying $\langle q, \cdot \rangle$ to (3.4) we find the relations

$$
\begin{aligned}
\dot{x} &= \frac{dx}{d\tau} &&= 1 + a(\tau, v) = \frac{1 - \langle \dot{q}(\tau), v \rangle}{1 + \langle q(\tau), \mathcal{F}(p(\tau) + v) - \dot{p} \rangle}, \\
\dot{v} &= \widehat{\mathcal{F}}(\tau, v) &&= (1 + a(\tau, v))\mathcal{F}(p(\tau) + v) - \mathcal{F}(p(\tau)), \quad \langle q(\tau), v(\tau) \rangle = 0.
\end{aligned}
\tag{3.5}
$$

To use the functional analytic setup developed above we have to return to a system defined on all of the space H. Therefore we define an isomorphism between (x, v) and $u \in H$ via

$$u(\tau) = (\tau - x(\tau))\dot{p}(\tau) + v(\tau) \in H, \tag{3.6}$$

which has the inverse $v = Q(\tau)u = u - \langle q, u \rangle \dot{p}$ and $\tau - x = \langle q, u \rangle$. In this way we reintroduce the Floquet multiplier 1, which would be lost by dropping the equation for x. After some elementary calculations (3.4) takes the form

$$\frac{1}{1 - \langle \dot{q}, u \rangle - \langle q, \dot{u} \rangle}\left(\dot{p} + Q\dot{u} + \dot{Q}u\right) = \mathcal{F}(p + Qu), \tag{3.7}$$

where $\dot{Q}(\tau)u = -\langle \dot{q}, u \rangle \dot{p} - \langle q, u \rangle \ddot{p}$.

Lemma 3.2

In the coordinate u, the system (3.3) takes the form

$$\dot{u} = \overline{\mathcal{F}}(\tau, u) = \frac{1 + \langle q, \dot{Q}u \rangle}{1 + \langle q, \mathcal{F}(p + Qu) - \dot{p} \rangle}\left(\mathcal{F}(p + Qu) - \dot{p}\right) + \langle q, u \rangle \ddot{p}, \tag{3.8}$$

232

and the linearization at $u = 0$ reads $D_u \overline{\mathcal{F}}(\tau, 0) = D_w \mathcal{F}(p(\tau))$.

Moreover, for every solution $u : (\tau_1, \tau_2) \to H$ of (3.8) and every $c \in \mathbb{R}$ the function $\widetilde{u}(\tau) = c\dot{p}(\tau) + u(\tau)$ is also a solution of (3.8).

Proof: The form (3.8) follows by simple calculations from (3.7). The assertion on the linear part is readily obtained from $\ddot{p} = D_w \mathcal{F}(p(\tau))\dot{p}$ which follows by differentiating $\dot{p} = \mathcal{F}(p(\tau))$.

The invariance under adding multiples of $\dot{p}$ is a consequence of the fact that x does not appear explicitly on the right–hand side of (3.5). $\square$

To equation (3.8) the center manifold theorem 3.1 is applicable as the linear part satisfies all the assumptions and the nonlinear part is periodic in τ and smooth in (τ, u).

Remark 3.3 Even when starting with a semilinear problem (3.3), the transformed system (3.8) is quasilinear. This is because of the scalar multiplication of $\mathcal{F}$ with the function $a(\tau, Qu)$. In contrast to this, the linear part $\widetilde{A}(\tau)$ still consists of a τ–independent dominant part and a τ–periodic lower order perturbations.

It is possible to avoid the appearence of a quasilinear system by choosing a slightly different coordinate systems, see e.g., [He80, Thm.9.2.2]. This involves Whitney's embedding theorem, and hence is not as constructive as our method.

Now we are able to construct a spatial center manifold around the given periodic orbit Γ_{per}, which contains all solutions $u : \mathbb{R} \to D(A)$ staying orbitally close to Γ_{per}.

Theorem 3.4
Let $p : \mathbb{R} \to D(A)$ be a solution of (3.3) with $p(x + 2\pi) = p(x)$ $\forall x$, such that $D_w \mathcal{F}(p(x)) = A + B(x)$ satisfies the assumptions (A1)–(A3). Further assume $\mathcal{F} \in C^{r+1}(D(A), H)$ for some $r \geq 1$. Let N be the number of Floquet multipliers on the unit circle (counting multiplicity), then there is an N–dimensional center manifold $\mathcal{M}_C \subset D(A)$ containing $\Gamma_{per} = \{ p(\tau) : \tau \in S^1 \}$. It has the following properties:

a) $\mathcal{M}_C$ is locally invariant with respect to the differential equation (3.3).

b) There is an $\varepsilon > 0$ such that all solutions $u : \mathbb{R} \to D(A)$ with $\mathrm{dist}_{D(A)}(u(x), \Gamma_{per}) < \varepsilon$, $\forall x \in \mathbb{R}$, lie in the spatial center manifold $\mathcal{M}_C$.

c) Let $P(\tau)$ be the projection constructed in (2.9), then the tangent space at $\mathcal{M}_C$ in $p(\tau)$ is given by $E(\tau) = \mathrm{Range}(P(\tau))$.

Proof: First we construct a center manifold for the system (3.8). For the linear system $\dot{u} = (A + B(\tau))u$ we find the center subspace from the projection $P(\tau)$ defined in (2.9). The range of $P(\tau)$ is N–dimensional and u is decomposed into $u = u_0 + u_1$ with $u_0(\tau) = P(\tau)u(\tau) \in E(\tau)$. Using Theorem 3.1 we find the $(N+1)$–dimensional center manifold

$$\overline{\mathcal{M}}_C = \{ (\tau, u_0 + u_1) \in \mathbb{R} \times D(A) \; : \; u_1 = h(\tau, u_0), \; \|u_0\| \leq \varepsilon \}.$$

The reduction function satisfies $h(\tau, u_0) = h(\tau + 2\pi, u_0) = \mathcal{O}(\|u_0\|^2)$ and $P(\tau)h(\tau, u_0) = 0$.

Note that $P(\tau)\dot{p}(\tau) = \dot{p}(\tau)$. Hence, $u_0(\tau) = \alpha\dot{p}(\tau) + v_0(\tau)$ and $u_1 = v_1$ where $v_0(\tau) + v_1(\tau) = v(\tau) = Q(\tau)u(\tau)$ as above. Now we define the manifold

$$\widehat{\mathcal{M}}_C = \{ (\tau, v_0 + v_1) \in S^1 \times D(A) \; : \; v_0 \in Q(\tau)P(\tau)H, \; \|v_0\| \leq \varepsilon, \; v_1 = h(\tau, v_0) \}$$

which is N–dimensional. Moreover, it is invariant with respect to (3.5) since for any solution u of (3.8) the function $v(\tau) = Q(\tau)u(\tau)$ solves (3.5).

The desired center manifold in the phase space $D(A)$ is now obtained by reversing the transformation $w = p(\tau) + v$:

$$\mathcal{M}_C = \{ w = p(\tau) + v \in D(A) \; : \; (\tau, v) \in \widehat{\mathcal{M}}_C \}.$$

By construction, we have found a locally invariant manifold for (3.3). The tangent space at $p(\tau)$ is equal to $\text{span}\{\dot{p}(\tau)\} \oplus Q(\tau)P(\tau)H = E(\tau)$. Furthermore, any solution $w = w(x) : \mathbb{R} \to D(A)$ of (3.3) which is orbitally close to Γ_{per} leads to a solution $v(\tau) = w(x(\tau)) - p(\tau)$ of (3.5). $v = v(\tau) : \mathbb{R} \to D(A)$ is small, bounded, and hence contained in $\overline{\mathcal{M}}_C$. Thus, by definition of $\mathcal{M}_C$, the solution $w = w(x)$ lies in $\mathcal{M}_C$. $\square$

Finally we want to study the structure of the reduced spatial system on the center manifold. It is finite dimensional and phrased in terms of the variable v_0 with $v_0(\tau) \in Q(\tau)P(\tau)H$:

$$\frac{d}{d\tau}x = 1 + \bar{a}(\tau, v_0), \qquad \frac{d}{d\tau}v_0 = \overline{f}(\tau, v_0), \tag{3.9}$$

where $\bar{a}(\tau, v_0) = a(\tau, v_0 + h(\tau, v_0))$ and $\overline{f}(\tau, v_0) = P(\tau)\widehat{\mathcal{F}}(\tau, v_0 + h(\tau)) + \dot{P}(\tau)[v_0 + h(\tau, v_0)]$. Further analysis is possible by introducing coordinates in the manifold $\{ (\tau, v_0) \; : \; v_0 \in E(\tau) \}$. A convenient way is to follow the ideas of [MS86, Io88] and to use the generalized eigenfunctions of the Floquet operator. In this way the linear part of the reduced problem becomes τ–independent, while the nonlinear terms are either 4π– or 2π–periodic (as a general reference with examples we refer to [IoA92]).

Assume that -1 is not a Floquet multiplier, then there are $N-1$ 2π–periodic functions (real and imaginary part of the generalized eigenfunctions) ϕ_k such that $E(\tau) = \mathrm{span}\{\phi_1(\tau), \ldots, \phi_{N-1}(\tau)\}$. Moreover, the coordinates $\alpha = (\alpha_1, \ldots, \alpha_{N-1})$ defined by

$$v_0(\tau) = \alpha(\tau) \cdot \phi(\tau) := \sum_{k=1}^{N-1} \alpha_k(\tau)\phi_k(\tau) \tag{3.10}$$

lead to the center manifold

$$\mathcal{M}_C = \{ w = p(\tau) + \alpha \cdot \phi(\tau) + h(\tau, \alpha) \ : \ \tau \in S^1, \ |\alpha| \leq \varepsilon \}$$

with $h(\tau, \alpha) = \mathcal{O}(|\alpha|^2)$ and the differential equation

$$\dot{x} = 1 + \widehat{n}(\tau, \alpha), \qquad \dot{\alpha} = \widehat{A}\alpha + \widehat{N}(\tau, \alpha), \tag{3.11}$$

where

$$\begin{aligned}
\widehat{n}(\tau, \alpha) &= \widehat{n}(\tau + 2\pi, \alpha) = \mathcal{O}(|\alpha|), \\
\widehat{N}(\tau, \alpha) &= \widehat{N}(\tau + 2\pi, \alpha) = \mathcal{O}(|\alpha|^2),
\end{aligned} \tag{3.12}$$

for $\alpha \to 0$. The eigenvalues of the matrix $\widehat{A}$ are exactly the critical Floquet exponents on the imaginary axis except for the trivial exponent 0.

If -1 is a Floquet multiplier then we have to allow for 4π–periodic functions, since the eigenfunctions ϕ_k associated to -1 satisfy $\phi_k(2\pi) = \sigma_k\phi_k(0)$ with $\sigma_k = \pm 1$. We define the matrix $S = \mathrm{diag}(\pm 1, \ldots, \pm 1)$ with the entry σ_k if the associated multiplier is -1 and with the entry $+1$ in all other cases. Now the functions ϕ_k in the ansatz (3.10) are only 4π–periodic but satisfy $\alpha \cdot \phi(\tau + 2\pi) = (S\alpha) \cdot \phi(\tau)$ for all α and τ. Inserting the ansatz into (3.9) we again find a system in the form of (3.11), but now the nonlinearities have the symmetry

$$\begin{aligned}
\widehat{n}(\tau, \alpha) &= \widehat{n}(\tau + 2\pi, S\alpha) = \mathcal{O}(|\alpha|), \\
\widehat{N}(\tau, \alpha) &= \widehat{N}(\tau + 2\pi, S\alpha) = \mathcal{O}(|\alpha|^2).
\end{aligned} \tag{3.13}$$

4   Reversible and Hamiltonian systems

In applications we often encounter systems which are reflection symmetric with respect to the axial direction x. When we rewrite the system as a differential equation with respect to the x–variable this leads to a reversible systems. This means that there is an involution $R : H \to H$ which anticommutes with the vector field:

$$R^2 = I, \qquad \mathcal{F}(Rw) = -R\mathcal{F}(w).$$

As a consequence $\widehat{w}(x) = Rw(-x)$ is a solution of (3.3) whenever $w(x)$ is one. This reversibility carries over to the center manifold if the basic periodic orbit $p(x)$ is *reversible*, that is $p(x) = Rp(x_0 - x)$, $\forall x \in \mathbb{R}$, and fixed x_0. Without loss of generality we may assume $x_0 = 0$ and find $p(0) = Rp(0)$ and $p(\pi) = Rp(-\pi) = Rp(\pi)$, by 2π–periodicity.

The linear part $\widetilde{A}(x) = D_w \mathcal{F}(p(x))$ satisfies $\widetilde{A}(x)R = -R\widetilde{A}(-x)$. This leads to the following consequences for the Floquet spectrum and the projection $P(x)$:

Lemma 4.1
a) If $\lambda \in \mathbb{C}$ is a Floquet exponent with eigenfunction $\phi(x)$ (that is $(e^{\lambda x}\phi)' = \widetilde{A}(x)e^{\lambda x}\phi$), then $-\lambda$ is a Floquet multiplier with eigenfunction $\widetilde{\phi}(x) = R\phi(-x)$.

b) The projection $P(x)$ onto the center space commutes with R: $P(x)R = RP(-x)$.

c) The function ϕ_k and ψ_k can be chosen even or odd, that is, $\phi_k(-x) = \pm R\phi_k(x)$ and $\psi_k(-x) = \pm R^*\psi_k(x)$.

Proof: We introduce the operator $L_{per} u = u' - \widetilde{A}(x)u$ on the space of periodic functions. With $(\widetilde{R}u)(x) = Ru(-x)$ we find the relation $L\widetilde{R} = -\widetilde{R}L$. Since $\widetilde{R}^{-1} = \widetilde{R}$ we have $-L = \widetilde{R}L\widetilde{R}^{-1}$, which implies the symmetry of the spectrum stated in a).

Moreover we find $-(L + \lambda)^{-1} = \widetilde{R}(L - \lambda)^{-1}\widetilde{R}$. Hence the Dunford projections $P_{(-r,r)}$ defined in Section 2.3 satisfies $P_{(-r,r)}\widetilde{R} = \widetilde{R}P_{(-r,r)}$. Letting $r \to \infty$ we find $P_{\mathbb{R}}\widetilde{R} = \widetilde{R}P_{\mathbb{R}}$. Since $P_{\mathbb{R}}$ and $\widetilde{R}$ act pointwise in x assertion b) follows.

Using the relation $P_{\mathbb{R}} = \widetilde{R}P_{\mathbb{R}}\widetilde{R}$ we obtain

$$\tfrac{1}{4}\sum_{k=1}^{N}\left(\langle u, \psi_k + \widetilde{R}^*\psi_k\rangle(\phi_k + \widetilde{R}\phi_k) + \langle u, \psi_k - \widetilde{R}^*\psi_k\rangle(\phi_k - \widetilde{R}\phi_k)\right)$$
$$= \tfrac{1}{4}(P_{\mathbb{R}}u + P_{\mathbb{R}}\widetilde{R}u + \widetilde{R}P_{\mathbb{R}}u + \widetilde{R}P_{\mathbb{R}}\widetilde{R}u + P_{\mathbb{R}}u - P_{\mathbb{R}}\widetilde{R}u - \widetilde{R}P_{\mathbb{R}}u + \widetilde{R}P_{\mathbb{R}}\widetilde{R}u) = P_{\mathbb{R}}u$$

However, since Range($P_{\mathbb{R}}$) is N–dimensional, half of the terms in the above sum must be always zero. $\qquad\square$

For the coordinate change $w = p(\tau) + v$ we choose $q = q(\tau)$ such that $q(\tau) = -R^*q(-\tau)$ where $\langle R^*w, u\rangle = \langle w, Ru\rangle$. Then, the system (3.8) again is reversible: $\overline{\mathcal{F}}(\tau, Ru) =$

$-R\overline{\mathcal{F}}(-\tau, u)$. Now the center manifold $\overline{\mathcal{M}}_C$ of (3.8) can be chosen such that it is reversible, that is $(\tau, u) \in \overline{\mathcal{M}}_C \Longleftrightarrow (-\tau, Ru) \in \overline{\mathcal{M}}_C$ (or equivalently $h(\tau, Ru_0) = Rh(-\tau, u_0)$), see [Mi86]. Using the definition of $\mathcal{M}_C \subset D(A)$ we find $w \in \mathcal{M}_C \Longleftrightarrow Rw \in \mathcal{M}_C$. Thus, the center manifold is reversible together with the reduced flow on the center manifold.

If we choose the basis ϕ_k according to Lemma 4.1(c), then the reduced action of the reversibility is given by $R_0\alpha = (\varepsilon_1\alpha_1, \ldots, \varepsilon\alpha_{N-1})$, where $|\varepsilon_k| = 1$ with $\phi_k(-x) = \varepsilon_k R\phi_k(x)$.

Remark 4.2 In general we can not prove that the dimension of the center manifold is even. The eveness is well known in the case of finite dimensions when the subspaces $H_+ = \{\, w \in H : Rw = w \,\}$ and $H_- = \{\, w \in H : Rw = -w \,\}$ have the same dimension. This statement has no counterpart when H_+ and H_- are infinite dimensional. (We may always add to a given system the reversible system $\gamma' = 0$.) However, the algebraic multiplicity of the Floquet multiplier -1 has to be even as in the finite–dimensional case. This can be concluded from the reduced system directly. Let $\Phi_0(x, s)$ be the fundamental matrix of the reduced linear problem with $\Phi_0(s, s) = I$. By reversibility we know $\Phi_0(-x, -s) = R_0\Phi_0(x, s)R_0$ in addition to the standard relations $\Phi_0(x + 2\pi, s + 2\pi) = \Phi_0(x, s) = \Phi_0(s, x)^{-1}$. As a consequence the Floquet matrix $C = \Phi_0(2\pi, 0)$ satisfies $C^{-1} = \Phi_0(0, 2\pi) = \Phi_0(-2\pi, 0) = R_0CR_0$. Since $\det \Phi_0(x, s) > 0$ and $\det R_0 = 1$ we find $\det C = 1$. Thus, the multiplicity of the eigenvalue -1 must be even, since all other eigenvalues are on the unit circle and either appear in pairs or are $+1$.

If the elliptic problem under consideration was derived as Euler–Lagrange equations from a variational problem we can write the differential equation with respect to x as a Hamiltonian system in the way described in [Mi91]. In the simplest case the symplectic structure ω is given in the form $\omega(u, v) = \langle \Omega u, v \rangle$, where Ω is a constant skew–symmetric invertible operator. The differential equation is then given as

$$w' = \mathcal{F}(w) = \Omega^{-1}\nabla H(w). \tag{4.1}$$

Again we are interested in solutions which are orbitally close to a given periodic solution. Without using the Hamiltonian structure at all, we construct the center manifold as above. However, having obtained an invariant finite dimensional submanifold, we ask how the flow on it can be described using Hamiltonian structures. First of all the

237

Hamiltonian function $H(w)$ provides a conserved quantity for the reduced system on the center manifold. Next we show that the center manifold around any given orbit is again a symplectic manifold. As a consequence, the reduced problem will again be Hamiltonian system on $\mathcal{M}_C$.

Lemma 4.3

Let $p(x)$ be a periodic solution of (4.1) and $Lu = u' - DF(p(x))u$ the linearization around p. Let $P(x)$ be the projection onto the critical Floquet exponent constructed in (2.9). Then the restriction of ω to $E(x) = \text{Range} P(x)$ is non–degenerate, i.e., if $u \in E(x)$ and $\omega(u,v) = 0 \ \forall v \in E(x)$ then $u = 0$. (As a consequence $\dim E(x)$ is even.)

Proof: We essentially apply the ideas from [Mi91, Lemma 3.1], but for convenience of the reader we repeat all the necessary steps. We introduce on $L_2(S^1, H)$ a symplectic form $\tilde\omega$ via

$$\tilde\omega(u(\cdot), v(\cdot)) = \int_0^{2\pi} \omega(u(x), v(x))\, dx.$$

Since $\tilde{A}(x) = DF(p(x)) = \Omega^{-1} D^2 H(p(x))$, where $D^2 H(w)$ is the symmetric second derivative, we find for all smooth periodic u and v the relation

$$\tilde\omega(Lu, v) = \int_0^{2\pi} \langle \Omega(u' - \Omega^{-1} D^2 Hu), v\rangle dx = \int_0^{2\pi} \langle \Omega u, -v' + \Omega^{-1} D^2 Hv\rangle dx = \tilde\omega(u, -Lv).$$

From this we find $\tilde\omega((L-\lambda)^{-1}u, v) = \tilde\omega(u, -(L+\lambda)^{-1}v)$ for all u, v. If we now define the projections $P_{(-n,n)}$, using the Dunford integral as in Section 2.3 with $\Gamma_{(-n,n)} \subset \mathcal{C}$ symmetric to the origin, we see that $\tilde\omega(P_{(-n,n)}u, v) = \tilde\omega(u, P_{(-n,n)}v)$ for all u, v. Hence, this remains true when going to the limit $P_{I\!R}$. We decompose $X = L_2(S^1, H)$ into $X_0 \oplus X_1$ with $X_j = \text{Range}(jI - P_{I\!R})$: For $u_j \in X_j$ we find

$$\tilde\omega(u_0, u_1) = \tilde\omega(P_{I\!R}u_0, u_1) = \tilde\omega(u_0, P_{I\!R}u_1) = 0.$$

This implies that $\tilde\omega$ restricted to X_0 is nondegenerate, i.e., $\tilde\omega(u_0, v_0) = 0 \ \forall v_0 \in X_0$ implies $u_0 = 0$.

Now we use the special form of $P_{I\!R} : u \mapsto P(x)u(x)$. It implies $\omega(P(x)u, v) = \omega(u, P(x)v)$ for all x, u, and v. Hence, for $u_0 \in E(x)$ we have $\omega(u_0, v) = 0 \ \forall v \in H$ if and only if $\omega(u_0, v_0) = 0 \forall v_0 \in E(x)$. Hence, the restriction to $E(x)$ is also nondegenerate. $\qquad\square$

The above Lemma guarantees that the symplectic form when restricted to the center manifold is nondegenerate along the periodic orbit. By continuity it is then also nondegenerate on a whole neighborhood. After possibly shrinking the center manifold we may henceforth assume that the restricted symplectic form is nondegenerate on the whole center manifold. By a general result (see e.g., [Mi91, Theorem 4.1]) the invariance of the center manifold implies that reduced vector field is exactly the Hamiltonian system defined by the reduced symplectic structure and the reduced Hamiltonian function. We define $\widehat{H}(\tau, \alpha) = H(p(\tau) + \phi(\tau) \cdot \alpha + h(\tau, \alpha))$ as the restriction of H to $\mathcal{M}_C$. The restriction $\widehat{\omega}$ is given by $\widehat{\omega}_{(\tau,\alpha)}((\delta\tau_1, \delta\alpha_1), (\delta\tau_2, \delta\alpha_2)) = \langle \widehat{\Omega}(\tau, \alpha) \binom{\delta\tau_1}{\delta\alpha_1}, \binom{\delta\tau_2}{\delta\alpha_2} \rangle$ with

$$\widehat{\Omega}(\tau, \alpha) = \begin{pmatrix} 0 & \widehat{\rho}(\tau, \alpha) \\ -\widehat{\rho}^T(\tau, \alpha) & \widetilde{\Omega}(\tau, \alpha) \end{pmatrix} \text{ where } \begin{cases} \widehat{\rho} = -(p' + \phi'\alpha + \partial_\tau h)^*\Omega(\phi + \partial_\alpha h), \\ \widetilde{\Omega} = (\phi + \partial_\alpha h)^*\Omega(\phi + \partial_\alpha h). \end{cases}$$

Now the reduced system has the form

$$\frac{d}{dx}\begin{pmatrix} \tau \\ \alpha \end{pmatrix} = \widehat{\Omega}(\tau, \alpha)^{-1}\nabla_{\tau,\alpha}\widehat{H}(\tau, \alpha).$$

Here we stay with τ as a dependent variable while x remains the independent variable.

Lemma 4.4
The coordinates α can be chosen such that $\widehat{\Omega}(\tau, 0)$ does not depend on τ.

Proof: With $\phi(x) = (\phi_1, \ldots, \phi_{N-1})$ as above we define $\psi(x) = (p'(x), \phi(x))$. It satisfies the relation

$$(\psi \cdot \beta)' = \widetilde{A}(x)\psi \cdot \beta - \psi \cdot B\beta, \qquad \widetilde{A}(x) = D\mathcal{F}(p(x)) = \Omega^{-1}D^2H(p(x)),$$

for every constant $\beta \in \mathbb{R}^N = \mathbb{R}^{2m}$. Here $B \in \mathcal{L}(\mathbb{R}^N, \mathbb{R}^N)$ is a constant matrix which has eigenvalues equal to the Floquet exponents on the imaginary axis and ψ is the periodic transformation making the reduced linear part constant, that is $\psi(x)e^{Bx}\beta$ solves the linear problem $u' = \widetilde{A}(x)u$. We may choose $\psi(x)$ such that the eigenvalues of B have imaginary part in $(-1/2, 1/2)$. Note that $\pm 1/2$ is explicitly excluded since it would correspond to a multiplier -1. But then we have to double the period and obtain an eigenvalue 0 for the matrix $\widetilde{B}$.

From the definition of $\widehat{\Omega}$ we have $\widehat{\Omega}(x, 0) = \psi(x)^T\Omega\psi(x)$. Differentiating with respect to x gives after some cancellation due to $\mathcal{F} = \Omega^{-1}D^2H$ the linear ODE

$$\frac{d}{dx}\widehat{\Omega}(x) = \widehat{\Omega}B + B^T\widehat{\Omega}.$$

It has the unique solution $\widehat{\Omega}(x) = e^{B^T x}\widehat{\Omega}(0)e^{Bx}$. However, every entry of e^{Bx} is a linear combination of terms of the form $x^m \cos(\sigma_j x + \delta)$ where $|\sigma_j| < 1/2$ since $i\sigma_j$ is an eigenvalue of B. Hence, entries of $e^{B^T x}\widehat{\Omega}(0)e^{Bx}$ consist of terms $x^m \cos((\sigma_j \pm \sigma_k)x + \delta)$. But since $\widehat{\Omega}$ is 2π–periodic it has to be constant. $\qquad\square$

Remark 4.5 If the eigenvalues σ of B are not chosen such that $\operatorname{Im}\sigma \in (-1/2, 1/2)$ then $\widehat{\Omega}(x)$ may not be constant. As an example consider

$$B = \begin{pmatrix} 0 & 0 \\ 0 & S \end{pmatrix}, \quad \widehat{\Omega}(0) = \begin{pmatrix} S & M \\ -M^T & S \end{pmatrix}, \quad \text{where } S = \begin{pmatrix} 0 & 1 \\ -1 & 0 \end{pmatrix}.$$

Then $\widehat{\Omega}(x) = e^{B^T x}\widehat{\Omega}(0)e^{Bx}$ is 2π–periodic and not constant whenever $M \neq 0$.

The linear part of the reduced system reads $\beta' = B\beta$. From $\widehat{\Omega}(0)B + B^T\Omega(0) = 0$ we conclude that $B = \widehat{\Omega}(0)^{-1}D$ where D is symmetric. Defining $H_2(\beta) = \frac{1}{2}\langle D\beta, \beta\rangle$ the linear part is the linear Hamiltonian system $\beta' = \widehat{\Omega}(0)^{-1}\nabla H_2(\beta)$. Using Darboux's theorem (see e.g. [Mi91]) we are even able to adjust the coordinates (τ, α) further such that $\widehat{\Omega}$ is constant and equals $\begin{pmatrix} \Omega_1 & 0 \\ 0 & \Omega_{N-1} \end{pmatrix}$ with $\Omega_m = \begin{pmatrix} 0 & -I_m \\ I_m & 0 \end{pmatrix}$, without destroying the property that the linearized flow is x–independent. This means that the Hamiltonian has the form

$$\widehat{H}(\tau, \alpha) = \alpha_1 + H_2(\alpha) + M(\tau, \alpha), \quad \text{where } M(\tau, \alpha) = \mathcal{O}(|\alpha|^3)$$

and $H_2(\alpha) = \frac{1}{2}\delta_1\alpha_1^2 + \alpha_1\langle\widetilde{\delta}, \widetilde{\alpha}\rangle + \frac{1}{2}\langle\widetilde{A}\widetilde{\alpha}, \widetilde{\alpha}\rangle$ is a quadratic form $(\widetilde{\alpha} = (\alpha_2, \ldots, \alpha_{N-1}))$. The associated Hamiltonian system reads

$$
\begin{aligned}
\tau' &= \partial_{\alpha_1}\widehat{H} &&= 1 + \langle\widetilde{\delta}, \widetilde{\alpha}\rangle + \mathcal{O}(|\alpha|^2), \\
\alpha_1' &= -\partial_\tau\widehat{H} &&= \mathcal{O}(|\alpha|^2), \\
\widetilde{\alpha} &= \Omega_{N-1}^{-1}\nabla_{\widetilde{\alpha}}H &&= \Omega_{N-1}^{-1}(\alpha_1\widetilde{\delta} + \widetilde{A}\widetilde{\alpha} + \mathcal{O}(|\alpha|^2)).
\end{aligned}
\tag{4.2}
$$

To this system we may apply normal form theory. In cases where the Floquet multipliers are real it is then possible to find a normal form where H does not depend on τ. As a consequence the variable α_1 will be a constant of the motion. This reduces the dimension of the system by 2. It only remains to study the Hamiltonian system for $\widetilde{\alpha}$.

In fact, such a reduction can be done for general systems (4.2) by restricting the system onto energy levels and using τ as new time variable. (We follow the arguments given

240

in [GH90, Sect. 4.8].) For small values h the equation $h = \widehat{H}(\tau, \alpha_1, \widetilde{\alpha})$ can be solved for $\alpha_1 = -K(\tau, \widetilde{\alpha}, h)$, since $\tau' = \partial_{\alpha_1}\widehat{H} \approx 1$. From $h = \widehat{H}(\tau, -K(\tau, \widetilde{\alpha}, h), \widetilde{\alpha})$ we find the relation $0 = -\partial_{\alpha_1}H\nabla_{\widetilde{\alpha}}K + \nabla_{\widetilde{\alpha}}H$. Thus, we arrive at

$$\dot{\widetilde{\alpha}} = \frac{d}{d\tau}\widetilde{\alpha} = \widetilde{\alpha}'/\tau' = \frac{1}{\partial_{\alpha_1}H}\Omega_{N-1}^{-1}\nabla_{\widetilde{\alpha}}H = \Omega_{N-1}^{-1}\nabla_{\widetilde{\alpha}}K(\tau, \widetilde{\alpha}, h).$$

Thus, on each energy level $\widehat{H} = h$ we have found a new Hamiltonian system for $\widetilde{\alpha} = \widetilde{\alpha}(\tau)$. The reduced Hamiltonian has the expansion

$$K(\tau, \widetilde{\alpha}, h) = -h + \frac{1}{2}\langle\widetilde{A}\widetilde{\alpha}, \widetilde{\alpha}\rangle + \frac{1}{2}\delta_1 h^2 + h\langle\widetilde{\delta}, \widetilde{\alpha}\rangle + \mathcal{O}(|(\widetilde{\alpha}, h)|^3).$$

5 Linearized instability

So far we have only studied elliptic systems on infinite cylinders. Often such systems occur as the stationary problem of an evolutionary problems such as parabolic or hyperbolic systems. Then it is desirable to analyze the stability properties of the steady states we have constructed above using spatial center manifold theory. In this section we develop a method to detect certain types of instabilites by using a second center manifold reduction applied to the linear spectral problem associated to a given stationary state which is orbitally close to the original periodic orbit $\Gamma_{per} = \{\, p(\tau) : \tau \in S^1 \,\} \subset H$.

The method is a generalization of the idea introduced in [BrM96, BrM95] where the principle of reduced instability was developed for bifurcations from a fixed point. In fact, as long as only spatially periodic solutions are concerned there is an even simpler method which is solely based on the Liapunov–Schmidt reduction, see [Mi95, Mi96a]. The main problem in the stability theory on unbounded cylindrical domains is the fact, that periodic states may be stable with respect to perturbations which have the same spatial periodicity but they are unstable with respect to perturbations which are quasiperiodic or periodic with a slightly changed period. This type of instability is called sideband instability, however, over the last thirty years only formal methods such as multiple scaling or amplitude equations where available to study this phenomenon. First mathematically rigorous methods were developed only recently: in [BrM96] and [Mi95] these methods are introduced by means of simple examples, in [BrM95] a rigorous proof of the Benjamin–Feir instability of periodic surface waves on a fluid layer of finite depth is given, and in [Mi96a] the sidedand instabilities of the Bénard convection rolls are established.

241

Here we want to generalize this method to the case of bifurcations from an arbitrary spatially periodic orbit. To this end consider a general evolutionary system $\mathcal{N}(\frac{\partial}{\partial t}, \frac{\partial}{\partial x}, \phi) = 0$, where t is the time, x the axial variable along the infinite cylinder, and all cross–sectional variables are hidden in the state variable $\phi \in E$ where E is a Hilbert space over the cross–section. We assume that the problem can be rewritten in spatial dynamics form

$$\partial_x w = \mathcal{G}(w, \partial_t w), \quad w \in H, \tag{5.1}$$

where typically w consists of a finite number of derivatives $\frac{\partial^{n+m}}{\partial x^n \partial t^m} \phi$.

The relation to the steady theory in Section 3 is that $\mathcal{G}(w, 0) = \mathcal{F}(w)$, and we again assume the existence a periodic steady state p which satisfies $p(x + 2\pi) = p(x)$ for all $x \in \mathbb{R}$. Assume we have found another steady state W on the spatial center manifold $\mathcal{M}_C$ around the periodic orbit Γ_{per}. Out interest is to analyze the stability properties of W. The linearization at W is

$$\partial_x z = D_w \mathcal{G}(W(x), 0)[z] + D_{(\partial_t w)} \mathcal{G}(W(x), 0) \partial_t z, \tag{5.2}$$

and the associated spectral problem is derived by seeking solutions $z(x, t) = e^{\lambda t} Z(x)$ with $\lambda \in \mathbb{C}$. This yields the linear elliptic problem

$$\partial_x Z = \Big(D_w \mathcal{G}(W(x), 0) + \lambda D_{(\partial_t w)} \mathcal{G}(W(x), 0) \Big) Z. \tag{5.3}$$

We call W linearly unstable when (5.2) has a spatially bounded solution which grows over all bounds for $t \to \infty$. We call W spectrally unstable when (5.3) has a solution (λ, Z) such that $\Re \lambda > 0$ and Z is nontrivial and bounded over $x \in \mathbb{R}$.

The general philosophy of the *principle of reduced instability* as introduced in [BrM96, BrM95, Mi95, Mi96a] is to assume small $|\lambda|$ and to consider (5.3) as a nonautonomous spatial dynamical systems to which the center–manifold reduction is applicable. However, for general non–periodic W it seems to be very difficult to do so, since it is not clear how to use the fact that W is orbitally close to Γ_{per}. In fact, even the construction of W using the center manifold $\mathcal{M}_C$ was done by the help of a local coordinate system around Γ_{per}. We will employ an analogous procedure for the time–dependent problem now.

As in Section 3 we let $w(x(\tau, t), t) = p(\tau) + v(\tau, t)$ with $\langle q(\tau), v(\tau, t) \rangle = 0$, where q was chosen such that $\langle q, \dot{p} \rangle = 1$ (recall $(\dot{\ }) = \partial_\tau (\ )$). The basic system (5.1) transforms into

$$\frac{1}{\dot{x}}(\dot{p} + \dot{v}) = \mathcal{G}\Big(p + v, \partial_t v - \partial_t x \frac{1}{\dot{x}}(\dot{p} + \dot{v}) \Big). \tag{5.4}$$

242

Denoting by $a = \widehat{\mathcal{G}}(w, w_1, \beta)$ the local solution of the implicit equation $a = \mathcal{G}(w, w_1 - \beta a)$ for small β, we obtain the spatial dynamical system

$$\frac{1}{\dot{x}}(\dot{p} + \dot{v}) = \widehat{\mathcal{G}}(p + v, \partial_t v, \partial_t x), \quad \langle q(\tau), v(\tau, t) \rangle = 0. \tag{5.5}$$

For the typical case $\mathcal{G}(w, \partial_t w) = \mathcal{F}(w) + M \partial_t w$ we find the simple formula $\widehat{\mathcal{G}}(w, w_1, \beta) = (I + \beta M)^{-1}(\mathcal{F}(w) + M w_1)$.

We let $u(\tau, t) = (\tau - x(\tau, t))\dot{p} + v(\tau, t)$ in order to obtain a partial differential for the unrestricted variable $u \in H$ rather than v with $\langle q, v \rangle = 0$. As in the steady case (cf. Lemma 3.2) we obtain

$$\dot{u} = \overline{\mathcal{G}}(\tau, u, \partial_t u) = \frac{1 + \langle q, \dot{Q}u \rangle}{1 + \langle q, \widehat{\mathcal{G}} - \dot{p} \rangle} \left(\widehat{\mathcal{G}} - \dot{p} \right) + \langle q, u \rangle \ddot{p}, \tag{5.6}$$

where $\widehat{\mathcal{G}} = \widehat{\mathcal{G}} \Big(p(\tau) + Q(\tau)u, Q(\tau)\partial_t u, -\langle q, \partial_t u \rangle \Big)$.

By construction, the steady part of (5.6) coincides with (3.8), that is $\overline{\mathcal{G}}(\tau, u, 0) = \overline{\mathcal{F}}(\tau, u)$. We now return back to the original steady solution W for (5.1) which has the form $u = U(\tau)$ as a steady solution of (5.6). The spectral stability of U can be studied by substituting $u(\tau, t) = U(\tau) + e^{\lambda t} Z(\tau)$ into (5.6) and then restricting to linear terms in Z:

$$\dot{Z} = \left(D_u \overline{\mathcal{G}}(\tau, U(\tau), 0) + \lambda D_{(\partial_t u)} \overline{\mathcal{G}}(\tau, U(\tau), 0) \right) Z, \tag{5.7}$$

where $D_u \overline{\mathcal{G}}(\tau, U(\tau), 0) = D_u \overline{\mathcal{F}}(\tau, U(\tau))$ since the steady parts coincide.

The second term $D_{(\partial_t u)} \overline{\mathcal{G}}(\tau, U(\tau), 0)$ can be calculated from the function $\mathcal{G}(w, \partial_t w)$ explicitly. In the simple case $\mathcal{G}(w, \partial_t w) = \mathcal{F}(w) + M \partial_t w$ we obtain

$$D_{(\partial_t u)} \overline{\mathcal{G}}(\tau, U(\tau), 0)Z = \frac{1 + \langle q, \dot{Q}U \rangle}{(1 + \langle q, \mathcal{F}(W) - \dot{p} \rangle)^2} \left((1 + \langle q, \mathcal{F}(W) - \dot{p} \rangle) M \widehat{Z} - \langle q, M \widehat{Z} \rangle \right)$$

where $W(\tau) = p(\tau) + Q(\tau)U(\tau)$ and $\widehat{Z} = Q(\tau)Z + \langle q, Z \rangle \mathcal{F}(W(\tau))$.

The main point about the above construction is that for small $|\lambda|$ and small bounded $U(\tau)$ the whole linear problem (5.7) can be considered as a small perturbation of the linear system obtained for $(\lambda, U) = (0, 0)$ but this system is exactly the basic linear problem studied in Section 2. Hence, the assumption for the construction of a spatial center manifold are fulfilled in this case also. Note that the periodicity is not an essential feature in the proof of Theorem 3.1. We only need the periodicity of the

243

leading linear part in order to define the spectral splitting. The terms included into the function $\mathcal{N}$ which has small Lipschitz constant may have an arbitrary dependence on τ, cf. [Mi88].

Thus, it is possible to reduce the spectral problem (5.7) to a linear ODE of the same dimension as the spatial center manifold:

$$\dot{z}_0 = G(\tau, \lambda) z_0. \tag{5.8}$$

Proceeding as in [BrM95, BrM96] it is possible to show that $G(\tau, 0)$ is exactly the linearization at $U_0(\tau)$ of the nonlinear ODE on the spatial center manifold of the steady problem. Moreover, the lowest order terms in λ can be expanded by using the invariance of the linear spatial center manifold. Having this tool in mind, it should be possible to generalize the existing stability theory for small amplitude spatially periodic solution to solutions with arbitrarily large amplitudes.

6 Examples in reaction–diffusion systems

We consider in this section travelling or standing waves of reaction–diffusion systems in infinite cylindrical domains.

$$\tilde{u}_t = D\Delta\tilde{u} + f(y, \tilde{u}), \qquad \text{in } (x, y) \in \Omega = \mathbb{R} \times \Sigma.$$

Here, $\tilde{u} \in \mathbb{R}^m$ is the vector of concentrations (including temperature) and D is a symmetric positive definite matrix containing the (cross–) diffusion rates. Travelling waves have the form $\tilde{u}(t, x, y) = u(x - ct, y)$ where c is the wave speed, and periodic solutions of this type are often called wave–trains. The function u has to satisfy the elliptic problem

$$u_{xx} + \Delta_\Sigma u + cD^{-1}u_x + D^{-1}f(y, u) = 0, \qquad \text{in } (x, y) \in \Omega = \mathbb{R} \times \Sigma.$$

We either prescribe Neumann or Dirichlet boundary conditions for the single components u_j. The system can be rewritten as a differential equation with respect to x in the following way:

$$\frac{d}{dx}\begin{pmatrix} u \\ v \end{pmatrix} = A\begin{pmatrix} u \\ v \end{pmatrix} + \begin{pmatrix} 0 \\ -D^{-1}(cv + f(y, u)) \end{pmatrix}, \quad \text{with } A = \begin{pmatrix} 0 & I \\ -\Delta_\Sigma & 0 \end{pmatrix}. \tag{6.1}$$

For $c = 0$ the system is reversible with respect to the reflection $R(u, v) = (u, -v)$. If moreover there is a potential for f, that is $f(y, u) = \nabla_u F(y, u)$, then (6.1) is a Hamiltonian system. With $w = Du_x$ the Hamiltonian function is

$$H(u, w) = \int_\Sigma \left(\frac{1}{2} w \cdot D^{-1} w - \frac{1}{2} |\nabla_\Sigma (D^{1/2} u)|^2 + F(y, u) \right) dy,$$

and the Hamiltonian system reads $u' = \partial_w H$, $w' = -\partial_u H$, see Section 4 and [Mi91] for the usage of this structure.

Our aim here is to show that the theory developed above can always be applied to systems of the above type. Moreover, we give an example where the Floquet theory can be carried out almost explicitly by using the theory for the scalar Hill problem.

Starting from a given periodic solution $p = p(x, y) = p(x+T, y)$, $T > 0$, the associated Floquet problem is given by

$$L_{per} U = \frac{d}{dx} U - \left[A + \begin{pmatrix} 0 & 0 \\ -D^{-1} \nabla_u f(y, p(x, y)) & -c D^{-1} \end{pmatrix} \right] U. \qquad (6.2)$$

Here the basic Hilbert space is $H = (H_0^1(\Sigma))^m \times (L_2(\Sigma))^m$ and $D(A) = (H^2(\Sigma) \cap H_0^1(\Sigma))^m \times (H_0^1(\Sigma))^m$ for the case of Dirichlet boundary conditions (with the usual changes when some components satisfy Neumann conditions).

In the first part of the following lemma we show that the operator L_{per} always has a discrete spectrum. This is due to the fact that the dominant part A can be controlled very well and that the x–periodic perturbation is of very low order. In the second part we ask how general a perturbation can be without losing the discrete spectrum.

For α, $\beta \in \mathbb{R}$ we introduce the Hilbert spaces

$$\mathcal{H}_{\alpha,\beta} = H_\#^\alpha((0, T), D(K^\beta)) \quad \text{with } K = (1 - \Delta_\Sigma)^{1/2}.$$

The norm is given by $\|u\|_{\alpha,\beta} = \left(\sum_{n \in \mathbb{Z}} \|u_n\|_{D(K^\beta)}^2 \right)^{1/2}$ where $u(x, \cdot) = \sum e^{i2\pi nx/T} u_n(\cdot)$. For α, $\beta > 0$ these spaces have the interpolation property $\|u\|_{\theta\alpha,(1-\theta)\beta} \leq C(\|u\|_{\alpha,0} + \|u\|_{0,\beta})$ for all $\theta \in (0, 1)$. Moreover, let $\mathcal{Y}_\gamma = \mathcal{H}_{\gamma,0} \cap \mathcal{H}_{0,\gamma}$ for $\gamma \geq 0$.

Lemma 6.1

a) Assume $f \in C^1(\mathbb{R}^m, \mathbb{R}^m)$, $p \in C^0(\overline{\Omega}, \mathbb{R}^m)$ and that D is diagonal. Then, the Floquet operator L_{per} defined in (6.2) has a discrete spectrum.

b) Consider the more general operator $\tilde{L}_{per} = d/dx - A - B$ from $D(\tilde{L}_{per}) = \mathcal{Y}_2 \times \mathcal{Y}_1$ into $\mathcal{Y}_1 \times \mathcal{Y}_0$. Assume there is an $\varepsilon > 0$ such that B is a bounded linear operator from

$(\mathcal{H}_{0,1-\varepsilon} \cap \mathcal{H}_{2-\varepsilon,0}) \times \mathcal{H}_{1-\varepsilon,-\varepsilon}$ into $\{0\} \times \mathcal{H}_{0,0}$. Then, the Floquet spectrum of $\widetilde{L}_{per}$ is discrete.

Remark 6.2 The result in part b) is sharp in the sense that for $\varepsilon = 0$ the conclusion does not hold in general. A counter example is given in the Appendix. In applications to the Navier–Stokes equations we arrive at Floquet operators of the form $u_{xx}+\Delta_\Sigma u+ a(x,y)u + b(x,y)u_x + c(x,y)\cdot\nabla_\Sigma u$. This again corresponds to the case $\varepsilon = 0$, where no positive results are known.

Proof: Without loss of generality we may assume that the period of p is 2π. We want to find $\sigma \in I\!\!R$ such that $\lambda = \sigma + i/2$, is in the resolvent set. By the theory of Section 2 it is sufficient to show that $0 = L_{per}U - \lambda U$ has only the trivial solution. With $U = (u,v)^T = (u, u_x - \lambda u)^T$ this is equivalent to

$$M_\lambda u + B(u, u_x - \lambda u) = 0, \quad \text{with } M_\lambda u = u_{xx} - 2\lambda u_x + \lambda^2 u + \Delta_\Sigma u + cD^{-1}(u_x - \lambda u).$$

We first show that M_λ is invertible for $\lambda = \sigma + i/2$. We use the fact that M_λ acts diagonally on the component u_j of u. Denote by b_j the standard basis in $I\!\!R^m$, then $M_\lambda(e^{inx}\phi_k b_j) = m_\lambda(n,k,j)e^{inx}\phi_k b_j$ whenever ϕ_k is an eigenfunction of $-\Delta_\Sigma$ with eigenvalue $\rho_k \geq 0$ and $m_\lambda(n,k,j) = (in - \lambda)^2 - \rho_k + c(in - \lambda)/d_j$. Note that we have found an orthogonal basis of eigenfunctions for the operator M_λ. Hence, the inverse exists and is bounded from $\mathcal{H}_{0,0}$ into $\mathcal{H}_{\alpha,\beta}$ if and only if

$$\nu(\sigma,\alpha,\beta) = \sup\left\{ \frac{(1 + |n|)^\alpha(1 + \rho_k)^{\beta/2}}{|m_{\sigma+i/2}(n,k,j)|} : n \in \mathbb{Z},\ k \in I\!\!N,\ j = 1,\ldots,m \right\}$$

is finite. From

$$|m_{\sigma+i/2}(n,k,j)|^2 = \left((n - 1/2)^2 + \rho_k + c\sigma/d_j - \sigma^2\right)^2 + (2\sigma - c/d_j)^2(n - 1/2)^2 \quad (6.3)$$

we easily find $\nu(\sigma,0,0) = \mathcal{O}(1/\sigma)$ for $\sigma \to \infty$.

To prove part a) we note that $u \mapsto B(u,0) = -D^{-1}\nabla_u f(y,p(\cdot))u$ is a bounded mapping from $\mathcal{H}_{0,0}$ into itself. Now, choose σ so large that $\|M_{\sigma+i/2}^{-1}B(u,0)\|_{0,0} \leq \frac{1}{2}\|u\|_{0,0}$ for all u. Assuming that u is a nontrivial solution of $M_{\sigma+i/2}u + B(u,0) = 0$ we arrive at a contradiction after applying $M_{\sigma+i/2}^{-1}$.

To prove part b) we assume for simplicity $c = 0$. Using (6.3) and substituting $a = (n - 1/2)^2$ yields

$$\nu^2(\sigma,\alpha,\beta) \leq C\tilde{\nu}(\sigma,\alpha,\beta)^2 = C\sup\left\{ \frac{a^\alpha(1 + \rho)^\beta}{(a + \rho - \sigma^2)^2 + 4a\sigma^2} : a \geq 1/4,\ \rho \geq 0 \right\}.$$

Elementary estimates show that $\tilde{\nu}$ is finite if and only if α, β, $\alpha + \beta \le 2$. Moreover,

$$\tilde{\nu}(\sigma, \alpha, \beta) = (C(\alpha, \beta) + o(1)_{\sigma \to \infty})\sigma^{\delta(\alpha, \beta)}, \quad \delta(\alpha, \beta) = \max\{-2, \alpha - 2, \beta - 1, \alpha + \beta - 2\}.$$

Here we have $\delta \le -1 - \varepsilon$ if and only if $\beta \le -\varepsilon$ and $\alpha \le 1 - \varepsilon$.

Denote by b the operator norm of $B = B(u, v)$ from $(\mathcal{H}_{0,1-\varepsilon} \cap \mathcal{H}_{2-\varepsilon,0}) \times \mathcal{H}_{1-\varepsilon,-\varepsilon}$ into $\{0\} \times \mathcal{H}_{0,0}$. Now choose σ large enough such that for all u and $w = M_{\sigma+i/2}u$ the estimate

$$\|w\|_{2-\varepsilon,0} + \|w\|_{0,1-\varepsilon} + \sigma\|w\|_{1-\varepsilon,-\varepsilon} \le \frac{1}{2b}\|u\|_{0,0}$$

holds. Note that the left–hand side is $\mathcal{O}(\sigma^{-\varepsilon})$ for $\sigma \to \infty$. As above we conclude that $M_{\sigma+i/2}u + B(u, u_x - (\sigma + i/2)u) = 0$ has no nontrivial solution. $\square$

We now study a simple example which allows us to carry through the Floquet theory and the center manifold reduction almost explicitly. We choose a problem with a periodic solution which is independent of the cross–sectional variable y. Then, the Floquet operator decouples with respect to an expansion in the eigenfunctions of the cross–sectional operator. To be specific, we consider the elliptic equation

$$u_{xx} + u_{yy} + f(u) = 0, \quad (x, y) \in \Omega = \mathbb{R} \times (0, \pi/\beta), \quad u_y = 0, \quad (x, y) \in \partial\Omega. \quad (6.4)$$

We trivially find periodic solutions which are y–independent, by solving the ODE $p'' + f(p) = 0$. We denote by p_α the unique solution with $p(0) = \alpha$ and $p'(0) = 0$. We assume that in a certain regime of α the solution p_α is periodic with period $T(\alpha)$ and that $p_\alpha''(0) < 0$.

We now want to study the linear Floquet problem around this solution: $\bar{u}_{xx} + \bar{u}_{yy} + f'(p(x))\bar{u} = 0$. We can take advantage of the fact, that Fourier expansion $\bar{u}(x, y) = \sum_{n=1}^{\infty} u_n(x) \sin(\alpha n y)$ leads to a decoupled set of ODEs:

$$u_n'' - (\beta n)^2 u_n + f'(p(x))u_n = 0, \quad n = 0, 1, 2, \ldots \quad (6.5)$$

This is Hill's problem, when $-(\beta n)^2$ is replaced by the spectral parameter λ:

$$u'' + (f'(p(x)) + \lambda)u = 0. \quad (6.6)$$

Let $v_{\alpha,\lambda}(x)$ be the unique solution with initial conditions $u(0) = 1$ and $u'(0) = 0$ and $w_{\alpha,\lambda}$ the unique solution with $u(0) = 0$ and $u'(0) = 1$. Then $\Phi_{\alpha,\lambda}(x) = \begin{pmatrix} v & w \\ v' & w' \end{pmatrix}$ is the fundamental solution. The eigenvalues of $C(\alpha, \lambda) = \Phi(T(\alpha))$ are the Floquet multipliers. Since v is even and w is odd and since $\Phi^{-1}(T) = \Phi(-T)$ with $\det \Phi = 1$, we find $D(\alpha, \lambda) = v_{\alpha,\lambda}(T(\alpha)) = w'_{\alpha,\lambda}(T(\alpha))$. The Floquet multipliers ρ of (6.6) lie on the unit circle if and only if $|D(\alpha, \lambda)| \leq 1$, in fact they satisfy $\rho^2 - 2D(\alpha, \lambda)\rho + 1 = 0$. The function D oscillates for $\lambda \to \infty$ such that all local maxima are ≥ 1 and all local minima ≤ -1. Moreover, there is a λ_0, such that $D > 1$ for $\lambda < \lambda_0$ and $D(\alpha, \lambda_0) = 1$. The values λ with $D(\alpha, \lambda) = 1$ correspond to the eigenvalues of the Hill operator $-u'' - f'(p_\alpha(x))u$ on the space of periodic functions. This follows since $\det \Phi = 1$ and $D = 1$ necessarily implies $v'(T) = 0$ or $w(T) = 0$. Hence, v or w are T–periodic. In our case we have $D(\alpha, 0) = 1$ since p'_α is an eigenfunction for the eigenvalue $\lambda = 0$. Since p' has exactly two zeros, it is clear from the Sturm–Liouville theory that $\lambda = 0$ is not the lowest eigenvalue. Ordering the eigenvalues $\lambda_0 < \lambda_1 \leq \lambda_2 \leq \ldots$ we find either $\lambda_1 = 0$ or $\lambda_2 = 0$, since λ_{2n-1} and λ_{2n} have eigenfunctions with exactly $2n$ zeros. Note that p'_α is the odd eigenfunction with lowest eigenvalue.

When $D = -1$ we obtain antisymmetric eigenfunctions. They have twice the period and satisfy $u(x + T(\alpha)) = -u(x)$. The associated eigenvalues are denoted by $\mu_1 \leq \mu_2 \leq \ldots$ It is well known that $\lambda_0 < \mu_1 \leq \mu_2 < \lambda_1 \leq \lambda_2 < \mu_3 \ldots$. Floquet multipliers are on the unit circle if and only if $\lambda \in [\lambda_0, \mu_1] \cup [\mu_2, \lambda_1] \cup \ldots$

We return to our elliptic problem and keep a fixed periodic solution p in mind. Without loss of generality we again assume that p has period 2π. We have to replace the spectral parameter λ by $-(\beta n)^2$. Hence only a finite number of Fourier modes can have critical Floquet multipliers, namely $-(\beta n)^2 \in [\lambda_0, \mu_0] \cup [\mu_1, \lambda_1]$. For small β of course arbitrarily many Fourier modes can be unstable. We only want to consider the case that besides of $n = 0$ exactly one other mode is critical, let us say $n = 1$. Thus, we obtain a four dimensional center manifold containing the one–parameter family of periodic solutions. Since the linear part is already decoupled, this will be the case for the reduced system.

We have to distinguish several different cases.

Non–real multiplier $\rho = e^{i2\pi\omega}$

The reduced system has the form

$$\dot{x} = 1 + \delta A + n(\tau, A, B, C), \quad \frac{d}{d\tau}\begin{pmatrix} A \\ B \\ C \end{pmatrix} = \begin{pmatrix} N_0(\tau, A, B, C) \\ \omega C + N_1(\tau, A, B, C) \\ -\omega B + N_2(\tau, A, B, C) \end{pmatrix}.$$

In the case of $\omega \notin \mathbb{Q}$ the normal form procedure together with the reversibility with respect to $(x, A, B, C) \rightarrow (-x, A, -B, C)$ leads to

$$n = p(A, B^2 + C^2) + \mathcal{O}(|(A, B, C)|^M), \quad N = \begin{pmatrix} 0 \\ CP_1(A, B^2 + C^2) \\ -BP_1(A, B^2 + C^2) \end{pmatrix} + \mathcal{O}(|(A, B, C)|^M),$$

where p and P_1 are polynomials of their arguments and $M \in \mathbb{N}$ can be chosen as large as one likes. Obviously, the normal form (after omitting the remainders $\mathcal{O}(\ldots)$) is completely integrable, since A and $B^2 + C^2$ are constant on solutions. General solutions are periodic in B and C which leads to invariant tori around the given periodic solution $A = \text{const}$, $B = C = 0$.

If $\omega = p/q$, with $p, q \in \mathbb{N}$ and relatively prime, the normal form will no longer lead to an autonomous system. For instance, $n(\tau, A, B, C)$ may contain the terms $\text{Re}(be^{il\tau}(B + iC)^n(B - iC)^m)$ with $ql + (n - m)p = 0$. We do not want to discuss this question further. The related subharmonic bifurcations are described in [Va90].

Floquet multiplier $\rho = +1$

This corresponds to the case $-\beta^2 = \lambda_0$ or $-\beta^2 = \lambda_1$, when $\lambda_2' = 0$. In both cases the associated 2π–periodic eigenfunction ϕ is an even function. Since $\partial_\lambda D(-\beta^2) \neq 0$ there is only one periodic eigenfunction. Hence, there is a nontrivial Jordan block in the Floquet matrix.

The reduced system reads

$$\dot{x} = 1 + \delta A + n(\tau, A, B, C), \quad \frac{d}{d\tau}\begin{pmatrix} A \\ B \\ C \end{pmatrix} = \begin{pmatrix} N_0(\tau, A, B, C) \\ C + N_1(\tau, A, B, C) \\ N_2(\tau, A, B, C) \end{pmatrix}.$$

Normal form theory leads to an autonomous system with

$$n = p(A,B) + \mathcal{O}(|(A,B,C)|^M), \quad N = \begin{pmatrix} P_0(A,B) \\ BP_1(A,B) \\ P_2(A,B) + CP_1(A,B) \end{pmatrix} + \mathcal{O}(|(A,B,C)|^M).$$

According to Lemma 4.1 we can choose coordinates such that the reduced reversibility operator is $R_0(A,B,C) = (A,B,-C)$. Hence, n and N_2 have to be even in C whereas N_0 and N_1 are odd. As a consequence we find $P_0 = P_1 \equiv 0$. Thus, the normal form reduces to

$$\dot{A} = 0, \quad \dot{B} = C, \quad \dot{C} = \ddot{B} = P_2(A,B).$$

Another interesting normal form is obtained when the reduced reversibility acts as $R_0(A,B,C) = (A,-B,C)$. Then, n and N_1 must be even in B and N_0 and N_2 must be odd in B. The normal form reads

$$\dot{x} = 1 + \delta A + p(A,B^2), \quad \dot{A} = BP_0(A,B^2),$$
$$\dot{B} = C + B^2 P_1(A,B^2), \quad \dot{C} = BP_2(A,B^2) + BCP_1(A,B^2).$$

It is not clear whether the system is integrable or not in the general case.

This case is not realizable in problem (6.4). However, if we add a nonlocal term we will find a suitable example:

$$u_{xx} + u_{yy} + \delta[u]\cos(\beta y) + f(u) = 0, \quad \text{where } [u] = \int_0^{\pi/\beta} u(y)\cos(\beta y)\, dy.$$

Thus, for $n \neq 1$ the Floquet equation (6.5) remain the same while for $n = 1$ the spectral parameter is replaced by $\lambda = \delta\pi/(2\beta) - \beta^2$. We may choose $\delta > 0$ so large that λ hits λ_3 or λ_4. One of these cases corresponds to an odd eigenfunction. In this special case the system is Hamiltionian and thus the normal form is integrable, since a conserved quantity exists.

The Floquet multiplier $\rho = -1$

If the Floquet multiplier -1 is present we find 4π–periodic coefficients in the reduced system. The linear part is the same as above, since Floquet multipliers -1 change into $+1 = (-1)^2$ when the period is doubled:

$$\dot{x} = 1 + \delta A + n(\tau, A, B, C), \quad \frac{d}{d\tau}\begin{pmatrix} A \\ B \\ C \end{pmatrix} = \begin{pmatrix} N_0(\tau, A, B, C) \\ C + N_1(\tau, A, B, C) \\ N_2(\tau, A, B, C) \end{pmatrix}.$$

We have the additional symmetry (with $S(A, B, C) = (A, -B, -C)$)

$$n(\tau + 2\pi, A, B, C) = n(\tau, S(A, B, C)), \quad N(\tau + 2\pi, A, B, C) = SN(\tau, S(A, B, C)).$$

The normal form is the same as above with the additional symmetry imposed by S:

$$\dot{A} = Q_0(A, B^2), \quad \dot{B} = C + BQ_1(A, B^2), \quad \dot{C} = BQ_2(A, B^2) + CQ_1(A, B^2).$$

With the reversibility we will restrict the normal form further. We have $R_0(A, B, C) \to$ $(A, \varepsilon B, -\varepsilon C)$ with $\varepsilon = \pm 1$ depending whether the eigenfunction is even or odd. In both cases we find $Q_0 = Q_1 = 0$ and the normal form reduces to

$$\dot{A} = 0, \quad \dot{B} = C, \quad \dot{C} = \ddot{B} = BQ_2(A, B^2).$$

Note that the linearization around the trivial branch of (periodic) solutions $(A, 0, 0)$ gives the eigenvalues 0 and $\pm\sqrt{Q(A, 0)}$. The latter should pass through 0, hence $\partial_A Q(0, 0) = q_1 \neq 0$ is a natural assumption. From this we see that $Q(A, B^2) = 0$ can be solved for $A = a(B^2)$ providing additional periodic solutions $(A, B, C) = (a(\beta^2), \beta^2, 0)$. However, using the symmetry S we have to identify $+\beta$ and $-\beta$ to one solution of double the period.

From the expansion $Q(A, B^2) = q_1 A + q_2 B^2 + \mathcal{O}(A^2 + B^4)$ we find four different types of phase diagrams depending on the signs of q_2 and $a_1 A$: the period doubling solutions exist for $q_2 q_1 A < 0$. Moreover, there are homoclinic solutions in the same regime: in one case they are asymptotic to the trivial periodic family, and in the other case to the period doubling solutions.

7 Hydrodynamical problems

7.1 Bifurcations from Stokes' waves

The Stokes problem is concerned with periodic traveling waves of an inviscid fluid in a two–dimensional channel with free surface. Let $\mathcal{S} = \{(x, y) \in \mathbb{R} : x \in \mathbb{R}, y \in (0, Y(x))\}$ be the physical flow domain with bottom $y = 0$ and free surface $y = Y(x)$ and let (u, v) be the velocity in (x, y)–direction.

$$u_x + v_y = 0, \quad u_y - v_x = 0, \quad \text{for} \quad (x, y) \in \mathcal{S};$$
$$v = 0, \quad \text{for} \quad y = 0; \tag{7.1}$$
$$v = Y'u, \quad \tfrac{1}{2}(u^2 + v^2) + gY = \text{const.} \quad \text{for} \quad y = Y(x).$$

In $\mathcal{S}$ we have incompressibility and irrotationality of the flow. Moreover, the flow is parallel to the boundary of $\mathcal{S}$. The last relation is Bernoulli's law.

In order to write the system as a differential equation with respect to x we have to transform the problem suitably. One way is given in [Mi88]; however this leads to a quasilinear problem and it is not clear whether the methods devoloped here can be applied. We use another formulation introduced in [AK89]. Note that the flow $\delta = \int_0^{Y(x)} u\, dx$ through each cross–section is independent of x due to incompressibility. We introduce the potential and stream function ϕ and ψ via $(u,v) = \delta(\phi_x, \phi_y) = \delta(\psi_y, -\psi_x)$. Obviously ψ is constant on the boundaries of $\mathcal{S}$, namely $\psi(x,0) = 0$ and $\psi(x, Y(x)) = 1$. Moreover, $\Phi(x + iy) = \phi(x,y) + i\psi(x,y)$ is a holomorphic function mapping $\mathcal{S}$ one–to–one onto the strip $\Omega = I\!\!R \times (0,1)$. Let χ be the inverse mapping, that is, $X = x + iy = \chi(\phi + i\psi)$.

Since $U = u - iv$ is a holomorphic function of $X = x + iy$, we see that $T : \Omega \to \mathbb{C}$ defined by $e^{T(\Phi)} = U(\chi(\Phi))$ again is holomorphic. Letting $\tau(\phi, \psi) + i\theta(\phi, \psi) = T(\phi + i\psi)$ with $\tau,\ \theta \in I\!\!R$ we find the equivalent system

$$\tau_\phi = \theta_\psi,\ \theta_\phi = -\tau_\psi, \quad \text{for}\ \ (\phi, \psi) \in \Omega;$$
$$\theta = 0 \quad \text{for}\ \ \psi = 0; \tag{7.2}$$
$$\tfrac{1}{2}e^{2\tau} + g\eta = \text{const.} \quad \text{for}\ \ \psi = 1.$$

Here $\eta(\phi)$ is defined through $\eta(\phi(x, Y(x))) = Y(x)$. Differentiating this relation with respect to x and using $Y'(x) = v/u = -\tan\theta(\phi(x, Y(x)), 1)$ results in

$$\eta_\phi = Y'/(\phi_x + \phi_y Y') = -\delta\tan\theta/(e^\tau(\cos\theta + \sin\theta\tan\theta)) = -\delta e^{-\tau}\sin\theta.$$

Thus, the boundary condition at $\psi = 1$ has now the form $\tau_\phi = -ge^{-2\tau}\eta_\phi = g\delta e^{-3\tau}\sin\theta$.

Using the auxiliary variable $a = \tau(1)$ we introduce the Hilbert space $H = L_2(0,1)^2 \times I\!\!R$ with elements $w = (\tau, \theta, a)$ and find the spatial dynamical system

$$\tfrac{d}{d\phi}w = Aw + \mathcal{M}(w), \quad \text{with}\ \ Aw = (\theta_\psi, -\tau_\psi, -\theta(1))^T,$$
$$\text{and}\ \ \mathcal{M}(\delta, w) = (0, 0, \theta(1) + \delta g e^{-3a}\sin\theta(1))^T. \tag{7.3}$$

Here A has the domain $D(A) = \{\, (\tau, \theta, a) \in H^1(0,1)^2 \times I\!\!R\ :\ \theta(0) = 0,\ \tau(1) = a\,\}$ and is selfadjoint with respect to the scalar product $\langle w, w_1 \rangle = \int_0^1 (\tau\bar{\tau}_1 + \theta\bar{\theta}_1)d\psi + a\bar{a}_1$. The resolvent is compact and satisfies the estimate

$$\|(A - i\xi)^{-1}f\| \le \frac{1}{|\xi|}\|f\| \quad \text{for all}\ f \in H,\ \xi \in I\!\!R \setminus \{0\}.$$

System (7.3) is reversible with respect to the involution $R : (\tau, \theta, a) \mapsto (\tau, -\theta, a)$, and it is semilinear, since the nonlinear term $\mathcal{M}$ is well-defined on the interpolation space $D(A^\beta)$ for $\beta \in (1/2, 1]$. Thus, the functional analysis of Sections 2 and 3 can be carried through, as soon as we are able to show that the Floquet spectrum is discrete. This is the contents of the following result.

Theorem 7.1
Let $p : \mathbb{R} \rightarrow D(A)$ be any continuous periodic function with period T, then the Floquet operator $L_{per}w = w_\phi - (A + D_w\mathcal{M}(p(x)))w$ has a discrete spectrum.

Proof: According to Section 2 it is sufficient to show that there exists at least one $\lambda \in \mathbb{C}$ such that $L_{per}w = w$ has only the trivial periodic solution.

We may assume $T = 2\pi$ and denote (ϕ, ψ) again by (x, y). The resolvent problem reads

$$\tau_x + \lambda\tau = \theta_y, \ \theta_x + \lambda\theta = -\tau_y, \ \text{for } y \in (0, 1),$$
$$\theta(x, 0) = 0, \ a_x + \lambda a = f(x), \ a(x) = \tau(x, 1), \tag{7.4}$$

where $f(x) = b(x)a(x) + c(x)\theta(x, 1)$ is the x-dependent periodic perturbation. We solve (7.4) for given f by expanding into Fourier series with respect to x. Using $\tau(1) = a$ we find

$$(\lambda + in)\tau_n = \partial_y\theta_n, \ (\lambda + in)\theta_n = -\partial_y\tau_n, \ \text{for } y \in (0, 1),$$
$$\theta_n(0) = 0, \ (\lambda + in)\tau_n(1) = f_n.$$

This yields the explicit solution

$$\tau_n(y) = \frac{\cos[(\lambda + in)y]}{(\lambda + in)\cos(\lambda + in)}f_n, \quad \theta_n(y) = \frac{\sin[(\lambda + in)y]}{(\lambda + in)\cos(\lambda + in)}f_n. \tag{7.5}$$

We now choose $\lambda = k\pi + i/2$, $k \in \mathbb{Z}$. Then the desired traces at $y = 1$ satisfy the estimates

$$|\tau_n(1)|, |\theta_n(1)| \leq C|f_n|/|\lambda + in| \leq \tilde{C}|f_n|/(|k| + |n + \tfrac{1}{2}|),$$

where $\tilde{C} = \sup\{|\tan(k\pi + i(n + \tfrac{1}{2}))| : k, n \in \mathbb{Z}\}$ which is finite due to periodicity and $\tan(\lambda + i(n + \tfrac{1}{2})) = i\tanh(n + \tfrac{1}{2} - i\lambda) \rightarrow i$ for $n \rightarrow \infty$.

Summing up this estimate over n, using Parseval's identity and $a = \tau(1)$, we obtain

$$\int_{S^1}(|a(x)|^2 + |\theta(x, 1)|^2)dx \leq \frac{C}{|k|}\int_{S^1}|f|^2 dx.$$

253

However, for a possible eigenfunction (τ, θ, a) of the Floquet operator the function f is given as $f(x) = b(x)a(x) + c(x)\theta(x, 1)$ with the prescribed bounded periodic functions b and c. Hence, independently of $\lambda = k\pi + i/2$ we find

$$\int_{S^1} |f|^2 dx \leq \left(\|b\|_\infty + \|c\|_\infty \right)^2 \int_{S^1} (|a(x)|^2 + |\theta(x, 1)|^2) dx.$$

However, for large enough $|k|$ the last two estimates imply $a = \theta(\cdot, 1) = f \equiv 0$. Now (7.5) shows that $w = (\tau, \theta, a)^T$ is trivial. $\qquad \square$

This results gives a rigorous basis to the work of C. Baesens and R. MacKay [BM92]. The basic periodic solutions are the so–called Stokes waves. They are symmetric (reversible) such that u and Y are even in x whereas v is odd. As is to be expected there exists a whole family of Stokes waves, and in [BM92] the movement of the Floquet multipliers is studied. An extensive numerical investigation indicates that, when moving along the one–parameter family, single pairs of Floquet multipliers move along the unit circle leaving and entering at ± 1.

In addition it can be shown that the differential equation can be understood as a Hamiltonian system, see [Mi91, BM92, Br94]. In light of the linearization technique of Section 5 it should be possible to generalize the recent proof of the Benjamin–Feir instability in [BrM95] from small amplitude Stokes waves to those of arbitrary amplitudes.

7.2 Navier–Stokes equations

We consider steady flows of a viscous fluid in a fixed infinite cylindrical domain $\Omega = \mathbb{R} \times \Sigma$. Of course, we could also allow for additional state variables like temperature or concentrations which are coupled to the velocity by convection, diffusion, and constitutive relations. However, for simplicity of the presentation we refrain from this generalization. The Navier–Stokes equations are

$$\begin{aligned} u_t + (u \cdot \nabla)u + \nabla p &= \nu \Delta u + f, \quad \nabla \cdot u = 0, \quad \text{in } \Omega, \\ u &= U_\partial \quad \text{on } \partial\Omega = \mathbb{R} \times \partial\Sigma. \end{aligned} \tag{7.6}$$

Here f and U_∂ should not depend on the axial variable x. Looking for steady traveling waves with speed d we have to replace u_t by $-du'$, where $'$ for ∂_x was used.

We write (7.6) as a differential equation with respect to x exactly as was done in [IoMD89]. We decompose u into the scalar component u_a parallel to the axis and u_c

parallel to the cross–section: $u = (u_a, u_c)$. The subscript c denotes cross–sectional components, also in the case of derivatives such as ∇_c and Δ_c. We let $v_c = \nu u_c'$ and $v_a = \nu u_a' - p$. Using incompressibility $(0 = \nabla \cdot u = u_a' + \nabla_c \cdot u_c)$ the pressure can be expressed through $w = (u_a, u_c, v_a, v_c)^T$ as $p = -\nu \nabla_c \cdot u_c - v_a$, and we find the spatial dynamical formulation

$$w' = \frac{d}{dx} w = \mathcal{F}(w) = \mathcal{A}_S w + \mathcal{M}(d, w), \tag{7.7}$$

with

$$\mathcal{A}_S w = \begin{pmatrix} -\nabla_c \cdot u_c \\ \frac{1}{\nu} v_c \\ -\nu \Delta_c u_a \\ -\nu \Delta_c u_c - \nabla_c (v_a + \nu \nabla_c \cdot u_c) \end{pmatrix}, \mathcal{M} = \begin{pmatrix} 0 \\ 0 \\ (u_c \cdot \nabla_c) u_a - (\nabla_c \cdot u_c)(u_a - d) - f_a \\ (u_c \cdot \nabla_c) u_c + (u_a - d) v_c - f_c \end{pmatrix}$$

To fit into the functional analytic setup described in Sections 2 and 3 we define the spaces (see also [IoMD89])

$$H = \{ w = (u, v) \in H_1(\Sigma)^3 \times L_2(\Sigma)^3 : u|_{\partial\Sigma} = 0 \},$$
$$D(\mathcal{A}_S) = \{ w = (u, v) \in H^2(\Sigma)^3 \times H^1(\Sigma)^3 : w \in H, \ v_c|_{\partial\Sigma} = \nabla_c u_c|_{\partial\Sigma} = 0 \}.$$

Note that the pressure p is eliminated completely and that no x–derivatives appear in $\mathcal{F}$. Moreover, v_a, which includes the pressure, only appears with derivatives, hence adding a constant to it does not change the solutions. From incompressibility we see that the mean flux $\int_\Sigma u_a dy$ through each cross–section is constant along solutions, if $\int_{\partial\Sigma} U_\partial \cdot n \, d\sigma = 0$. This always leads to a double Floquet exponent 0, which, however, can be suppressed by projecting out the mean value of v_a and by restricting the mean flux to a given constant.

The system is reversible with $R(u_a, u_c, v_a, v_c)^T = (-u_a, u_c, v_a, -v_c)^T$ if $d = f_a = (U_\partial)_a = 0$. If we work with prescribed mean flux only the case $\int_\Sigma u_a dy = 0$ maintains reversibility. For simplicity we restrict ourself further on to the case of zero wave speed d and zero mean flux $\int_\Sigma u_a dy = 0$.

We assume the existence of a reversible periodic orbit w_p with $w_p(x + T) = p(x) = Rw_p(-x)$, e.g., the Taylor vortices in the Taylor–Couette problem. The ansatz $w = w_p(\tau) + \widehat{w}$ and $\widetilde{w} = \widehat{w} + \alpha w_p'(\tau)$ then leads, as described in Section 3 to the differential equation $\frac{d}{d\tau} \widetilde{w} = \dot{\widetilde{w}} = \widetilde{\mathcal{F}}(\tau, \widetilde{w})$, where $\widetilde{w} = 0$ corresponds to the periodic solution w_p of (7.7). The vector fields $\widetilde{\mathcal{F}}$ is T–periodic in τ and a smooth mapping from $\mathbb{R} \times D(\mathcal{A}_S)$

255

into H. The linearization at $\widetilde{w} = 0$ has the form

$$\widetilde{A}(\tau) = \mathcal{A}_S + B(\tau) \quad \text{with } B(\tau) = D_w \mathcal{M}(w_p(\tau)).$$

From $B(\tau)\widetilde{w} = (0, \ 0, \ (u_c \cdot \nabla_c)\widetilde{u}_a + (\widetilde{u}_c \cdot \nabla_c)u_a - (\nabla_c \cdot u_c)\widetilde{u}_a - (\nabla_c \cdot \widetilde{u}_c)u_a, \ (u_c \cdot \nabla_c)\widetilde{u}_c + (\widetilde{u}_c \cdot \nabla_c)u_a + u_a \widetilde{v}_a + \widetilde{u}_a v_c)^T$ we see that $B(\tau)$ is a bounded linear operator from H into H. This follows since $w_p = (u_a, u_c, v_a, v_c)$ is in $D(\mathcal{A}_S)$ and each term is a product $\widehat{a}\,\widehat{b}$ where either $\widehat{a} \in L_2(\Sigma)$ and $\widehat{b} \in H^2(\Sigma)$ or $\widehat{a}, \ \widehat{b} \in H^1(\Sigma)$. In both cases the product is well defined as element of $L_2(\Sigma)$ (Σ has dimension 1 or 2).

Altogether we find that the abstract methods developed above are applicable to the Navier–Stokes system written in the form (7.7). In [IoMD89] the estimate of the resolvent of $\mathcal{A}_S$ was established. Hence the assumptions (A1), (A2), and (A4) are satisfied. Yet, up to now we were not able to show (A3) in general, that is, the Floquet exponents are discrete. Assuming that (A3) holds does not seem to be a stringent assumption for the following reasons. First we know that the condition is satisfied generically according to Theorem 2.1. Second we are interested in bifurcations at the threshold of instability. As long as a periodic solution is stable there are no Floquet exponents on the imaginary axis, except for the trivial ones. Thus, discreteness of the spectrum is guaranteed. Moving the parameters of the system we expect that only finitely many exponents reach the axis. It seems rather unlikely that the first exponents want to meet the imaginary axis exactly at the same time when the spectrum changes from discrete to continuous.

Periodic patterns typically appear from a trivial state homogeneous by a instability with non–zero wavelength. Restricting the analysis to a fixed period $T = 2\pi/k$ the bifurcation of spatially periodic solutions occurs at the Reynold number $Re = r(T)$. Normalizing the solutions to be reversible we have a two–parameter family of periodic solutions $w = W(Re, T)$ in a region $\{ (Re, T) : T \in (T_1, T_2), \ Re \in (r(T), r(T) + \delta) \}$. In Bénard's problem and also in Taylor–Couette flow the periodic solutions W as well as the function $r(T)$ depend smoothly on Re and T. Often $r(T)$ has a nondegenerate minimum at some $T_c > 0$. In most experiments one sees for $Re > r(T_c)$ periodic solution with periods close to T_c. However, not all periodic solutions $W(Re, T)$ with $r(T) \leq Re$ can be observed, this is due to the so–called Eckhaus instability. We will come back to this phenomenon in the next subsection.

Often there is a period T_0 with $T_0 < T_c < 2T_0$ such that $r(T_0) = r(2T_0)$ and that $r(T_0)$ is not far above $r(T_c)$. Then, for $0 < Re - r(T_0)$ small there bifurcate solutions of period T_0 and $2T_0$. This case is investigated in [BO86] for the Bénard problem.

Calculating the Floquet multipliers of the periodic solution with period close to T_0 we find that they are close to -1. Thus, in the (Re, T)–plane there is a curve $T = \widehat{T}(Re)$ with $T_0 = \widehat{T}(r(T_0))$ such that the periodic solution $w = W(Re, \widehat{T}(Re))$ has -1 has a Floquet multiplier of multiplicity 2 (by reversibility). For generic points on this curve, we expect one Jordan block of length 2 associated with the multiplier -1. Doing a center manifold analysis around one of these periodic orbits we find a four–dimensional center manifold. The reduced system is exactly of the form given in Section 6. As shown there we have period doubling and homoclinic solutions as well as subharmonic bifurcations.

7.3 Eckhaus instability: Four–fold Floquet multiplier 1

As one particular example in Taylor–Couette flow, we examine the so–called Eckhaus instability ([Ec65]) more closely. Above we have introduced the family $w = W(Re, T)$ of symmteric periodic solutions. For the Taylor–Couette problem the existence of this family can be established rigorously, at least close to the threshold of instability, i.e., $Re \approx r(T_c)$, see [IoA92] and the references therein. However, not all the existing Taylor vortex flows are stable, only those with "intermediate" periods. The instability of the vortex structures with too large or too small periods is called Eckaus instability, since this phenomenon was analized first by Eckhaus in [Ec65]. We want to explain here how this instability is related to the movement of the spatial Floquet multipliers associated to the periodic Taylor vortices. In fact, a pair of real (spatial) Floquet multipliers meets exactly at the Eckhaus instability point and then moves along the unit circle. This leads to steady bifurcations with very large spatial wavelength. At the point the Floquet multipliers meet in zero they joint to trivial multipliers sitting there already and form a single Jordan block of length 4. An analysis of the associated ODE is presented in [IoP93], and the analogous phenomenon in the nonlinear Schrödinger equation is studied in [Br92].

To support our claims concerning the Navier–Stokes equations we look at the bifurcation of the Taylor vortices from the x–independent Couette flow w_C. Letting $w = w_C + u$ eqn. (7.7) can be written as

$$u' = (\mathcal{A}_S + \mathcal{A}_C)u + \mathcal{N}(u),$$

where $\mathcal{A}_C u = D_w \mathcal{M}(w_C)$ is the convective part associated to the Couette solution and $\mathcal{N}(u) = \mathcal{M}(0, w_C + u) - \mathcal{M}(w_C) - \mathcal{A}_C u = \mathcal{O}(\|u\|^2)$.

In [IoMD89] to this autonomous system center manifold theory was applied in order to show that all small bounded solutions can be found by studying a finite dimensional ODE. For the case corresponding to the bifurcation of Taylor vortices, the associated reduced ODE is four dimensional and can be written in complex form as follows:

$$A' = ik_0 A + B + f(\mu, A, \overline{A}, B, \overline{B}), \quad B' = ik_0 B + g(\mu, A, \overline{A}, B, \overline{B}), \qquad (7.8)$$

where $f, g = \mathcal{O}(|\mu|(|A| + |B|) + |A|^2 + |B|^2)$ and

$$f(\mu, \overline{A}, A, -\overline{B}, -B) = -\overline{f(\mu, A, \overline{A}, B, \overline{B})}, \quad g(\mu, \overline{A}, A, -\overline{B}, -B) = \overline{g(\mu, A, \overline{A}, B, \overline{B})},$$

due to reversibility $(x, A, B) \rightarrow (-x, \overline{A}, -\overline{B})$. The parameter μ is an external parameter (e.g., $\mu = Re - r(T_c)$) which drives the system from the regime of stable Couette flow ($\mu < 0$) into unstable Couette flow ($\mu > 0$). The value $k_0 > 0$ is the critical wavelength which becomes unstable first. Note that this ODE captures all small bounded solutions of the problem. This includes the bifurcating Taylor vortices, but also all solutions bifurcating from the Taylor vortices (subharmonics etc.).

Using normal form theory the system can be transformed into the following equivalent system

$$
\begin{aligned}
A' &= ik_0 A + B + iAp(\mu, |A|^2, \operatorname{Im} \overline{A}B) + \mathcal{O}(|A|^M + |B|^M), \\
B' &= ik_0 B + iBp(\mu, |A|^2, \operatorname{Im} \overline{A}B) + Aq(\mu, |A|^2, \operatorname{Im} \overline{A}B) + \mathcal{O}(|A|^M + |B|^M).
\end{aligned}
$$
$$(7.9)$$

Here p and q are real polynomials in their last two arguments with μ dependent coefficients. As explained in [IoMD89] we may assume $q(\mu, u, v) = -q_1\mu + q_2 u + q_3 v + \mathcal{O}(\mu^2 + u^2 + v^2)$ with q_1, $q_2 > 0$. Scaling μ, A, and B appropriately we even have $q_1 = q_2 = 1$.

Neglecting the higher order terms, we can look for periodic solutions of the form $A = \alpha e^{ik(x-\gamma)}$ and $B = i\alpha\beta e^{ik(x-\gamma)}$, with $\alpha, \beta, \gamma \in \mathbb{R}$. They represent the Taylor vortices. This leads to the system

$$k - k_0 = \beta + p(\mu, \alpha^2, \alpha^2\beta), \quad \beta(k - k_0) = \beta p(\mu, \alpha^2, \alpha^2\beta) - q(\mu, \alpha^2, \alpha^2\beta).$$

The first relation can be solved as $\beta = k - k_0 + \mathcal{O}(|\mu| + \alpha^2)$ and inserted into the second one. Using the expansion of q we are able to solve for α^2 in terms of μ and $k - k_0$:

$$\alpha^2 = \mu - (k - k_0)^2 + \mathcal{O}(|\mu||k - k_0| + \mu^2 + |k - k_0|^3).$$

258

Thus, periodic solutions exist for $\mu > 0$ only, and in the limited range of wavelengths given by $|k - k_0| < \sqrt{\mu} + \mathcal{O}(|\mu|)$.

Since the following analysis is robust with respect to the addition of higher order terms, we assume for simplicity that $p \equiv 0$ and $q(\mu, u, v) = -\mu + v$. Then, the periodic solutions are given exactly by

$$A = \kappa e^{i(k_0+\delta)(x+\gamma)}, \quad B = i\delta\kappa e^{i(k_0+\delta)(x+\gamma)}, \quad \text{with } \kappa = \sqrt{\mu - \delta^2}$$

We set the phase shift $\gamma = 0$. The parameter δ varies from $(-\sqrt{\mu}, \sqrt{\mu})$.

We now calculate the Floquet multipliers for the periodic solutions of this family. To calculate the linearization around the period orbits we switch to the real form of the problem. With $A = w_1 + iw_2$ and $B = w_3 + iw_4$ we find for $w \in \mathbb{R}^4$ the system

$$w' = \begin{pmatrix} 0 & -k_0 & 1 & 0 \\ k_0 & 0 & 0 & 1 \\ 0 & 0 & 0 & -k_0 \\ 0 & 0 & k_0 & 0 \end{pmatrix} w + \begin{pmatrix} 0 \\ 0 \\ w_1(w_1^2 + w_2^2 - \mu) \\ w_2(w_1^2 + w_2^2 - \mu) \end{pmatrix}.$$

The periodic solution is $p(x) = (\kappa \cos, \kappa \sin, -\delta\kappa \sin, \delta\kappa \cos)^T$, where "sin" and "cos" have the argument $(k_0 + \delta)x$. The linearization reads

$$u' = \widetilde{A}(x)u = \begin{pmatrix} 0 & -k_0 & 1 & 0 \\ k_0 & 0 & 0 & 1 \\ 3p_1^2 + p_2^2 - \mu & 2p_1p_2 & 0 & -k_0 \\ 2p_1p_2 & p_1^2 + 3p_2^2 - \mu & k_0 & 0 \end{pmatrix} u.$$

However, there is a periodic change of coordintes rendering the system constant. It is given by $u(x) = D(x)v(x)$ with $D(x) = \begin{pmatrix} E(x) & 0 \\ 0 & E(x) \end{pmatrix}$ where $E(x) = \begin{pmatrix} \cos & -\sin \\ \sin & \cos \end{pmatrix}$. We find

$$v' = D^{-1}(x)(A(x)D - D'(x))v = \widehat{A}v = \begin{pmatrix} 0 & \delta & 1 & 0 \\ -\delta & 0 & 0 & 1 \\ 2\mu - 3\delta^2 & 0 & 0 & \delta \\ 0 & -\delta^2 & -\delta & 0 \end{pmatrix} v.$$

From $\det(\widehat{A} - \lambda) = \lambda^2(\lambda^2 + 6\delta^2 - 2\mu)$ we conclude that the Floquet exponents are 0 (double) and $\pm 2\pi\sqrt{\mu - 3\delta^2}/(k_0 + \delta)$. Thus, for $|\delta| < \sqrt{\mu/3}$ (that is $\kappa = |A|$ large)

259

there are no nontrivial Floquet exponents on the imaginary axis, for $\delta^2 = \mu/3$ we have a four–fold exponent 0, and for $\delta^2 \in (\mu/3, \mu)$ critical exponents are present.

We want to investigate the case $\delta^2 = \mu/3$ in more detail. Since $\widehat{A}$ has only a one–dimensional kernel its Jordan normal form consists of one block of length 4. Thus, we are led to study the system

$$
\begin{aligned}
\dot{x} = dx/d\tau = 1 + A + n(\tau, A, B, C), \quad \dot{A} = B + N_0(\tau, A, B, C), \\
\dot{B} = C + N_1(\tau, A, B, C), \quad \dot{C} = N_2(\tau, A, B, C).
\end{aligned}
\tag{7.10}
$$

The same system also is expected to appear in fully developed Taylor vortices, when points on the Eckhaus instability curve are investigated. We are in the reversible case with $R_0(A, B, C) = (A, -B, C)$. Hence, n and N_1 are odd in (τ, B), and N_0, N_2 are even. According to [IoP93] the normal form reads

$$
\dot{A} = B + \mathcal{O}, \quad \dot{B} = C + AP(A, B^2 - 2AC) + \mathcal{O}, \quad \dot{C} = BP(A, B^2 - 2AC) + \mathcal{O}.
$$

Neglecting the higher order terms we find the family of periodic solutions $(A, B, C) = (\alpha, 0, \gamma(\alpha))$ with $\gamma + P(\alpha, -2\alpha\gamma) = 0$.

Moreover, the functions $I = B^2 - 2AC$ and $J = C - \int_0^A P(s, B^2 - 2AC)\, ds$ are first integrals. Hence, $C = J + AQ(A, B^2 - 2AJ)$ where Q is analytic (or polynomial after neglecting higher order terms). Thus, we are left with

$$
\ddot{A} = \dot{B} = J + A\widetilde{P}(A, \dot{A}^2 - 2AJ) + \mathcal{O}.
$$

The relevant nonlinear term of $\widetilde{P}$ can be found from $P(A, I) = p_1 A + \mathcal{O}(A^2 + |I|)$. Using $J = C - \frac{1}{2}p_1 A^2 + \mathcal{O}(|A|(A^2 + |I|))$ we obtain $C = J + \frac{1}{2}A^2 + \mathcal{O}(|A|(A^2 + |B^2 - 2AJ|))$ and $\widetilde{P}(A, \widetilde{I}) = \frac{3}{2}A + \mathcal{O}(A^2 + |\widetilde{I}|)$.

We assume $p_1 \neq 0$ which is generically the case. Hence, for J with $Jp_1 < 0$ there are two fixed points: one a saddle ($p_1 A > 0$) and the other a center ($p_1 A < 0$). Attached to the saddle there is a homoclinic solution (recall the symmetry with respect to the A–axis). For $J \to 0$ the homoclinic orbit shrinks into the origin. For $Jp_1 > 0$ no bounded solutions exist.

The question which of these solutions persist for the full problem (7.10) is treated in [IoP93]. It follows that reversible homoclinic orbits as well as reversible periodic orbits (subharmonics) persist. Furthermore the theory of [IoL90] predicts that two–dimensional invariant tori with quasi–periodic flow exist for (7.10). We do not pursue these questions further. We only mention that the homoclinic solutions may be considered as (weak) defects of the perfect periodic pattern. The solutions converge at both infinities to periodic states but differ in the middle.

Appendix: A counter example to Floquet theory

We construct an elliptic problem of the type

$$u_{xx} + u_{yy} + A(x)u + C(x)u_x = 0, \quad (x, y) \in \mathbb{R} \times (0, \pi).$$

Here A and B are operators which depend periodically on $x \in \mathbb{R}$. The aim is to construct A and B in such way that the problem has a solution u which decays faster than exponential for $x \to \pm\infty$.

We expand u in a Fourier series with respect to y:

$$u(x, y) = \sum_{n=1}^{\infty} u_n(x) \sin(ny),$$

and let $U = (u_n)_{n \in \mathbb{N}} \in \ell_2$. Then our problem in equivalent to

$$u_n'' - n^2 u_n + A_n(x)U + C_n(x)U' = 0 \quad \forall n \in \mathbb{N}.$$

We define first the solution which decays faster than exponential and then show that an associated periodic problem exists which has the given solution.

We choose a cut–off funtion $\chi \in C^\infty(\mathbb{R}, [0, 1])$ with $\chi(t) = 0$ for $|t| > 1$, $\chi(t) = 1$ for $|t| < 1/2$, and $t\chi'(t) \le 0$ for all $t \in \mathbb{R}$. Now we let $\chi_m(t) = \chi(t - m)$ for $m \in \mathbb{Z}$ and define

$$u_n(x) = \alpha_n(\chi_n(x)e^{-nx} + \chi_{1-n}(x)e^{nx}),$$

where the sequence $\alpha_n \ge 0$ will be determined later. Note that each u_n has compact support, and for $x \in [m, m+1]$, $m \in \mathbb{N}$, we have $\|U(x)\|_2 \le (\alpha_m + \alpha_{m+1})e^{-m^2}$. We have

$$\rho_n(x) = u_n'' - n^2 u_n = \alpha_n([\chi_n'' - 2n\chi_n']e^{-nx} + [\chi_{1-n}'' + 2n\chi_{1-n}']e^{nx}).$$

To arrive at a linear homogeneous problem, we have to express the remainder ρ_n through the variables $U = (u_m)$ again.

The main idea is to separate the periodicity interval $[0, 2]$ in two parts: in $[0, 1]$ there is interaction only between the modes $2m$ and $2m + 1$ for all $m \in \mathbb{N}$ whereas in $[1, 2]$ interaction takes place between the modes $2m - 1$ and $2m$. Thus, u_n is zero on $[0, n - 1]$, then stimulated on the interval $[n - 1, n]$ by interaction with u_{n-1}. On $[n, n+1]$ the variable u_n is forced down to zero by the variable u_{n+1}.

In particular, we look for the interaction in the form

$$\rho_n(x) = a_n(x)u_n(x) + b_n(x)u_{n+1}(x) + c_n(x)u_n'(x) + d_n(x)u_{n-1}'(x),$$

where $d_1(x) \equiv 0$. For $x \in [n-1, n]$ we assume $b_n(x) = c_n(x) = 0$ and for $x \in [n, n+1]$ we assume $a_n(x) = d_n(x) = 0$. Hence, a_1 is smooth and well–defined through $a_1(x) = \rho_1(x)/u_1(x)$ on $[0,1]$ and 0 on $[1,2]$.

Using the periodicity we are able to determine all the other coefficients. Let $x \in [n, n+1]$, then $x - 2n \in [-n, 1-n]$ and $\rho(x-2n) = b_n(x)u_{n+1}(x-2n) + c_n(x)u'_n(x-2n)$. Together with $\rho_n = b_n u_{n+1} + c_n u'_n$ this defines a linear system for (b_n, c_n). The determinant is

$$\Delta_n(x) = u_{n+1}(x)u'_n(x-2n) - u_{n+1}(x-2n)u'_n(x) = \alpha_n \alpha_{n+1} e^{-n^2 - n} \delta_n(x)$$

with $\delta_n(x) = \chi_{n+1}(\chi'_{n+1} + n\chi_n)e^{n-x} - \chi_n(\chi'_n - n\chi_n)e^{x-n} \geq n/e$ for $x \in [n, n+1]$. For the last estimate note $-\chi'_n, \chi'_{n+1} \geq 0$ on $[n, n+1]$, $\chi_{n+1} = 1$ on $[n, n+1/2]$, and $\chi_n = 1$ on $[n+1/2, n+1]$.

Since b_n and c_n are given via

$$b_n(x) = \frac{1}{\Delta_n(x)}(u'_n(x-2n)\rho_n(x) - u'_n(x)\rho_n(x-2n)),$$
$$c_n(x) = \frac{1}{\Delta_n(x)}(u_{n+1}(x)\rho_n(x-2n) - u_{n+1}(x-2n)\rho_n(x)),$$

we find, by elementary estimates, $|b_n(x)| \leq M\alpha_n n e^n/\alpha_{n+1}$ and $|c_n(x)| \leq M$, where fixed constant M independent of n. Recalling $\rho_n(\pm n) = \rho_n(1 \pm n) = 0$, it is clear that b_n and c_n can be continued to continuous functions which are 0 on $[n+1, n+2]$ and periodic with period 2.

For a_m and d_m, $m \geq 2$, we have the relations

$$a_{n+1}(x) = \frac{1}{\Delta_n(x)}(u'_n(x-2n)\rho_{n+1}(x) - u'_n(x)\rho_{n+1}(x-2n)),$$
$$d_{n+1}(x) = \frac{1}{\Delta_n(x)}(u_{n+1}(x)\rho_{n+1}(x-2n) - u_{n+1}(x-2n)\rho_{n+1}(x)),$$

and the estimates $|a_{n+1}(x)| \leq Mn$ and $|d_{n+1}(x)| \leq M\alpha_{n+1}e^{-n}/\alpha_n$.

Using the choice $\alpha_n = e^{-n^2/2}$ we find $|a_n|, |b_n| \leq Mn$ and $|c_n|, |d_n| \leq M$. Thus, the operator $\tilde{A}(x)$ with $(\tilde{A}(x)U)_n = a_n(x)u_n + b_n(x)u_{n+1}$ is bounded from ℓ_2^1 into ℓ_2. Similarly, $(\tilde{C}(x)U)_n = c_n(x)u_n + d_n(x)u_{n-1}$ defines a bounded operator from ℓ_2 into itself. In fact, a more careful analysis shows $\tilde{A} \in C_\#^k(\mathbb{R}, \mathcal{L}(\ell_2^1, \ell_1))$ and $\tilde{C} \in C_\#^k(\mathbb{R}, \mathcal{L}(\ell_2, \ell_1))$ for each $k \in \mathbb{N}_0$.

Doing the inverse Fourier expansion, we find the associated operators

$$A(\cdot) \in C_\#^k(\mathbb{R}, \mathcal{L}(H_0^1(0,\pi), L_2(0,\pi))) \quad \text{and} \quad C(\cdot) \in C_\#^k(\mathbb{R}, \mathcal{L}(L_2(0,\pi), L_2(0,\pi))).$$

Thus, we have found a periodic lower order perturbation of the Dirichlet problem on the strip which possesses a C^∞–solution $u(x,y) = \sum u_n(x)\sin(ny)$ which decays faster than exponential for $|x| \to \infty$ (together with all the derivatives).

References

[AGH88] D. Armbruster, J. Guckenheimer and P. Holmes. Heteroclinic cycles and modulated travelling waves in systems with $O(2)$ symmetry. Physica D **29**, 257–282, 1988.

[AK89] C. J. Amick and K. Kirchgässner. A theory of solitary water waves in the presence of surface tension. Arch. Rat. Mech. Anal. **105**, 1–49, 1989.

[An92] S. Angenent. A variational interpretation of Melnikov's function and exponentially small separatrix splitting. Preprint, 1992.

[AR67] R. Abraham and J. Robbin. *Transversal Mappings and Flows.* Benjamin Inc., Amsterdam, 1967.

[Ar72] V. I. Arnol'd. Lectures on bifurcations and versal systems. Russ. Math. Surv. **27**, 54–123, 1972.

[Ar83] V. I. Arnol'd. *Geometrical Methods in the Theory of Ordinary Differential Equations.* Springer-Verlag, New York, 1983.

[Ar93] V. I. Arnol'd (ed.). *Dynamical Systems V. Theory of Bifurcations and Catastrophes.* Enc. Math. Sciences **5**, Springer-Verlag, Berlin, 1992.

[AS74] V. S. Afraimovich and L. P. Shilnikov. On attainable transitions from Morse-Smale systems to systems with many periodic points. Math. USSR Izvestija **8**, 1235–1270, 1974.

[AS91] V. S. Afraimovich and M. A. Shereshevsky. The Hausdorff dimension of attractors appearing by saddle-node bifurcations. Int. J. Bif. Chaos **1**, 309–325, 1991.

[AW75] D. G. Aronson and H. F. Weinberger, (1975), Nonlinear diffusion in population genetics, combustion and nerve propagation. In Lecture Notes in Mathematics, Vol. **446**, 5-49, Springer-Verlag, New-York.

[AY78] J. C. Alexander and J. A. Yorke. Global bifurcation of periodic orbits. Amer. J. Math. **100**, 263–292, 1978.

[Be80] L. A. Belyakov. Bifurcation set in a system with homoclinic saddle curve. Mat. Zam. **28**, 911–922, 1980.

[Be84] L. A. Belyakov. Bifurcation of systems with homoclinic curve of a saddle-focus with saddle quantity zero. Mat. Zam. **36**, 838–843, 1984.

[Be90] W.-J. Beyn. The numerical computation of connecting orbits in dynamical systems. IMA J. Numer. Anal. **9**, 379–405, 1990.

[BLN86] H. Brand, P. Lomdahl, and A. C. Newell. Evolution of the order parameter in situtations with broken rotational symmetry. Phys. Lett. A **118**, 67, 1986.

[BLR92] H. Berestycki, B. Larrouturou and J.- M. Roquejoffre. Stability of travelling fronts in a model for flame propagation, part 1: linear stability. Arch. Rat. Mech. Anal. **117**, 97-117, 1992.

[BM92] C. Baesens and R.S. MacKay. Uniformly travelling water waves from a dynamical systems viewpoint: some insight into bifurcations from the Stokes family. J. Fluid Mechanics, **241**, 333–347, 1992.

[Bo76] R. Bogdanov. Versal deformations of a singularity of a vector field on the plane in the case of zero eigenvalues. Russ. (1976), Engl.: Sel. Mat. Sov. **1**, 389–421, 1981.

[Bo81] R. Bogdanov. Bifurcation of the limit cycle of a family of plane vector fields. Russ. (1976), Engl.: Sel. Mat. Sov. **1**, 373–387, 1981.

[BO86] F.H. Busse and A.C. Or. Subharmonic and asymmetric convection rolls. J. Applied Math. Physics (ZAMP), **37**, 608–623, 1986.

[Br83] M. Bramson. Convergence of solutions of the Kolmogorov equation to travelling waves. Mém. Amer. Math. Soc. **285**, 1983.

[Br92] T.J. Bridges. Hamiltonian bifurcations of the spatial structure for coupled nonlinear Schrödinger equations. Physica **D 57**, 375–394, 1992.

[Br94] T.J. Bridges. Hamiltonian spatial structure for three–dimensional water waves in a moving frame of reference. J. Nonlinear Science **4**, 221–251, 1994.

[BrM95] T.J. Bridges and A. Mielke. A proof of the Benjamin–Feir instability. Archive Rational Mech. Analysis **133**, 145–198, 1995.

[BrM96] T.J. Bridges and A. Mielke. Linear instability of spatially–periodic states via Hamiltonian center–manifold techniques. Math. Nachr. **179**, 5–25, 1996.

[BS90] A. K. Bajaj and P. R. Sethna. Effect of symmetry-breaking perturbations on flow induced oscillations in tubes. Preprint, 1990.

[Bu72] T. Burak. On semigroups generated by restrictions of elliptic operators to invariant subspaces. Israel J. of Math. **12**, 79-93, 1972.

[By78] V. V. Bykov. On the structure of a neighborhood of a separatrix contour with a saddle-focus (Russ.). Meth. Qual. Th. Diff. Eq. **133**, 3–32, (Gorki, 1978).

[By80] V. V. Bykov. On bifurcations of dynamical systems which are close to systems with a separatrix contour (Russ.). Meth. Qual. Th. Diff. Eq. **224**, 44–72, (Gorki, 1980).

[By88] V. V. Bykov. On nontrivial hyperbolic sets arising from a contour which consists of saddle separatrices (Russ.). Meth. Qual. Bif. Th. **120**, 22–32, (Gorki, 1988).

[Ca81] J. Carr. *Applications of Centre Manifold Theory.* Appl. Math. Sci. **35**, Springer-Verlag, New-York 1981.

[CD89] S.-N. Chow and B. Deng. Bifurcation of a unique periodic orbit from a homoclinic orbit in infinite-dimensional systems. Trans. Amer. Math. Soc. **312**, 539–587, 1989.

[CDF90] S.-N. Chow, B. Deng and B. Fiedler. Homoclinic bifurcation at resonant eigenvalues. J. Dyn. Diff. Eq. **2**, 177–244, 1990.

[CDT90] S.-N. Chow, B. Deng and D. Terman. The bifurcation of a homoclinic orbit and periodic orbits from two heteroclinic orbits. SIAM J. Math. Anal. **21**, 179–204, 1990.

[CDT91] S.-N. Chow, B. Deng and D. Terman. The bifurcation of a homoclinic orbit from two heteroclinic orbits – a topological approach. Applic. Analysis **42**, 275–300, 1991.

[CE93] P. Collet and J. P. Eckmann. The time dependent amplitude equation for the Swift-Hohenberg problem. Comm. Math. Physics, **132**, 139-153, 1993.

[CES90] V. Coti Zelati, I. Ekeland and E. Séré. A variational approach to homoclinic orbits in Hamiltonian systems. Math. Ann. **288**, 133–160, 1990.

[CFT85] P. Coullet, S. Fauve, and E. Tirapegui. Large scale instability of nonlinear standing waves. J. Physique Lett., **46**, 787, 1985.

[Ch61] S. Chandrasekhar. *Hydrodynamic and Hydromagnetic Stability.* Oxford University Press, 1961.

[CH82] S.-N. Chow and J. K. Hale. *Methods of Bifurcation Theory.* Springer-Verlag, New York, 1982.

[Ch93] P. Chossat. Forced reflectional symmetry breaking of an $O(2)$-symmetric homoclinic cycle. Nonlinearity **6**, 723–731, 1993.

[CK72] S. C. Chikwendu and J. Kevorkian. A perturbation method for hyperbolic equations with small nonlinearities. SIAM J. Appl. Math.,**22**, 235–258, 1972.

[CK88] J. D. Crawford and E. Knobloch. Classification and unfolding of degenerate Hopf bifurcations with O(2) symmetry: No distinguished parameter. Physica **D31**, 1–48, 1988.

[CK91] J. D. Crawford and E. Knobloch. Symmetry and symmetry-breaking bifurcations in fluid dynamics. Ann. Rev. Fluid Mech.**23**, 341–387, 1991.

[CK94] A. Champneys and Y. Kuznetsov. Numerical detection and continuation of codimension-two homoclinic bifurcations. Int. J. Bif. Chaos **4**, 795–822, 1994.

[CL90] S.-N. Chow and X.-B. Lin. Bifurcation of a homoclinic orbit with a saddle-node equilibrium. Diff. Integr. Eq. **3**, 435–466, 1990.

[ClK94] T. Clune and E. Knobloch. *Pattern selection in three–dimensional magnetoconvection.* Physica, **D74**, 151–176, 1994.

[CM78] S.-N. Chow and J. Mallet-Paret. The Fuller index and global Hopf bifurcation. J. Diff. Eq. **29**, 66–85, 1978.

[CMY83] S.-N. Chow, J. Mallet-Paret and J. Yorke. A periodic orbit index which is a bifurcation invariant. In [Pa83], 109–131.

[Co78] C. C. Conley. *Isolated Invariant Sets and the Morse Index.* CBMS Notes **38**, AMS, Providence, 1978.

[CoR91] V. Coti Zelati and P. H. Rabinowitz. Homoclinic orbits for second order Hamiltonian systems possessing superquadratic potentials. J. Amer. Math. Soc. **4**, 693–727, 1991.

[CoR92] V. Coti Zelati and P. H. Rabinowitz. Homoclinic type solutions for a semilinear elliptic PDE on IR^n. Comm. Pure. Appl. Math. **45**, 1217–1269, 1992.

[Cr86] M. C. Cross. Traveling and standing waves in binary fluid convection in finite geometries. Phys. Rev. Lett. **57**, 2935, 1986.

[Cr88] M. C. Cross. Structure of nonlinear traveling-wave states in finite geometries. Phys. Rev. A **38**, 3593, 1988.

[CrK92] M. C. Cross and E. Y. Kuo. One dimensional structure near a Hopf bifurcation at finite wavenumber. Physica **D59**, 90–120, 1992.

[CW89] V. Croquette and H. Williams. Nonlinear competition between waves on convective rolls. Phys. Rev. A **39**, 2765, 1989.

[DaG87] G. Dangelmayr and J. Guckenheimer. On a four parameter family of planar vector fields. Arch. Rat. Mech. Anal. **97**, 321–352, 1987.

[DaK86] G. Dangelmayr and E. Knobloch. Interaction between standing and traveling waves and steady states in magnetoconvection. Phys. Lett. A **117**, 394, 1986.

[DaK87a] G. Dangelmayr and E. Knobloch. On the Hopf bifurcation with broken O(2) symmetry. In W. Güttinger and G. Dangelmayr, editors, *The Physics of Structure Formation. Theory and Simulation*, volume 37 of *Springer Series in Synergetics*, pages 387–393, Berlin, 1987. Springer-Verlag.

[DaK87b] G. Dangelmayr and E. Knobloch. The Takens-Bogdanov bifurcation with O(2)-symmetry. Phil. Trans. Roy. Soc. A **322**, 243–279, 1987.

[DaK89] G. Dangelmayr and E. Knobloch. Hopf bifurcation in reaction diffusion equations with broken translational symmetry. In L. Kaitai, J. Marsden, M. Golubitsky, and G. Iooss, editors, *Bifurcation Theory and its Numerical Analysis*, pages 162–170. Xian Jiaoting University Press, 1989.

[DaK90] G. Dangelmayr and E. Knobloch. Dynamics of slowly traveling wave trains in finite geometry. In F. H. Busse and L. Kramer, editors, *Nonlinear Evolution of Spatio-Temporal Structures in Dissipative Continuous Systems*, pages 399–410. Plenum Press, 1990.

[DaK91] G. Dangelmayr and E. Knobloch. Hopf bifurcation with broken circular symmetry. Nonlinearity **4**, 399–427, 1991.

[DaKW91] G. Dangelmayr, E. Knobloch, and M. Wegelin. Dynamics of travelling waves in finite containers. Europhysics Letters **1**(8), 723–729, 1991.

[DaRG93] G. Dangelmayr, J. D. Rodriguez, and W. Güttinger. Dynamics of waves in extended systems. Lectures in Applied Mathematics **29**, 145-161, 1993.

[DaWK91] G. Dangelmayr, M. Wegelin, and E. Knobloch. Traveling wave convection in finite containers. Eur. J. Mech. B/Fluids, **10**(2-Suppl.), 125–130, 1991.

[De90] B. Deng. Homoclinic bifurcations with nonhyperbolic equilibria. SIAM J. Math. Anal. **21**, 693–720, 1990.

[De76] R. Devaney. Homoclinic orbits in Hamiltonian systems. J. Diff. Eq. **21**, 431–438, 1976.

[De77] R. Devaney. Blue sky catastrophes in reversible and Hamiltonian systems. Ind. Univ. Math. J. **26**, 247–263, 1977.

[De91a] B. Deng. The bifurcations of countable connections from a twisted heteroclinic loop. SIAM J. Math. Anal. **22** (1991), 653–679.

[De91b] B. Deng. The existence of infinitely many travelling front and back waves in the FitzHugh-Nagumo equations. SIAM J. Math. Anal. **22**1631–1650, 1991.

[DGJM91] G. Dyławerski, K. Gęba, J. Jodel and W. Marzantowicz. S^1-equivariant degree and the Fuller index. Ann. Pol. Math. **52**, 243–280, 1991.

[DGSS95] B. Dionne, M. Golubitsky, M. Silber, and I. Stewart. Time-periodic spatially-periodic planforms in euclidean equivariant PDE. Preprint, 1995.

[DKT87] A. E. Deane, E. Knobloch, and J. Toomre. Traveling waves and chaos in thermosolutal convection. Phys. Rev. A **36**, 2862, 1987.

[DoF89] E. J. Doedel and M. J. Friedman. Numerical computation of heteroclinic orbits. J. Comp. Appl. Math. **26**, 155–170, 1989.

[DoK85] E. J. Doedel and J. P. Kernevez. Software for continuation problems in ordinary differential equations with applications. CALTECH 1985.

[DR81] P. G. Drazin and W. H. Reid. *Hydrodynamic Stability.* Cambridge University Press, 1981.

[DRS91] F. Dumortier, R. Roussarie, J. Sotomayor and H. Żoładek. *Bifurcations of Planar Vector Fields.* Springer-Verlag, Berlin 1991.

[Ec65] W. Eckhaus. *Studies in Nonlinear Stability Theory.* Springer Tracts in Natural Phil. Vol. **6**, Springer–Verlag, 1965.

[Ec75] W. Eckhaus. New approach to the asymptotic theory of nonlinear oscillations and wave propagation. Journal of Mathematical Analysis and Applications **49**, 575–611, 1975.

[EFF82] J. Evans, N. Fenichel and J. A. Feroe. Double impulse solutions in nerve axon equations. SIAM J. Appl. Math. **42**, 219–234, 1982.

[Ek90] I. Ekeland. *Convexity Methods in Hamiltonian Mechanics.* Springer-Verlag, Berlin, 1990.

[EW94] J.-P. Eckmann and C.E. Wayne. The non-linear stability of front solutions for parabolic partial differential equations. Comm. Math. Phys. **161**, 323-334, 1994.

[Fe82] J. A. Feroe. Existence and stability of multiple impulse solutions of a nerve axon equation. SIAM J. Appl. Math. **42**, 235–246, 1982.

[Fi37] R. A. Fisher. The advance of advantageous genes. *Ann. of Eugenics* **7**, 355-369, 1937.

[Fi84] G. Fischer. Zentrumsmannigfaltigkeiten bei elliptischen Differentialgleichungen. Math. Nachr. **115**, 137–157, 1984.

[Fi85] B. Fiedler. An index for global Hopf bifurcation in parabolic systems. J. reine angew. Math. **359**, 1–36, 1985.

[Fi86a] B. Fiedler. Global Hopf bifurcation for Volterra integral equations. SIAM J. Math. Anal. **17**, 911–932, 1986.

[Fi86b] B. Fiedler. Global Hopf bifurcation of two-parameter flows. Arch. Rat. Mech. Anal. **94**, 59–81, 1986.

[Fi88] B. Fiedler. *Global Bifurcation of Periodic Solutions with Symmetry.* Springer-Verlag, Berlin, 1988.

[FiS88] B. Fiedler and J. Scheurle. *Discretization of homoclinic orbits and invisible chaos.* Memoir AMS, Providence, in press, 1996.

[FiT96] B. Fiedler & D. Turaev. Coalescence of reversible homoclinic orbits causes elliptic resonance. to appear: Int. J. Bif. Chaos, 1996.

[FMS88] J. Fineberg, E. Moses, and V. Steinberg. Spatially and temporally modulated traveling-wave pattern in convecting binary mixtures. Phys. Rev. Lett. **61**, 838, 1988.

[Ga87] J.-M. Gambaudo. *Ordre, désordre, et frontière des systèmes Morse-Smale.* These, Nice, 1987.

[Ga94] Th. Gallay. Local stability of critical fronts in non-linear parabolic partial differential equations. Nonlinearity **7**, 741-764, 1994.

[GH90] J. Guckenheimer and P. Holmes. *Nonlinear Oscillations, Dynamical Systems, and Bifurcations of Vector Fields.* Springer-Verlag, New York, 1990 (second edition).

[GJK93] R. Gardner, C.K.R.T. Jones and T. Kapitula. Stability of travelling waves for non-convex scalar viscous conservation laws. Comm. Pure Appl. Math. **XLVI**, 505-526, 1993.

[GKW94] K. Gęba, W. Krawcewicz and J. Wu. An equivariant degree with applications to symmetric bifurcation problems I: construction of the degree. Proc. London Math. Soc. **69**, 377–398, 1994.

[GSS88] M. Golubitsky, I. Stewart, and D. G. Schaeffer. *Singularities and Groups in Bifurcation Theory. Volume II*, volume 69 of *Applied Mathematical Sciences.* Springer, 1988.

[Gu86] J. Guckenheimer. A codimension two bifurcation with circular symmetry. Contemporary Mathematics **56**, 175–184, 1986.

[Ha82] S. P. Hastings. Single and multiple pulse waves for the FitzHugh-Nagumo equations. SIAM J. Appl. Math. **42**, 247–260, 1982.

[Ha88] J. K. Hale. *Asymptotic Behavior of Dissipative Systems*. Math. Surv. **25**, AMS, Providence 1988.

[Har88] J. Harrison. C^2 counterexamples to the Seifert conjecture. Topology **27**, 249–278, 1988.

[He80] D. Henry. *Geometric Theory of Semilinear Parabolic Equations*. Lecture Notes in Mathematics, Vol. **840**, Springer-Verlag, New-York, 1980.

[Hi76] M. W. Hirsch. *Differential Topology*. Springer-Verlag, New York, 1976.

[HL95] Hairer and C. Lubich. The lifespan of backward error analysis for numerical integrators. Preprint, 1995.

[HMR92] P. J. Holmes, A. Mielke and O. O'Reilly. Cascades of homoclinic orbits to, and chaos near, a Hamiltonian saddle-center. J. Dyn. Diff. Eq. **4**, 95–126, 1992.

[Ho95] A. J. Homburg. Homoclinic intermittency. Fields Institute Communications 4, 191–200, 1995.

[HSY93] B. R. Hunt, T. Sauer and J. A. Yorke. Prevalence: a translation invariant "almost everywhere" on infinite dimensional spaces. Bull. Amer. Math. Soc. **28**, 306–307, 1993.

[HW79] P. de Hoog and R. Weiss. The numerical solution of boundary value problems with an essential singularity. SIAM J. Numer. Anal. **16**, 637–669, 1979.

[HW90] H. Hofer and K. Wysocki. First order elliptic systems and the existence of homoclinic orbits in Hamiltonian systems. Math. Ann. **288**, 483–503, 1990.

[HY83] J. Harrison and J. A. Yorke. Flows on S^3 and $\mathbb{R}^3$ without periodic orbits. In [Pa83], 401–407.

[IMV89] J. Ize, I. Massabò and V. Vignoli. *Degree theory for equivariant maps I.* Trans. AMS **315**, 433–510, 1989.

[IMV92] J. Ize, I. Massabò and V. Vignoli. Degree theory for equivariant maps: the general S^1-action. Mem. Amer. Math. Soc. **100**, x+179pp., 1992.

[Io88] G. Iooss. Global characterization of the normal form for a vector field near a closed orbit. *J. Diff. Eqns.* **76**, 47–76, 1988.

[IoA92] G. Iooss and M. Adelmeyer. *Topics in Bifurcation Theory and Applications*. Advanced Series in Nonlinear Dynamics Vol. 3, World Scientific, 1992.

[IoK92] G. Iooss and K. Kirchgässner. Water waves for small surface tension: an approach via normal form. Proc. Roy. Soc. Edinburgh **122 A**, 267-299, 1992.

[IoL90] G. Iooss and J. Los. Bifurcation of spatially quasi–periodic solutions in hydrodynamic stability problems. Nonlinearity **3**, 851–871, 1990.

[IoM91] G. Iooss and A. Mielke. Bifurcating time-periodic solutions of Navier-Stokes equations in infinite cylinders. J. Nonlinear Science **1**, 107-146, 1991.

[IoMD89] G. Iooss, A. Mielke and Y. Demay. Theory of steady Ginzburg–Landau equation in hydrodynamic stability problems. Europ. J. Mech. B/Fluids **3**, 229–268, 1989.

[IoP93] G. Iooss and M.–C. Pérouème. Perturbed homoclinic solutions in reversible 1:1 resonant vector fields. J. Diff. Eqns. **102**, 62–88, 1993.

[Iz76] J. Ize. *Bifurcation Theory for Fredholm Operators.* AMS memoir **174**, Providence, 1976.

[JPBCRK92] B. Janiaud, A. Pumir, D. Bensimon, V. Croquette, H. Richter, and L. Kramer. The Eckhaus instability for travling waves. Physica **D55**, 269–286, 1992.

[Ka76] T. Kato. *Perturbation Theory for Linear Operators.* Springer-Verlag, New-York, 1976.

[Ka95] T. Kapitula, (1994), On the stability of travelling waves in weighted L^∞ spaces, to appear in *J. Differential Equations.*

[KC81] J. Kevorkian and J. D. Cole. *Perturbation Methods in Applied Mathematics.* Springer, 1981.

[KD90] E. Knobloch and J. DeLuca. Amplitude equations for traveling wave convection. Nonlinearity **3**, 975–980, 1990.

[Ki82] K. Kirchgässner. Wave-solutions of reversible systems and applications. J. Differential Equations **45**, 113-127, 1982.

[Ki92] K. Kirchgässner. On the nonlinear dynamics of travelling fronts. J. Differential Equations **96**, 256-278, 1992.

[KKLN93] A. I. Khibnik, Y. A. Kuznetsov, V. V. Levitin and E. V. Nikolaev. Continuation techniques and iterative software for bifurcation analysis of ODEs and iterated maps. Phys. D **62**, 360–371, 1993.

[KKO93] M. Kisaka, H. Kokubu and H. Oka. Bifurcations to n-homoclinic orbits and n-periodic orbits in vector fields. J. Diff. Equ. **5**, 305–357, 1993.

[KM83] M. Kubiček and M. Marek. *Computational Methods in Bifurcation Theory and Dissipative Structures.* Springer-Verlag, New York, 1983.

[KM91] M. Krupa and I. Melbourne. Asymptotic stability of heteroclinic cycles in systems with symmetry. Preprint, 1991.

[Kn86a] E. Knobloch. On the degenerate Hopf bifurcation with O(2)-symmetry. Contemporary Mathematics **56**, 175–184, 1986.

[Kn86b] E. Knobloch. Oscillatory convection in binary mixtures. Phys. Rev. A **34**, 1538–1549, 1986.

[Kn92] E. Knobloch. Nonlocal amplitude equations. In S. Kai, editor, *Pattern Formation in Complex Dissipative Systems*. World Scientific, 1992.

[KnP81] E. Knobloch and M. R. E. Proctor. Nonlinear periodic convection in double-diffusive systems. J. Fluid Mech. **108**, 291–316, 1981.

[KnP92] E. Knobloch and R. Pierce. Spiral vortices in finite cylinders. In C. D. Andereck and F. Hayot, editors, *Ordered and Turbulent Patterns in Taylor–Couette Flow*, Plenum Press, 1992.

[KnWD81] E. Knobloch, N. Weiss, and L. N. DaCosta. Oscillatory and steady convection in a magnetic field. J. Fluid Mech. **113**, 153–186, 1981.

[Ko88] H. Kokubu. Homoclinic and heteroclinic bifurcations of vector fields. Japan J. Appl. Math. **5**, 455–501, 1988.

[Ko91] H. Kokubu. *Heteroclinic bifurcations associated with different saddle indices.* Collection: Dynamical Systems and related topics (Nagoya, 1990) 236–260, Adv. Ser. Dyn. Syst. **9** (1991), World Sci. Publ., River Edge, N.J. 1991.

[Ko93] H. Kokubu. A construction of three dimensional vector fields which have a codimension two heteroclinic loop at Glendinning-Sparrow T-point. Z. Angew. Math. Phys. **44** (1993), 510–536.

[KoS88] P. Kolodner and C. M. Surko. Weakly nonlinear traveling wave convection. Phys. Rev. Lett. **61**, 842, 1988.

[KoSW89] P. Kolodner, C. M. Surko, and H. Williams. Dynamics of traveling waves near the onset of convection in binary fluid mixtures. Physica **D37**, 319–333, 1989.

[KPP37] A. Kolmogorov, I. Petrovsky and N. Piscounov. Etude de l'équation de la diffusion avec croissance de la quantité de matière et son application à un problème biologique. Moscow Univ. Math. Bull. **1**, 1-25 1937.

[KSM92] P. Kirrmann, G. Schneider and A. Mielke. The validity of modulation equations for extended systems with cubic nonlinearities. *Proc. Royal Soc. Edinburgh* **A122**, 85-91, 1992.

[Ku82] P.A. Kuchment. Floquet theory for partial differential equations. Birkhäuser Verlag 1993. See also: Russ. Math. Surveys, **37**:4, 1–60, 1982.

[Ku90] Y. A. Kuznetsov. Computation of invariant manifold bifurcations. In D. Roose et al., eds., *Continuation and Bifurcations: Numerical Techniques and Applications*, 183–195. Kluwer, Netherlands, 1990.

[Ku95] Y. A. Kuznetsov. *Elements of Applied Bifurcation Theory.* Springer Verlag, New York, 1995.

271

[Le51] E. Leontovich. On the generation of limit cycles from separatrices (Russ.). Dokl. Akad. Nauk **78**, 641–644, 1951.

[Le80] M. Lentini and H. B. Keller. Boundary value problems on semi-infinite intervals and their numerical solution. SIAM J. Numer. Anal. **17**, 577–604, 1980.

[Li90] X.-B. Lin. Using Melnikov's method to solve Shilnikov's problems. Proc. Roy. Soc. Edinburgh **116A**, 295–325, 1990.

[LK96] A.S. Landsberg and E. Knobloch. Oscillatory bifurcation with broken translation symmetry, Phys. Rev. E **53**, 3579–3600, 1996; and: Oscillatory doubly diffusive convection in a finite container, Phys. Rev. E **53**, 3601–3609, 1996.

[LM68] J.–L. Lions and E. Magenes. *Problèmes aux limites non homogènes et applications, Vol. 1.* Dunod, Paris, 1968.

[Lu82] V. I. Lukyanov. Bifurcations of dynamical systems with a saddle-node separatrix loop. Diff Eq. **18**, 1049–1059, 1982.

[LZ91] W-G. Li and Z-F Zhang. The "blue sky catastrophe" on closed surfaces. Collection: Dynamical systems and related topics (Nagoya, 1990), 316–332, Adv. Ser. Dyn. Syst., **9** (1991), (ed.) K. Shiraiwa, World Sci. Publishing, River Edge, N.J.

[Ma90] P. Manneville. *Dissipative Structures and Weak Turbulence.* Academic Press, 1990.

[MC76] J. E. Marsden and M. McCracken. *The Hopf Bifurcation and its Applications.* Springer, 1976.

[Me80] V.S. Medvedev. A new type of bifurcations on manifolds. Math. Sb. **113(155)**, 487–492, 496, 1980.

[Me85] M. Medved. The unfoldings of a germ of vector fields in the plane with a singularity of codimension 3. Czech. Math. J. **35**, 1–42, 1985.

[Mi86] A. Mielke. A reduction principle for nonautonomous systems in infinite-dimensional spaces. J. Diff. Eqns. **65**, 68–88, 1986.

[Mi87] A. Mielke. Über maximale L^p-Regularität für Differentialgleichungen in Banach– und Hilbert–Räumen. Math. Annalen **277**, 51–66, 1987.

[Mi88] A. Mielke. Reduction of quasilinear elliptic equations in cylindrical domains with applications. Math. Meth. Appl. Sci. **10**, 51–66, 1988.

[Mi90] A. Mielke. Normal hyperbolicity of center manifolds and Saint–Venant's principle. Archive Rational Mech. Analysis **110**, 353–372, 1990.

[Mi91] A. Mielke. *Hamiltonian and Lagrangian Flows on Center Manifolds with Applications to Elliptic Variational Problems.* Lecture Notes in Mathematics Vol. **1489**, Springer–Verlag, 1991.

[Mi92] A. Mielke. Reduction of PDEs on domains with several unbounded directions: a first step towards modulation equations. ZAMP **43**, 449–470, 1992.

[Mi94a] A. Mielke. Essential manifolds for an elliptic problem in an infinite strip. J. Diff. Eqns. **110**, 322–355, 1994.

[Mi94b] A. Mielke. Floquet theory for, and bifurcations from spatially periodic patterns. Tatra Mountains Math. Publ. **4**, 153–158, 1994.

[Mi95] A. Mielke. A new approach to sideband instabilities using the principle of reduced instability. In "*Nonlinear Dynamics and Pattern Formation in the Natural Environment*, A. Doelman & A. van Harten (eds). Pitman Research Notes in Math. Vol. **335**, 1995." Pages 206–222.

[Mi96a] A. Mielke. Mathematical analysis of sideband instabilites with application to Rayleigh–Bénard convection. J. Nonlinear Science; submitted.

[Mi96b] A. Mielke. The complex Ginzburg–Landau equation on unbounded domains — sharp estimates and attractors —. Preprint 27/96 Universität Hannover, 1996. Submitted to Nonlinearity.

[MR93] P. C. Matthews and A. M. Rucklidge. Traveling and standing waves in magneto-convection. Proc. R. Soc. London Ser. A **441**, 649–658, 1993.

[MR95] J.-F. Mallordy and J.-M. Roquejoffre. A parabolic equation of the KPP type in higher dimensions. SIAM J. Math. Anal. **26**, 1-20, 1995.

[MS86] W. Meiske and K.R. Schneider. Existence, persistence and structure of integral manifolds in the neighbourhood of a periodic solution of autonomous differential systems. Časopis pro pěstování matem. **111**, 304–313, 1986.

[MV92] B. J. Matkowsky and V. Volpert. Coupled nonlocal complex Ginzburg Landau equations in gasless combustion. Physica,**D54**, 203–219, 1992.

[MV96] C. Martel and J. M. Vega. Finite size effects near the onset of the oscillatory instability. Nonlinearity, to appear.

[MW89] J. Mawhin and M. Willem. *Critical Point Theory and Hamiltonian Systems*. Springer-Verlag, New York, 1989.

[MY82] J. Mallet-Paret and J. A. Yorke. Snakes: oriented families of periodic orbits, their sources, sinks, and continuation. J. Diff. Eq. **43**, 419–450, 1982.

[Na86] W. Nagata. Symmetric Hopf bifurcations and magnetoconvection. Contemporary Mathematics **56**, 237–266, 1986.

[ND88] M. Neveling and G. Dangelmayr. Bifurcation analysis of interacting stationary modes in thermohaline convection. Phys. Rev. A **38**, 2536, 1988.

[NLHGD87] M. Neveling, D. Lang, P. Haug, W. Güttinger, and G. Dangelmayr. Interactions of stationary modes in systems with two and three degrees of freedom. In W. Güttinger and G. Dangelmayr, editors, *The Physics of Structure Formation. Theory and Simulation*, volume 37 of *Springer Series in Synergetics*, pages 153–165. Springer-Verlag, Berlin, 1987.

[No82] V.P. Nozdracheva. Bifurcations of a structurally unstable separatrix loop. Differential. Eqs. **18**, 1098–1104, 1982.

[No85] V.P. Nozdracheva. Bifurcations of a singular cycle with two separatrices. Collection: Integral and differential equations and approximate solutions (Russian), 107–124, Kalmytsk. Gos. Univ., Elista, 1985.

[NW69] A. C. Newell and J. A. Whitehead. Finite bandwidth, finite amplitude convection. J. Fluid Mech. **38**, 279, 1969.

[Pa83] J. Palis, jr., ed. *Geometric Dynamics*. Springer-Verlag, New York, 1983.

[PaM82] J. Palis and W. de Melo. (*Geometric Theory of Dynamical Systems, An Introduction*. Springer-Verlag, New-York, 1982.

[Pa84] K. J. Palmer. Exponential dichotomies and transversal homoclinic points. J. Diff. Eq. **55**, 225–256, 1984.

[Pe92] M.-C. Pérouéme. Bifurcations d'orbites homoclines dans les systèmes reversibles. Thesis, Nice 1992.

[PiK95] R. D. Pierce and E. Knobloch. Asymptotically exact Zakharov equations and the stability of water waves with bimodal spectra. Physica **D81**, 341–373, 1995.

[PiW95] R. D. Pierce and C. E. Wayne. On the validity of mean-field amplitude equations for counterpropagating wavetrains. Nonlinearity **8**, 769–779, 1995.

[Po1892] H. Poincaré. *Méthodes Nouvelles de la Mécanique Céleste I*. Gauthiers-Villars, Paris, 1892.

[PW67] M. H. Protter and H. F. Weinberger. *Maximum Principles in Differential Equations*. Prentice-Hall, Englewood Cliffs, N.J., 1967.

[Ra86] P. H. Rabinowitz. *Minimax Methods in Critical Point Theory with Applications to Differential Equations*. CBMS Notes **65**, AMS, Providence, 1986.

[Rb81] C. Robinson. Differentiability of the stable foliation for the model Lorenz equations. In D. Rand & L. Young, eds., *Dynamical Systems and Turbulence*. Springer-Verlag, Berlin, 1981.

[RB83] W. C. Rheinboldt and J. W. Burkardt. A locally parametrized continuation process. ACM Trans. Math. Software **9**, 215–235, 1983.

[Re79] J.W. Reyn. Generation of limit cycles from separatrix polynons in the phase plane. In collection: Geometrical approaches to differential equations. Lect. Notes Math., (ed.) R. Martini, **810**, 264–289, Springer Verlag, Berlin, 1979.

[RK95] G. Raugel and K. Kirchgässner. Long-time asymptotics of perturbed fronts of minimal speed for a KPP system. Manuscript, Universität Stuttgart, 1995.

[Ro92] J.-M. Roquejoffre. Stability of travelling fronts in a model for flame propagation, part 2: nonlinear stability. *Arch. Rat. Mech. Anal.* **117**, 119-153, 1992.

[RS80] M. Reed and B. Simon. *Methods of Mathematical Physics III, Scattering Theory.* Academic Press, 1980.

[RSW86] M. Roberts, J. W. Swift, and D. H. Wagner. The Hopf bifurcation on a hexagonal lattice. Contemporary Mathematics **56**, 283–318, 1986.

[Sa76] D. H. Sattinger. On the stability of waves of nonlinear parabolic systems. Adv. Math. **22**, 312-355, 1976.

[Sa77] D. H. Sattinger. Weighted norms for the stability of travelling waves. J. Differential Equations **25**, 130-144, 1977.

[Sa92] B. Sandstede. Personal communication. 1992.

[Sa95] B. Sandstede. Center manifolds for homoclinic solutions. Preprint, 1995.

[Sc87a] S. Schecter. The saddle-node separatrix-loop bifurcation. SIAM J. Math. Anal. **18**, 1142–1156, 1987.

[Sc87b] S. Schecter. Melnikov's method at a saddle-node and the dynamics of a forced Josephson junction. SIAM J. Math. Anal. **18**, 1699–1715, 1987.

[Sc91] B. Scarpellini. Center manifolds of infinite dimensions I: Main results and applications. J. Applied Math. Physics (ZAMP) **42**, 1–32, 1991. Center manifolds of infinite dimensions II: Proofs of the main results. J. Applied Math. Physics (ZAMP) **42**, 280–314, 1991.

[SC92] A. Scheel and P. Chossat. Bifurcation d'orbites périodiques à partir d'un cycle homocline symétrique. C. R. Acad. Sci. Paris, Ser. I **314**, 49–54, 1992.

[Sc94] A. Scheel. Existence of fast travelling waves for some parabolic equations -A dynamical systems approach- Serie A. Mathematik Preprint **A 17/94**, Freie Universität Berlin, Germany, 1994.

[Sch94a] G. Schneider. A new estimate for the Ginzburg-Landau approximation on the real axis. J. Nonlinear Science 4, 23-34, 1994.

[Sch94b] G. Schneider. Error estimates for the Ginzburg-Landau approximation. Z. angew. Math. Physik (ZAMP) **45**, 433-457, 1994.

[Se69] L. A. Segel. Distant side-walls cause slow amplitude modulation of cellular convection. J. Fluid Mech. **38**, 203, 1969.

[Se88] R. Seydel. *From Equilibrium to Chaos.* Elsevier, New York, 1988.

[SFMR89] V. Steinberg, J. Fineberg, E. Moses, and I. Rehberg. Pattern selection and transition to turbulence in propagating waves. Physica **D37**, 359–383, 1989.

[Sh92] M. V. Shashkov. On the bifurcations of separatrix contours on two-dimensional surfaxes I. Russ. (1989),Engl. Selecta Math. Sov., **11**, 341–353, 1992.

[Sh94] M. V. Shashkov. On the bifurcations of separatrix contours on two-dimensional surfaxes II. Russ. (1990), Engl. Selecta Math. Sov., Vol. 13, **2**, 175–182, 1994.

[ShT96] M. V. Shashkov and Turaev. On the complex bifurcation set for sa system with simple dynamics. to appear in Int. J. Bif. Chaos, 1996.

[Shi62] L. P. Shilnikov. Some cases of generation of periodic motions in an n-dimensional space. Soviet Math. Dokl. **3**, 394–397, 1962.

[Shi65] L. P. Shilnikov. A case of the existence of a countable number of periodic motions. Soviet Math. Dokl. **6**, 163–166, 1965.

[Shi66] L. P. Shilnikov. On the generation of a periodic motion from a trajectory which leaves and re-enters a saddle-saddle state of equilibrium. Soviet Math. Dokl. **7**, 1155–1158, 1966.

[Shi67] L. P. Shilnikov. The existence of a denumerable set of periodic motions in four-dimensional space in an extended neighborhood of a saddle-focus. Soviet Math. Dokl. **8**, 54–57, 1967.

[Shi68] L. P. Shilnikov. On the generation of a periodic motion from trajectories doubly asymptotic to an equilibrium state of saddle type. Math. USSR Sbornik **6**, 427–437, 1968.

[Shi69] L. P. Shilnikov. On a new type of bifurcation of multidimensional dynamical systems. Soviet Math. Dokl. **10**, 1368–1371, 1969.

[Shi70] L. P. Shilnikov. A contribution to the problem of the structure of an extended neighborhood of a rough equilibrium state of saddle-focus type. Math. USSR Sbornik **10**, 91–102, 1970.

[SK91] M. Silber and E. Knobloch. Hopf bifurcation on a square lattice. Nonlinearity **4**, 1063–1106, 1991.

[Sp82] C. Sparrow. *The Lorenz Equations: Bifurcations, Chaos, and Strange Attractors.* Springer-Verlag, New York, 1982.

[SRK92] M. Silber, H. Riecke, and L. Kramer. Symmetry-breaking Hopf bifurcation in anisotropic systems. Physica **D61**, 260–278, 1992.

[St90] M. Struwe. *Variational Methods.* Springer-Verlag, Berlin, 1990.

[Sw74] P. A. Schweitzer. Counterexamples to the Seifert conjecture and opening closed leaves of foliations. Ann. of Math. **100**, 386–400, 1974.

[SW89] W. Schöpf and W. Zimmermann. Multicritical behaviour in binary fluid convection. Europhysics Letters **8**, 41, 1989.

[Ta74] F. Takens. Singularities of vector fields. Publ. Math. IHES **43**, 47–100, 1974.

[Tr84] C. Tresser. About some theorems by L.P. Shilnikov. Ann. Inst. H. Poincaré **40**, 441–461, 1984.

[Tu85] D. V. Turaev. Bifurcations of two-dimensional dynamical systems that are close to a system with two separatrix loops. Russ. Uspekhi Mat. Nauk., **40**, 203–204, 1985.

[Tu86] D. V. Turaev. Bifurcations of a homoclinic "figure eight" saddle with a negative saddle value. Russ. Dokl. Akad. Nauk SSSR., **290**, 1301–1304, 1986.

[Tu88] D. V. Turaev. Bifurcations of a homoclinic "figure eight" of a multidimensional saddle. Russ. Math.-Surv., **43**, 264–265, 1988.

[Tu96] D. V. Turaev. On dimension of semi-local bifurcational problems Bifurcation and Chaos, to appear.

[Va90] A. Vanderbauwhede. Subharmonic branching in reversible systems. SIAM J. Appl. Anal., **21**, 954–979, 1990.

[Va91] A. Vanderbauwhede. Branching of periodic solutions in time–reversible systems. In *"Geometry and Analysis in Nonlinear Dynamics"*, H. Broer & F. Takens, eds, Longman, 1991.

[VaF92] A. Vanderbauwhede and B. Fiedler. Homoclinic period blow-up in reversible and conservative systems. Z. angew. Math. Phys. **43**, 292–318, 1992.

[VaV87] S.A. van Gils and A. Vanderbauwhede. Center manifolds and contractions on a scale of Banach spaces. J. Funct. Anal., **72**, 209–224, 1987.

[Ve93] J. M. Vega. On the amplitude equations arising at the onset of the oscillatory instability in pattern formation. SIAM J. Math. Anal. **24**, 603–617, 1993.

[VH91] A. van Harten. On the validity of Ginzburg-Landau's equation. J. Nonlinear Science **1**, 397-422, 1991.

[Wa89] H.-O. Walther. *Hyperbolic periodic solutions, heteroclinic connections and transversal homoclinic points in autonomous differential delay equations.* AMS Memoir **402**. AMS, Providence, 1989.

[We77] N. O. Weiss. Magnetic fields and convection. In E. A. Spiegel and J. P. Zahn, editors, *Problems of Stellar Convection*, pages 176–187. Springer, 1977.

[We93] M. Wegelin. *Nichtlineare Dynamik raumzeitlicher Muster in hierarchischen Systemen.* Dissertation, Universität Tübingen, Institut für Informationsverarbeitung, 1993.

[Wh68] G. T. Whyburn. Topological Analysis. Princeton Univ. Press, Princeton, 1968.

[Wi88] S. Wiggins. *Global Bifurcations and Chaos – Analytical Methods.* Springer-Verlag, New York, 1988.

[WKPS85] R. W. Walden, P. Kolodner, A. Passner and C. M. Surko. Traveling waves and chaos in convection in binary fluid mixtures. Phys. Rev. Lett. **55**, 496, 1985.

[Ya87] E. Yanagida. Branching of double pulse solutions from single pulse solutions in nerve axon equations. J. Diff. Eq. **66**, 243–262, 1987.